Lecture Notes in Physics

Volume 857

For further volumes:
http://www.springer.com/series/5304

The Lecture Notes in Physics

The series Lecture Notes in Physics (LNP), founded in 1969, reports new developments in physics research and teaching—quickly and informally, but with a high quality and the explicit aim to summarize and communicate current knowledge in an accessible way. Books published in this series are conceived as bridging material between advanced graduate textbooks and the forefront of research and to serve three purposes:

- to be a compact and modern up-to-date source of reference on a well-defined topic
- to serve as an accessible introduction to the field to postgraduate students and nonspecialist researchers from related areas
- to be a source of advanced teaching material for specialized seminars, courses and schools

Both monographs and multi-author volumes will be considered for publication. Edited volumes should, however, consist of a very limited number of contributions only. Proceedings will not be considered for LNP.

Volumes published in LNP are disseminated both in print and in electronic formats, the electronic archive being available at springerlink.com. The series content is indexed, abstracted and referenced by many abstracting and information services, bibliographic networks, subscription agencies, library networks, and consortia.

Proposals should be sent to a member of the Editorial Board, or directly to the managing editor at Springer:

Christian Caron
Springer Heidelberg
Physics Editorial Department I
Tiergartenstrasse 17
69121 Heidelberg/Germany
christian.caron@springer.com

Jean-Pierre Rozelot · Coralie Neiner

Editors

The Environments
of the Sun and
the Stars

 Springer

Editors

Jean-Pierre Rozelot
Observatoire de la Côte d'Azur
CNRS-Lagrange Dpt.
Université Nice-Sophia-Antipolis
Nice
France

Coralie Neiner
Observatoire de Paris-Meudon
CNRS-LESIA Dpt.
Université Paris-Diderot
Meudon
France

ISSN 0075-8450   ISSN 1616-6361 (electronic)
Lecture Notes in Physics
ISBN 978-3-642-30647-1   ISBN 978-3-642-30648-8 (eBook)
DOI 10.1007/978-3-642-30648-8
Springer Heidelberg New York Dordrecht London

Library of Congress Control Number: 2012946192

Printed on acid-free paper

Springer is part of Springer Science+Business Media (www.springer.com)

Preface

This book is the third of a series about the physics of the Sun and the stars bringing together the knowledge from the two communities working on these objects. The focus of this book is placed on the environments of the Sun and the stars, with a particular emphasis on recent observations and techniques, and on the interactions between the environment and the central object.

The General Overlook of this Book is as Followed The first three chapters of the book focus on the Sun, a star that can and has been studied in great detail. In the recent years many space missions provided us with important information about shock waves and discontinuities in the natural plasmas of the solar system. The solar wind, i.e., the particles escaping from the Sun via flares, prominence ejection, and coronal mass ejection, impact on the Earth magnetic environment with cascading effects on the Earth atmosphere. This allows us to study the Sun from the magnetic activity of the Earth. For example, from the changing correlation between sunspot and geomagnetic activity, the long-term variations in the components of the solar magnetic field can be estimated. In addition, one can study the link between the solar cycle and the planetary tidal forces.

Chapter 4 proposes a physical picture of tides in planetary systems and binary stellar systems. Tidal interactions are a key mechanism to understand the dynamics and evolution of celestial objects. The following chapter presents the case of massive binary stars, which can be studied thanks to the advent of interferometric techniques.

Interferometry also allows us to study the very close environments of young stars, as presented in Chap. 6. For example, interferometric observations directly probe the location of the dust and gas in the circumstellar disks and allow us to detect and characterize close companions. A more complete review about accretion disks is presented in the following chapter. With Chap. 8 we move on from young stars to main sequence stars, discuss mass loss through the winds of hot massive stars and how these winds interact with rotation and magnetic fields leading to stellar spin-down and large-scale disk-like structures. The next chapter presents the magnetic field and convection in the cool supergiant Betelgeuse. Finally, Chap. 10 focuses on evolved objects, such as bipolar nebulae, post-AGB stars or novae, and discuss in particular the difference between the torus and disk environment.

The audience targeted by this book consists of researchers, PhD students, post-docs, and also all scientists seeking a complementary culture or evolving toward new research topics.

This book is based on tutorials and discussions on the same topic held at a CNRS school in Roscoff (France) in 2011, which has allowed us to give a progress report on the very last solar developments (structure of the solar core for example) and stellar developments (CoRoT results, new stellar models) for a better understanding of environments in general. Let us remind that two previous books titled "The Rotation of the Sun and Stars" (LNP 765) and "The Pulsations of the Sun and the Stars" (LNP 832) resulted from two previous CNRS schools held in Obernai (France) in 2007 and Saint-Flour (France) in 2008. We hope that this new book about the "Environments of the Sun and the Stars" will provide an interesting sequel for the reader.

The editors sincerely thank the authors for the great quality of their contributions published here.

C. Neiner
J.P. Rozelot

Contents

Part I
The Sun as a Star: Its Environment

Chapter 1
Discontinuities and Turbulence in the Solar Wind

Laurence Rezeau

Abstract Shock waves and discontinuities are observed in the natural plasmas of the Solar System as in many other systems. The recent years have seen many space missions that have given lots of information about these phenomena. These observations make possible a detailed analysis of shocks and discontinuities. A summary of how their main properties can be deduced from MHD equations is presented. The in situ observations have also evidenced the presence of turbulence behind the bow shock and also in the solar wind itself. As the plasma is collisionless, the role of this turbulence is likely to be important in diffusion and reconnection phenomena.

1.1 Shock Waves and Discontinuities

1.1.1 Observations of Discontinuities

Shock waves are observed ahead of supersonic planes in air. They are characterized by a sharp drop in velocity and a sharp jump in all physical quantities, and they behave as a boundary separating two regions. Similar discontinuities are also observed in plasmas. In a laboratory plasma a shock is generated when a laser pulse impacts a target. In astrophysics, shocks are seen on some remote sensing observations (in accretion disks, associated with high-velocity jets, in supernova remnants, etc.), and they are observed in space plasmas thanks to in situ measurements. The solar wind creates a shock ahead of the planets, and it has been observed for instance ahead of the Earth by the four CLUSTER spacecraft. Two examples of such observations are shown on Fig. 1.1, showing a sharp variation of the magnetic field at the crossing of the shock (the magnitude of the field is multiplied by a factor around 3). The four spacecraft are a few thousand kilometers apart, and they observe very similar signatures in some cases (top part of the figure) or very different (low part) showing that the structure of the shock can be more or less simple. It can be seen also that the signal in these regions is more or less turbulent.

L. Rezeau (✉)
Laboratoire de Physique des Plasmas, LPP, UPMC, CNRS, Ecole Polytechnique, 4 avenue de Neptune, 94107 Saint-Maur des Fossés, France
e-mail: laurence.rezeau@upmc.fr

J.-P. Rozelot, C. Neiner (eds.), *The Environments of the Sun and the Stars*, Lecture Notes in Physics 857,
DOI 10.1007/978-3-642-30648-8_1, © Springer-Verlag Berlin Heidelberg 2013

Fig. 1.1 Magnetic field observed by CLUSTER. The four spacecraft cross the shock at different times

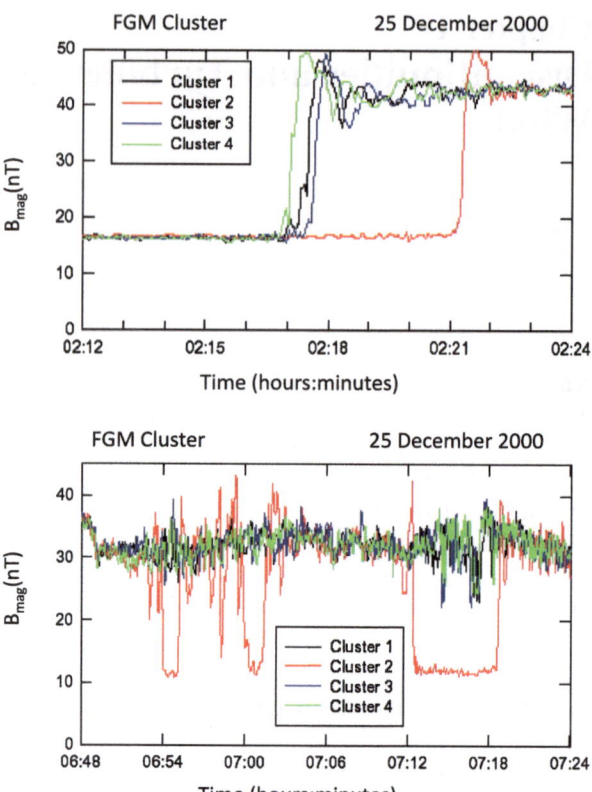

1.1.2 Jump Conditions at the Boundary

We assume that the boundary is one-dimensional, that a frame where it is stationary exists, and that all the physical parameters are constant on each side of the boundary. With these assumptions, the jump of any physical quantity a is defined as $\Delta a = a_2 - a_1$, where the indices 1 and 2 are related to upstream and downstream values of a. To obtain the jump conditions of all parameters, the MHD equations are integrated across the boundary. The simplest equation to integrate is $\nabla.(\mathbf{B}) = 0$, and it leads to the conservation of the normal component of the magnetic field across the boundary. To make the integration of the other equations simple, they have to be written in a conservative form, i.e. as $\partial_t a + \nabla.\mathbf{b} = 0$. The integration along the normal direction with the assumption of a one-dimensional stationary problem leads to the conservation of the normal component of $\mathbf{b}$ across the boundary. The integration of all the MHD equations leads to a system of equations called the Rankine–Hugoniot equations. These equations write as follows:

Conservation of mass: $\rho_2 v_{n2} = \rho_1 v_{n1} = \Phi_m$.

Conservation of momentum:

$$\rho_2 v_{n2} \mathbf{v}_2 + \left(p_2 + \frac{B_2^2}{2\mu_0} \right) \mathbf{n} - \frac{B_n \mathbf{B}_2}{\mu_0} = \cdots = \Phi_i .$$

Conservation of the total energy (fields and particles):

$$\frac{1}{2} \rho_2 v_{n2} v_2^2 + \frac{5}{2} p_2 v_{n2} - \frac{1}{\mu_0} \left(B_n \mathbf{B}_2.\mathbf{v}_2 - B_2^2 v_{n2} \right) = \cdots = \Phi_e .$$

Ohm's law: $v_{n2} \mathbf{B}_{T2} - B_{n2} \mathbf{v}_{T2} = v_{n1} \mathbf{B}_{T1} - B_{n1} \mathbf{v}_{T1} = \mathbf{n} \times \mathbf{E}_T .$

The different solutions of these equations cannot be seen at first sight, and when it is possible, it is easier to move to the de Hoffman–Teller frame (dHT frame). It is a frame in translation along the boundary which exists when the normal component of the magnetic field is nonzero. The dHT frame is the frame where $\mathbf{B}$ and $\mathbf{v}$ are parallel and therefore in which the tangential electric field, $\mathbf{E}_T$, is zero.

1.1.3 Different Kinds of Discontinuities

In the de HT frame the last equation is equal to zero and then allows a simplification of the conservation of momentum to

$$\left(v_n - \frac{B_n^2}{\mu_0 \Phi_m} \right) \mathbf{B}_T = \left(v_{n1} - \frac{B_n^2}{\mu_0 \Phi_m} \right) \mathbf{B}_{T1} .$$

This equation evidences two very different kinds of solutions. When the parenthesis is equal to zero on one side, it is equal to zero on the other side (because of the conservation of mass), and then there is no relation between the magnetic fields on both sides. In the other case where the parenthesis is nonzero, the two tangential magnetic fields are parallel, and the boundary is called a coplanar discontinuity. This means that at the crossing of the boundary, the magnitude of the magnetic field may change, but not its direction (Fig. 1.2).

In the case of the noncoplanar discontinuity, v_n is conserved, and therefore the plasma density is also conserved (because of the conservation of mass). The discontinuity is therefore not a shock because it induces no compression and no slowing of the plasma. These conservations induce the conservation of the pressure and of the modulus of the magnetic field. Therefore we only have a rotation of the magnetic field with no change in modulus. The discontinuity is therefore called a "rotational discontinuity." It can be shown easily that in the dHT frame, we also have the Walén relation $v = \frac{B}{\sqrt{\mu_0 \rho}} = V_A$, that is, the velocity of the plasma is equal to the Alfven velocity. This is a test that may be used to identify the rotational discontinuity in the data.

Let us study now the coplanar case which is the shock case. The elimination of variables gives a relation between the downstream normal velocity of the plasma and the upstream velocity (Fig. 1.3). The curve crosses the $v_{n2} = v_{n1}$ line in three values: these are the values obtained when one solves the MHD equations in the small

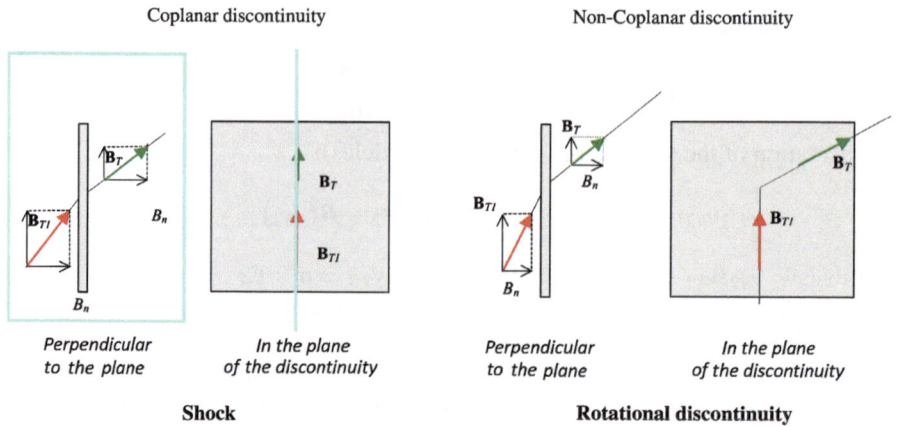

Fig. 1.2 Behavior of the magnetic field when crossing the different types of discontinuities

Fig. 1.3 The velocity downstream of the shock as a function of the upstream velocity (normal components)

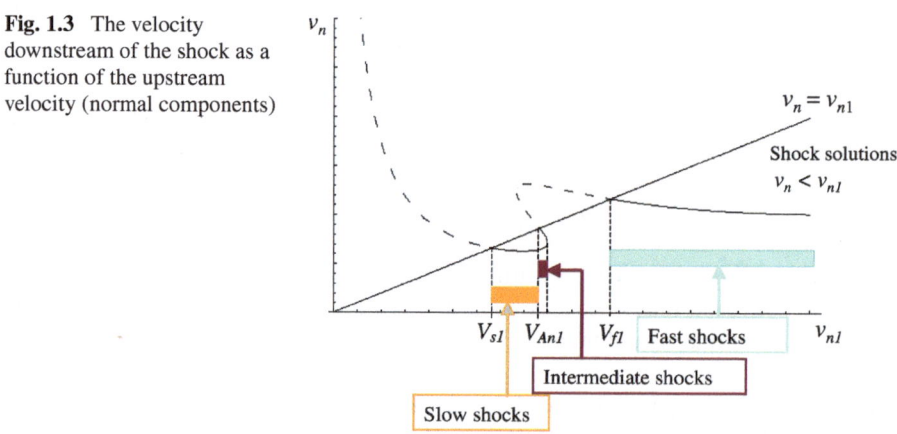

perturbation case, that is, the phase velocities of the three MHD wave modes, slow, Alfven, and fast modes. These values delimit the three kinds of solution that may exist: the slow shock, the intermediate shock, and the fast shock. All of them induce a compression and a slowing of the plasma. The only difference is the behavior of the tangential magnetic field: in the case of the fast shock it increases, in the other cases it decreases, without changing sign in the slow shock case and with a reversal in the intermediate shock case (see a summary on Fig. 1.4).

One more case has to be added to the different kinds of discontinuities: the tangential discontinuity which happens when the normal component of the magnetic field is equal to zero, together with the normal component of the velocity. In that case the dHT frame does not exist, and there is no connection between the two sides of the boundary. The only quantity which is conserved across the discontinuity is the total pressure (kinetic + magnetic), and the tangential magnetic fields are independent on both sides.

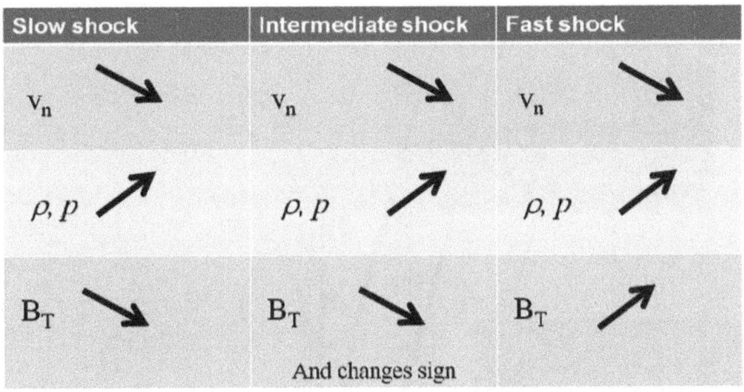

Fig. 1.4 Evolution of the physical parameters when crossing the different types of shocks

Slow shock	Intermediate shock	Fast shock
V_n ↘	V_n ↘	V_n ↘
ρ, p ↗	ρ, p ↗	ρ, p ↗
B_T ↘	B_T ↘	B_T ↗
	And changes sign	

What About the Boundaries Around the Planets? Having made this study of the different kinds of discontinuities that may exist in a plasma, let us go back to the observations in the Solar System. As already mentioned, the velocity of the solar wind is so high that a shock is observed ahead of the Earth. In fact it is the same for all the planets, although there is a difference depending on the existence of a planetary magnetic field or not, and all of them have been more or less explored by space missions. When the planet is not magnetized (Mars, Venus, and Pluto), the planet itself is the obstacle on the way of the solar wind, and the shock is very near to planet surface (less than one planet radius) [14]. The other planets, as the Earth, have an intrinsic magnetic field, and it is this magnetic field which is the obstacle for the solar wind (Mercury is an exception with a very small magnetic field; it will be explored in details by the future missions Messenger and Bepi Colombo). The magnetic field is modified by the solar wind pressure, defining a magnetosphere, where the field is compressed in the direction of the Sun and elongated in a tail on the opposite direction (Fig. 1.5). The noticeable phenomenon is that there is not only one boundary ahead of the magnetosphere, but two. The shock is observed first, and a magnetopause is following. The shock allows the slowing and compression of the plasma but with little modification of the solar wind magnetic field. The magnetopause is the boundary between this magnetic field (imposed by the Sun) and the planetary magnetic field (imposed by the internal dynamo of the planet).

The physical nature of these two boundaries has been studied in great details in the case of the Earth where many missions have given lots of data. As can be seen on Fig. 1.1, together with the shock crossing a very strong increase in the magnetic field is observed. The order of magnitude of the Alfven velocity is around 50 km s^{-1}, whereas the solar wind velocity is around 400–800 km s^{-1}. These two observations indicate that with no doubt, the shock is always a fast shock. But a detailed study evidences that the structure of the shock is more complicated than expected in MHD theory. First, there is a backward influence of the shock on the solar wind: some particles are reflected on the shock and make what is called a foreshock. Second, the detailed physics depends on the geometry of the shock with respect to the direction of the solar wind magnetic field. The nice "clean" signatures of the shock crossing (Fig. 1.1, top panel) are observed in the case where the shock is quasi-perpendicular (i.e., the shock normal is quasi-perpendicular to the incident magnetic field). In the other case where the shock is quasi-parallel (Fig. 1.1, lower

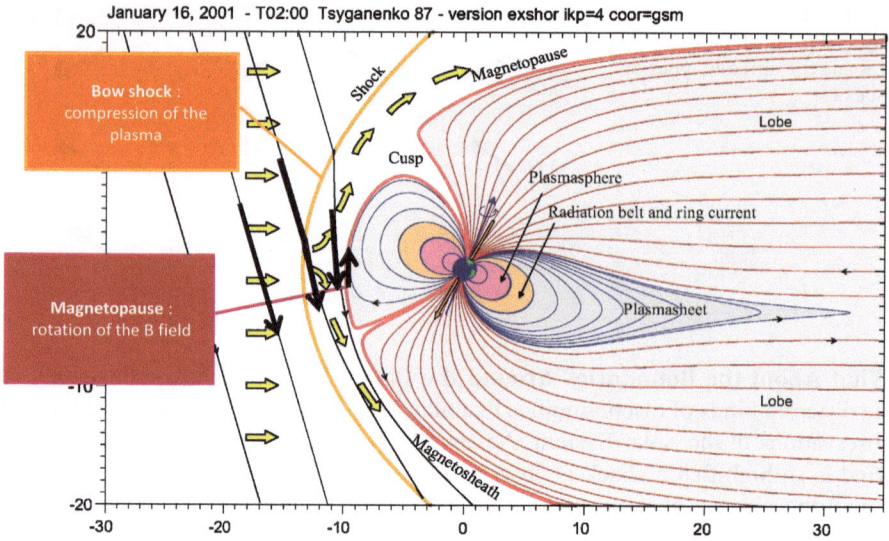

January 16, 2001 - T02:00 Tsyganenko 87 - version exshor ikp=4 coor=gsm

Fig. 1.5 Magnetic field lines in the magnetosphere. The solar wind blows from the left

panel), the structure is much more complicated and difficult to understand. Both cases may happen at the same time, depending on where the observations are made (on the morning or evening side of the shock). An internal structure of the shock has also been identified with a "foot", a "ramp" (sharp variation), and an "overshoot" on the downstream side of the shock [5]. The three-dimensional measurements made by CLUSTER have made possible a detailed analysis of the parameters of the shock (direction of the normal, velocity, thickness, etc.). It has also evidenced that in the quasi-perpendicular case the ramp is very thin (at the electron scale), and the foot is rather at the ion scale. It has also been shown that the shock is not really stationary, the foot grows and finally replaces the shock itself, driving a cyclic self-reformation of the shock [4].

The second question is about the nature of the magnetopause. As the magnetic field in the solar wind is independent of the magnetic field generated by the Earth, the discontinuity is not coplanar, and it is not a shock. Therefore it may be a tangential discontinuity, a rotational discontinuity, or something more complicated. The difference between the two cases is whether $B_n = 0$ or not. The physical consequences are very different in the two cases: in the first case the magnetosphere is closed, in the second case it "open." As the particles follow the field lines, in one case they may enter the magnetosphere easily. The diagnosis of this property can be made through minimum variance analysis [6], but it is always very difficult because B_n is small and there is a problem with the precision. Another difficulty comes with the assumption that the discontinuity is one-dimensional, and it is not obvious that is always the case (bubbles may be present in the boundary) and the stationarity. In some cases the Walén relation may be tested to prove that the magnetopause is a rotational discontinuity, but in many cases no clear conclusion can be reached.

1.2 Turbulence in the Boundary Regions and in the Solar Wind

1.2.1 Evidence of Turbulence in the Boundary Regions

Turbulence in a hydrodynamics view means fluctuations of the velocity field of the fluid. In space plasmas the velocity measurements are deduced from the particle function distributions measurements. Therefore they allow the measurement of the velocity fluctuations at low frequencies but not at high frequencies. On the other hand, the field measurements are made with a very high time resolution, because they aim at measuring high-frequency waves. Especially for the magnetic field captors, the fluxgate magnetometers give information about the large scale variations and the search-coil magnetometers to the small scales with a continuity between the two instruments. On the GEOS2 spacecraft, for instance, in 1978, a high level of magnetic turbulence was observed first in the magnetosheath, the region which immediately downstream of the shock, and second at the magnetopause itself [8]. The level of magnetic fluctuations is between 100 and 1000 times higher in the magnetosheath with respect to the level in the magnetosphere, and it is around 10 times higher within the magnetopause. These observations were made each time a spacecraft crossed these boundaries. These studies also evidenced that the fluctuations have a power law spectrum very similar to those observed in hydrodynamics turbulence, with a spectrum steeper in the magnetosheath (around $f^{-2.8}$) than in the magnetopause ($f^{-2.5}$) [9].

Why is the study of this magnetic turbulence interesting in space plasmas? It may first be a direct consequence of the shock itself, as it is behind a hydrodynamics shock. But it may also play an active role in the cross-scale coupling inducing diffusion in the plasma, and transports of heat or momentum. This is important since these plasmas are collisionless and therefore have no "classical" diffusion at all and therefore not any "normal" transport phenomena.

1.2.2 Cluster and Multipoint Measurements

The results already mentioned were obtained with one spacecraft measurements and with no possibility of distinguishing space effects from time effects. Consequently, the only spectra that were possible to calculate were frequency spectra (Fourier transform of the observed time series), with no possibility of getting any information about the scale spectra, except in the very special case of the solar wind turbulence which is carried by a very fast wind, and one may assume that the Doppler shift is dominant and that the frequency spectrum is the wave number spectrum along the velocity field. The ESA CLUSTER (group of four spacecraft with identical instruments onboard) was dedicated to remove this spatio-temporal ambiguity [3]. The method developed for analyzing the wave-number structure of the fluctuations is the k-filtering. The basic idea of the method is that for a monochromatic wave, the correlation between the same component of the field measured on two spacecraft gives access to the wavenumber component along the line relating the two spacecraft. With four spacecraft measuring the three components of the field, it is possible to get much more information, even in the case of a nonmonochromatic wavenumber spectrum, through all the possible correlations. The method was developed by Pinçon and Lefeuvre [7] using a maximum-likelihood nonlinear method. It was first applied on the CLUSTER data by Sahraoui et al. [10]. For each frequency observed on the spacecraft, f_s, a three-dimensional k spectrum is obtained, and the full spectrum is finally obtained by integrating over f_s. To give significant results, the k-filtering needs the signal to be sufficiently homogeneous and stationary in the period considered. The range of scales accessible to the analysis is limited to the inter-spacecraft distance (aliasing effect). Taking into account these limits, the method has given significant results about the turbulence in the magnetosheath: the spectra appear to be significantly anisotropic with respect to the magnetic field and to the magnetopause normal [11]. They evidence a power law spectrum in $k^{-8/3}$ along the direction of the velocity in the scale range 150–1800 km. They also show that the magnetic energy is distributed over the low frequency modes and among them the zero frequency mirror mode. CLUSTER brought another interesting result which is the correlation between the level of turbulence at the magnetopause and the solar wind pressure, evidencing the correlation between upstream activity and the fluctuations observed in the boundary [1]. CLUSTER was not designed to study the turbulence in the solar wind, but its magnetic sensors happen to have a sensitivity which is high enough to measure these fluctuations on a very broad frequency range (Fig. 1.6).

1.2.3 Our Present Knowledge and Questions

The classical view of turbulence cascade comes from hydrodynamics: power is injected in the system at large scales, and it is dissipated at a small scales. In between energy cascades with $k^{-5/3}$ power law in the inertial range, where the dominant feature is nonlinear coupling between eddies. Is the situation similar in the solar wind? Similarly, there is energy injected at large scale, this scale is of the order of 50 000 km (a typical size of the magnetosphere). But as it is a magnetized plasma, there are more characteristic scales than in an ordinary fluid: a breaking of the scale invariance can be expected at the ions scales (ion Larmor radius or inertial length) which are of the order of a few 100 km, and another breaking can be expected at the electrons scales of the order of 1 km. There is another big difference, the fact that there are no collisions and therefore no breaking of the spectrum at a dissipation scale related to collisional phenomena. The use of k-filtering has allowed one to reconstruct the wave-number spectra and to compare the experimental dispersion relations to theoretical ones. The first results obtained in the magnetosheath [2] have shown that mirror modes could be identified at large scales and other Hall-MHD modes at smaller scales, thus giving the indication that nonlinear coupling of linear

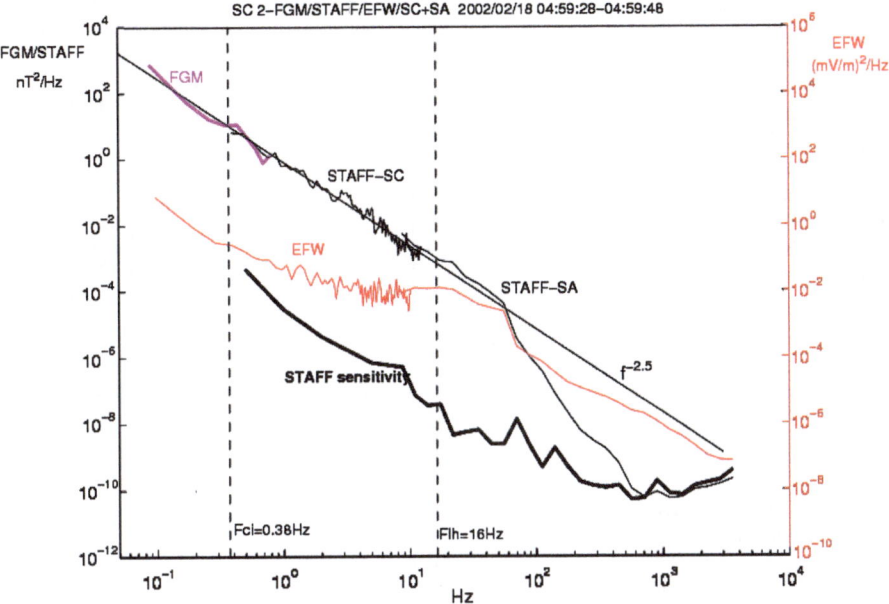

Fig. 1.6 Spectra obtained onboard CLUSTER with the magnetic sensors: FGM in the low-frequency range, STAFF-SC and SA for the high-frequency part, and with the electric sensors: EFW. With STAFF-SA, the spectra are computed onboard and transmitted with a lower resolution

modes was observed (weak turbulence). Further studies have been performed in the solar wind when Cluster orbit was in the right direction. Figure 1.7 shows the magnetic fluctuations spectra. It clearly evidences breakings of the spectrum at the ion scale and the electron scale [13]. In the range where k-filtering is possible, it shows that turbulence is perpendicular to the DC magnetic field and quasi-stationary, but not axisymmetric. It also evidences that it satisfies the kinetic Alfven mode dispersion relation. The smaller scale where electron dissipation occurs was not accessible because the spacecraft separation was too large.

Another important point to take into account is the information contained in the phases of the signal. It is well known in image processing that the phase is crucial in the existence of structures in a picture. Fourier phases contain more information on the coherence of signal than the amplitudes of the spectrum, but they are less tractable because of the folding of the values between 0 and 2π. One possible way of obtaining this information is to compare the original signal (O) with the surrogate signal obtained by reconstructing the signal from the original Fourier amplitudes and randomized phases (R), or with a coherent signal built in the same way but with a coherent phase (C). To know whether the strong fluctuations in the signal are due to phase coherence or not, the three signals are not compared directly but through their structure functions which are defined as

$$S(q, \tau) = \left\langle \left| x(t) - x(t + \tau) \right|^q \right\rangle,$$

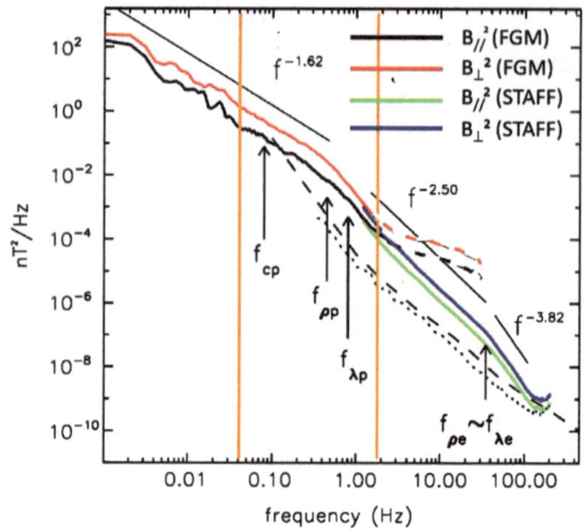

Fig. 1.7 Spectra obtained
from the FGM (for the
low-frequency range) and
STAFF (for the
high-frequency range)
instruments. The ion and
electrons typical frequencies
are calculated using the
Doppler shift. The range
where k-filtering is possible is
limited by vertical lines

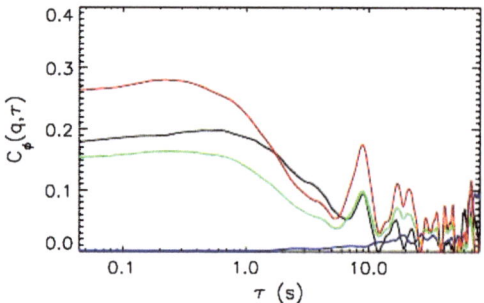

Fig. 1.8 Coherence indices
obtained for magnetic
fluctuations observed in the
magnetosheath. The different
colors refer to different values
of q ($1 = black$, $2 = blue$,
$3 = green$, $4 = red$)

where the angular brackets denote a time average. Then a coherence index is computed:

$$C(q,\tau) = \left(\frac{|S_O(q,\tau) - S_R(q,\tau)|}{|S_O(q,\tau) - S_R(q,\tau)| + |S_O(q,\tau) - S_C(q,\tau)|} \right)^{1/q}.$$

When the original signal has a similar structure function as the randomized signal, the index goes to 0, and when it is similar to structure function of the coherent signal, it goes to 1. Figure 1.8 [12] shows the coherence indices obtained for magnetic fluctuations observed in the magnetosheath. The case $q = 2$ is special because $C(2,t)$ should be exactly equal to 0 since the different signals have the same Fourier amplitudes, and it is therefore an estimate of the quality of the procedure. We clearly observe a significant coherence at small scales and a drop at large scales. This suggests the presence of coherent structures at scales that are of the order of the proton gyroradius, which are likely to be mirror structures.

As a conclusion, we can say that the Cluster four point measurements, together with the development of sophisticated signal processing tools, have allowed significant progresses in the understanding of turbulence in the solar wind and behind the earth bow shock. But there is a limitation in the scales that can be explored, which is due to separation of the spacecraft. Next concepts of missions should include multiscale separations to access electron and ion scales at the same time.

Acknowledgements   The author thanks Christian Mazelle, Gérard Belmont, and Fouad Sahraoui for their contribution to the content of this chapter and CNRS for the invitation to the school devoted on the topic "From solar environment to stellar environment" held in Roscoff (F).

References

1. Attié, D., Rezeau, L., Belmont, G., Cornilleau-Wehrlin, N., Lucek, E.: Power of magnetopause low-frequency waves: a statistical study. J. Geophys. Res. **113**, A07213 (2008)
2. Belmont, G., Sahraoui, F., Rezeau, L.: Measuring and understanding space turbulence. Adv. Space Res. **37**(8), 1503–1515 (2006)
3. Cornilleau-Wehrlin, N., Chanteur, G., Perraut, S., Rezeau, L., Robert, P., Roux, A., de Villedary, C., Canu, P., Maksimovic, M., de Conchy, Y., Hubert, D., Lacombe, C., Lefeuvre, F., Parrot, M., Pinçon, J.L., Décréau, P.M.E., Harvey, C.C., Louarn, Ph., Kofman, W., Santolik, O., Alleyne, H.St.C., Roth, M. (STAFF team): Ann. Geophys. **21**, 437–456 (2003). First results obtained by the Cluster STAFF experiment
4. Lembège, B., Savoini, P.: Nonstationarity of a two-dimensional quasiperpendicular supercritical collisionless shock by self-reformation. Phys. Fluids **4**, 11 (1992)
5. Mazelle, C., Lembège, B., Morgenthaler, A., Meziane, K., Horbury, T.S., Genot, V., Lucek, E.A., Dandouras, I.: Self-reformation of the quasi-perpendicular shock: CLUSTER observations. In: Twelfth International Solar Wind Conference, Saint-Malo, France. AIP Conf. Proc., vol. 1216 (2009)
6. Paschmann, G., Daly, P.W.: Multi-spacecraft analysis methods revisited. ISSI Scientific Report SR-008 (2008)
7. Pinçon, J.L., Lefeuvre, F.: The application of the generalized Capon method to the analysis of a turbulent field in space plasma: experimental constraints. J. Atmos. Terr. Phys. **54**, 1237–1247 (1992)
8. Rezeau, L., Perraut, S., Roux, A.: Electromagnetic fluctuations in the vicinity of the magnetopause. Geophys. Res. Lett. **13**, 1093–1096 (1986)
9. Rezeau, L., Morane, A., Perraut, S., Roux, A., Schmidt, R.: Characterization of Alfvenic fluctuations in the magnetopause boundary layer. J. Geophys. Res. **94**(A1), 101–110 (1989)
10. Sahraoui, F., Pinçon, J.L., Belmont, G., Rezeau, L., Cornilleau-Wehrlin, N., Robert, P., Mellul, L., Bosqued, J.M., Balogh, A., Canu, P., Chanteur, G.: ULF wave identification in the magnetosheath: k-filtering technique applied to cluster II data. J. Geophys. Res. **108**(A9), 1335 (2003)
11. Sahraoui, F., Belmont, G., Pinçon, J.L., Rezeau, L., Balogh, A., Robert, P., Cornilleau-Wehrlin, N.: Magnetic turbulent spectra in the magnetosheath: new insights. Ann. Geophys. **22**, 2283–2288 (2004)
12. Sahraoui, F.: Diagnosis of magnetic structures and intermittency in space-plasma turbulence using the technique of surrogate data. Phys. Rev. E **78**, 026402 (2008)
13. Sahraoui, F., Goldstein, M.L., Robert, P., Khotyaintsev, Y.V.: Evidence of a cascade and dissipation of solar-wind turbulence at the electron gyroscale. Phys. Rev. Lett. **102**, 231102 (2009)
14. Vignes, D., Mazelle, C., Rème, H., Acuña, M.H., Connerney, J.E.P., Lin, R.P., Mitchell, D.L., Cloutier, P., Crider, D.H., Ness, N.F.: The solar wind interaction with Mars: location and shapes of the BS and the MPB from the observations of the MAG/ER experiment onboard MGS. Geophys. Res. Lett. **27**, 49 (2000)

Chapter 2
Nature and Variability of Plasmas Ejected by the Sun

J.-C. Vial

Abstract The Sun not only emits radiation in the whole electromagnetic spectrum but also sends in the interplanetary medium plasmas of different natures (energy, continuous, or episodic flows, etc.) which contribute to its (small) mass loss. The escaping material when properly oriented may impact on the Earth magnetic environment with cascading effects on the Earth atmosphere. The continuous flow known as the solar wind is actually made of two categories, slow and fast winds. We discuss their properties, sources, and the mechanisms at work through the two types of models (fluid and particles). We describe the sporadic mass losses for the three main typical events: flares, prominence ejection, and coronal mass ejection. We discuss a possible unifying scenario which takes into account these three manifestations of magnetic disruption. We also extend the investigation to the whole heliosphere. Our conclusion proposes a few goals concerning the diagnostic and the understanding of the plasmas ejected by the Sun, along with the space missions which could provide some answers.

2.1 Introduction

We first recall some properties of the Sun from its internal structure to the outer heliosphere. We provide some basic parameters of the outer atmosphere which allow us to define a quiet Sun coronal model. In the second part, we derive that the corona cannot be in hydrostatic equilibrium and that necessarily a wind blows. In the third part, we establish the Sun permanent loss rate and characterize the two kinds of solar wind. We also raise the (open) issue of the "sources" of the fast and slow winds. In the fourth part, we compare the pros and cons of the fluid models with the kinetic/exospheric models. In the fifth part, we treat the activity-related solar plasma losses, an issue with has a strong impact on Space Weather. We show how flares, eruptive prominences (EPs), and Coronal Mass Ejections (CMEs) are closely related phenomena whose extension concerns the whole heliosphere, which

J.-C. Vial (✉)
Institut d'Astrophysique Spatiale, CNRS/Université Paris XI, Bâtiment 121, 91405 Orsay Cédex, France
e-mail: jean-claude.vial@ias.u-psud.fr

J.-P. Rozelot, C. Neiner (eds.), *The Environments of the Sun and the Stars*,
Lecture Notes in Physics 857,
DOI 10.1007/978-3-642-30648-8_2, © Springer-Verlag Berlin Heidelberg 2013

Fig. 2.1 Cartoon showing a
cut into the solar atmosphere

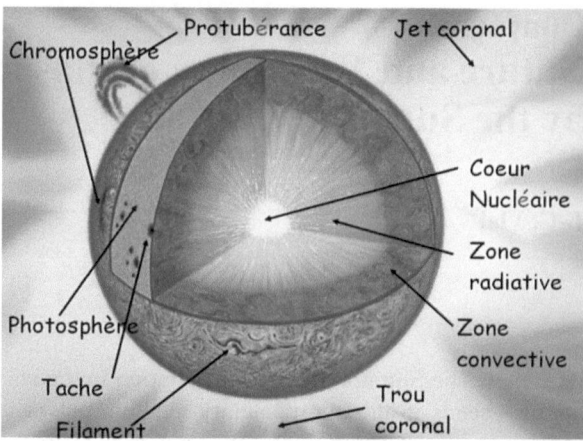

Table 2.1 Main parameters allowing the identification of the Sun

Parameter	Value
Age	4.5 Gy
Radius	696 000 km
Mass	1.99×10^{30} kg
Composition	90 % H and 10 % He
Average density	1 410 kg m^{-3}
Surface gravity	274 m s^{-2}
Escape velocity	618 km s^{-1}
Luminosity	3.9×10^{23} kW
Surface temperature	5780 K
Color temperature	6200 K
Sideral rotation period	25 d at equator; 31 d at poles, which means a differential rotation
Average mass loss	10^9 kg s^{-1} or 10^{-14} Ms year^{-1}

is shortly described in the sixth part. We finally conclude in addressing the current
and future work on the above-mentioned issues.

2.2 Some Properties of the Sun

The properties of this G star are summarized in Table 2.1.

Its general structuring is shown in Fig. 2.1, where one easily distinguishes the
three internal layers below the surface: the core where thermonuclear reactions take
place, the radiative zone where energy is transported through γ photons, and the
convective zone where convection transports the blocked energy. The layers above

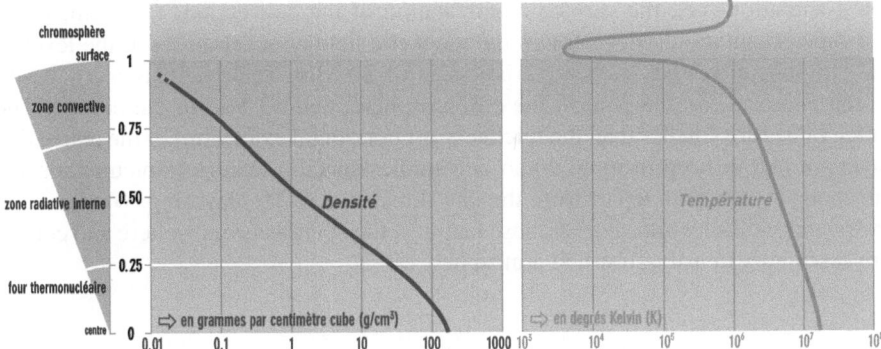

Fig. 2.2 The two figures display the run of log(density) (*left-hand side*, in g cm^{-3}) and temperature (*right-hand side*, in K) versus (*in ordinates*) the distance to Sun center normalized to the solar radius. The temperature variation has been extended above the surface in order to show the high coronal temperature

Fig. 2.3 Close-up of the variation of temperature and density below and above the solar surface. Courtesy E. Marsch

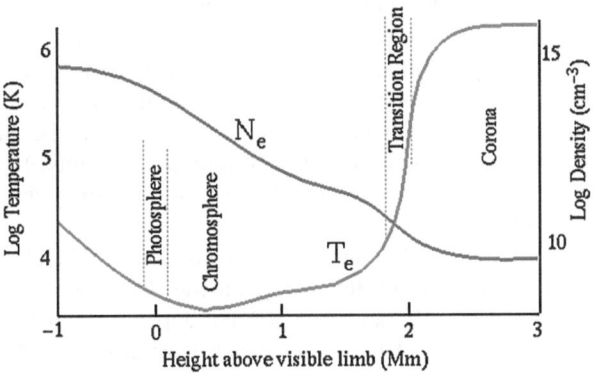

are the photosphere where the visible photons come from, the rather inhomogeneous chromosphere, and a strongly heterogeneous corona.

Average density and temperature values are displayed in Fig. 2.2, essentially in the solar interior. Huge decreases of density and temperature from the solar core to the surface are noticeable, but an increase of temperature has been sketched above the surface. Actually, a close look at the regions slightly below and above the surface (Fig. 2.3) clearly shows a strong temperature jump (of the order of 10^6 K) from the top of the chromosphere to an outer layer located a few hundred km above. This jump characterizes a region called Chromosphere Corona Transition Region or CCTR. This is the well-known coronal paradox which raises the issue of the stealthy heating of the corona.

Before returning to this issue, it is useful to provide a few figures concerning the spatio-temporal scales which are relevant in the outer atmosphere. We know that the solar radius is 700 000 km hereafter expressed as 700 Mm. The typical size of the granulation (i.e., the surface manifestation of the convection) is 1 Mm, while the supergranulation (which traces the magnetic field concentrations at the level of the chromosphere) has a characteristic size of 30 Mm. Scale heights are 0.1 Mm in the photosphere, 0.3 Mm in the chromosphere, and 50 Mm in the corona. The latter values are smaller than the thickness (or horizontal extension) of the respective layers, a fact to keep in mind when one studies specific features (spicules, fibrils, sunspots, etc.) which depart from the one-dimensional (1D) layering: whatever are the ranges, altitudes and sizes of these features, the overall chromosphere and corona are essentially density-stratified atmospheres.

2.2.1 An Unexplained Solar Region: The Corona

The plasma regime of solar material is essentially collisional at the exception of the corona, which expands into the interplanetary medium. This regime is often characterized by the Knudsen parameter (K = ratio of the mean free path (mfp) to the scale height). The electron mfp in the corona ranges from 5 to 500 km depending on the electron density (10^{16}–10^{14} m^{-3} or 10^{10}–10^8 cm^{-3}) of the region and altitude. Since the coronal scale height is about 50 Mm (in the low corona), $K \ll 1$ in the low corona. However, this parameter should be taken with much caution since it assumes a fully ionized atmosphere, which may not be the case in the very low corona where material can be partially ionized. Actually, the frontier between collision-dominated and noncollisional plasmas in the high chromosphere and low corona is still debated.

  Other physical quantities in the corona are: the electron–ion collision frequency (from 7 to 700 Hz depending on the density), the electron cyclotron frequency (between 3×10^6 and 3×10^8 Hz depending on the magnitude of the magnetic field B: 10^{-4} to 10^{-2} T), electron thermal speed (3900 km s^{-1}), the sound speed (166 km s^{-1}), the Alfvén speed (200 to 2000 km s^{-1}, depending on the magnetic field and density).

  In most of the corona (at the exception of prominences), the photon mean free paths, although wavelength dependent, are much larger than the scale height in the corona. From a comparison (Table 2.2) of the convective, thermal, and magnetic energy densities, at the bottom of the convection zone and at the photosphere, it can be seen that the magnetic energy emerges as a major parameter at the surface and in the outer atmosphere. More precisely, the respective thermal and magnetic parameters can be compared in the photosphere, the chromosphere, and the corona (Table 2.3), along with the associated energies and their thermal to magnetic ratio, the β parameter. It can be seen that for the chromosphere and the corona, a large range of values of the magnetic field is provided.

  Actually, the magnetic field is more and more heterogeneous in the outer atmosphere from both spatial and temporal standpoints: one finds open (especially at the poles) versus closed (in active regions) fields; and the magnetic field configuration and magnitude change from a period of minimum to maximum activity, as

Table 2.2 A comparison of kinetic, thermal, and magnetic energies at two locations in the solar sphere: bottom of the convective zone and surface. From [34]

Location	$0.7\,R_\odot$	$1.0\,R_\odot$
$\rho v^2/2$	5×10^5	1.5×10^2
$\rho k_g T/\mu$	7×10^{12}	1.5×10^4
$B^2/2\mu_0$	4×10^7 (presumed)	$(0.4\text{--}4) \times 10^4$ (measured)

Table 2.3 Variation with altitude (in fraction of the solar radius) of the temperature, scale height, sound speed, magnetic field, and plasma β. From [34]

	Photosphere	Upper chromosphere	Lower corona	Corona
Height ($R_\odot$)	0.0	$(2\text{--}5) \times 10^{-3}$	$10^{-2}\text{--}10^{-1}$	10^{-1}
Temperature T (K)	6×10^3	10^4	10^6	10^6
Scale height (m)	1.5×10^5	5×10^5	5×10^7	10^8
Sound speed (m s^{-1})	0.8×10^4	1.2×10^4	1.5×10^5	1.5×10^5
Magnetic field amplitude (T)	0.1 (strong)	$(2\text{--}10) \times 10^{-4}$?	$(2\text{--}10) \times 10^{-4}$?	10^{-4}?
Ratio of pressure to magnetic forces	~ 1	~ 1	<1	<1

Fig. 2.4 The real low solar corona, quiet (*left*) and active (*right*) as seen by the EIT imager on SOHO. *Left image*: the quiet Sun at the minimum of activity. Note the general poloidal field structuring. *Right image*: the active Sun at the maximum of activity. Note the general torodoidal fields structuring the plasma

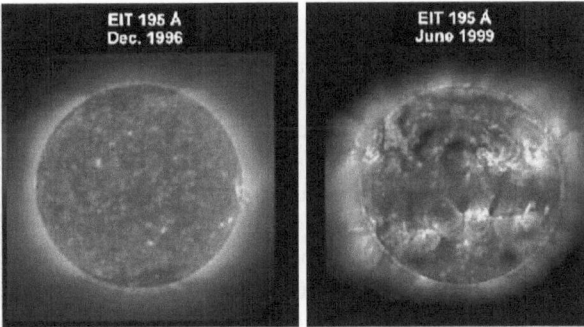

evidenced in the EUV images of Fig. 2.4. In a period of minimum of activity, the magnetic field is radial at the pole (location of decreased density, regions called coronal holes), and its overall structure is poloidal (left). Other areas have locally closed field lines. In a period of maximum of activity (right), one sees a number of local dipoles above and below the equator.

With the concept of large variations of the magnetic field, in terms of space (different structures) and time (solar activity), we now return to the radial variation of the β parameter (Fig. 2.5). The figure not only displays very large spatial (in the horizontal plane) and temporal variations of the plasma β, but evidences a systematic radial trend: between the convective zone and the high corona (where β is lower than 1), there is a regime in the high chromosphere and the low corona where β is much lower than 1. This means that the magnetic field "freezes" the

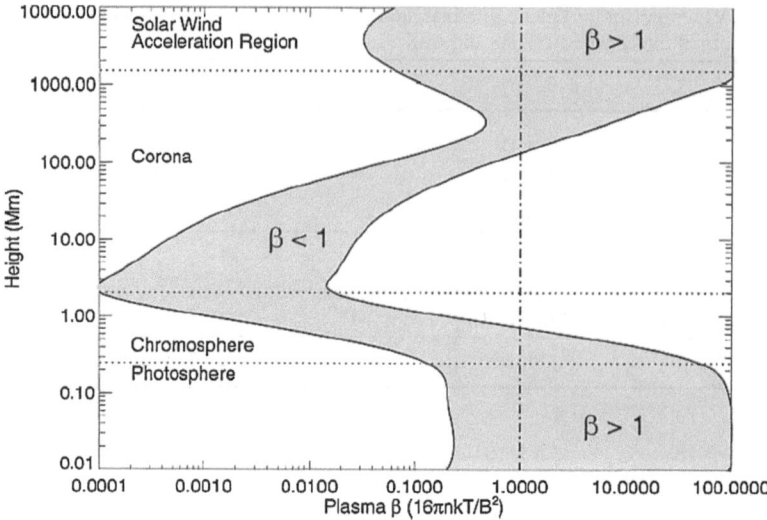

Fig. 2.5 Range of variations of the plasma β (*abscissa*) versus height in the solar atmosphere (*ordinate*). Note that at a given altitude the parameter may vary by more than two orders of magnitude, depending on the structure. The *left and right hand-side curves* correspond to a magnetic field of 0.25 and 0.01 T, respectively ([3] and Courtesy of G. Allen Gary)

plasma, contrary to what happens in the other regions. The situation is also complicated by the fact that the plasma is completely neutral in the lower layers (including the chromosphere), then partially ionized in the transition region between chromosphere and corona (CCTR) to become totally ionized in the corona. It also changes from a very collisional nature in the deep layers (up to the chromosphere) to a noncollisionality in the corona. This change occurs in the regions where β is lower than 1 and the plasma is partially ionized. This means that the plasma is not only dependent of the solar structure studied (we will not go through the whole solar zoo) but also on the degree of filamentation of the plasma which is thought to be beyond the limit of present observational capabilities (about 0.1 arcsec or 70 km at the Sun). For instance, it could happen that a plasma thought to be fully ionized and noncollisional is actually concentrated in small patches where densities are high enough to allow partial ionization and high collisionality. Until now, we focused on the "low corona" (say a hundred Mm above the solar surface) which is accessible to sophisticated remote-sensing techniques (X-ray, EUV, UV imaging, and spectroscopy). Farther out (up to a few solar radii), white-light coronagraphy and UV spectro-coronagraphy (on SOHO and STEREO) have provided a wealth of information, not only on the electron density but also on the electron, proton, and other species temperatures parallel and perpendicular to the magnetic field, to be discussed later.

From these various data, one can try to define a quiet Sun corona (Fig. 2.6, which displays average values of the electron density and temperature up to 10 solar radii

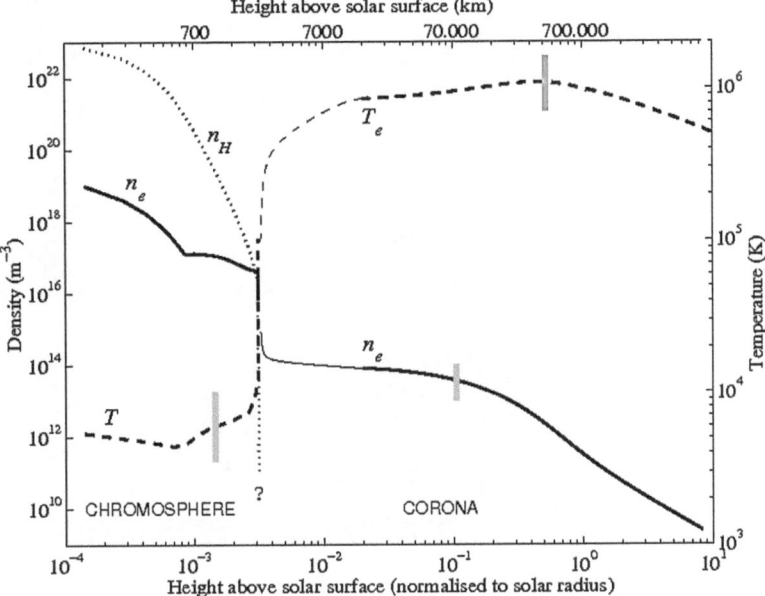

Fig. 2.6 A model of the quiet Sun corona extending up to about 10 solar radii. The altitude marked by an abrupt decrease of density and increase of temperature corresponds to the highly heterogeneous CCTR. From [35]

above the solar surface). The main feature of this figure is the increase of (electron) temperature with altitude up to about 2×10^6 K, some plateau, and a decrease above about one solar radius. With the advent of the STEREO mission, the two Heliospheric Imagers (HI) have been providing images much farther, which allow one to detect moving features in their fields of view. Above about 60 solar radii, the essential information comes from in situ measurements which directly provide most physical quantities (including the full magnetic field vector) but, obviously, in very localized space-time domains, much smaller than the size of the expanding solar structures.

As far as the active corona is concerned, it is still more difficult to build an average model. An example is given by Fig. 2.7 (see also Fig. 2.4, right), where eclipse and coronagraphic images, obtained in the increased phase of activity of Cycle 23, have been adjusted and show radial streamers all around the surface along with a CME in the North-East quadrant (recall that East is on the left). The loci of increased intensity correspond to increased electron density (along the line-of-sight) and also trace the magnetic field lines. These lines are mostly open above about 5 solar radii and consequently suggest the presence of outward flows.

Fig. 2.7 Composite of two
white-light images obtained
on 11 August 1999: the
internal one during an eclipse,
the external one with
C2/LASCO/SOHO. Courtesy
S. Koutchmy (Institut
d'Astrophysique de Paris (F))

2.2.2 A Simple Derivation of the Coronal Temperature

We first write the energy balance equation between the radiative losses W_R and the
conductive flux with the standard notation (q is the conduction factor):

$$\frac{q}{r^2}\frac{d}{dr}\left(r^2 T^{5/2}\frac{dT}{dr}\right) = W_R.$$ (2.1)

W_R is given by the equation

$$W_R = n^2 F(T) W\ m^{-3}$$ (2.2)

with n the density and the $F(T)$ variation of radiative losses with temperature as
shown in Fig. 2.8, easily parameterized with a set of power laws.

Since the radiative losses are proportional to the square of the density which
strongly decreases with altitude (as $n \propto R^{-1.96}$, [17]), they can be considered as
negligible. Equation (2.1) becomes

$$\frac{d}{dr}\left(r^2 T^{5/2}\frac{dT}{dr}\right) = 0,$$ (2.3)

which becomes the equation

$$\frac{d}{dr}\left(T^{5/2}\right) = \frac{\text{Constant}}{r^2}.$$ (2.4)

With the reasonable assumption that the temperature goes to 0 at infinity,
Eq. (2.4) is easily integrated into the equation

$$T \propto r^{-2/7} \quad \text{or} \quad T(r) = T_{R_0}\left(\frac{R_0}{r}\right)^{2/7},$$ (2.5)

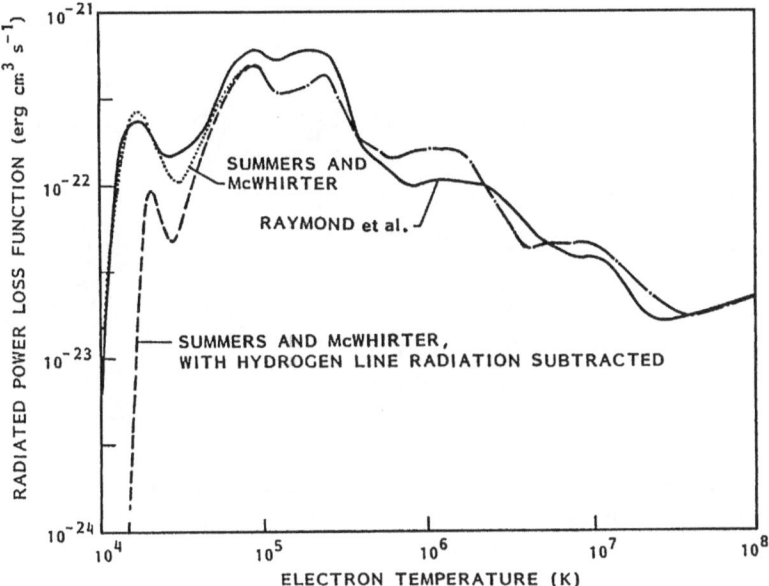

Fig. 2.8 Radiative loss function (in log) vs. temperature (in log) from various authors. The different behaviors at relatively low temperatures (*continuous* vs. *broken lines*) show the importance of H losses in the CCTR and the low corona

where R_0 is the solar radius.

We can now see if hydrostatic equilibrium stands. We write the hydrostatic equilibrium as

$$\frac{dp}{dr} = -GM_0\frac{\rho}{r^2}, \tag{2.6}$$

where G is the gravity constant, M_0 the mass of the Sun, and p the pressure which also follows the gas law $p = 2nkT$.

Then Eq. (2.5) allows us to derive the pressure as a function of radial distance:

$$p = p_o \exp\left(7GM_0r_0\left(\left(\frac{R_o}{r}\right)^{5/7} - 1\right)\Big/10p_oR_o\right). \tag{2.7}$$

All quantities with subscript $_o$ are taken at the solar surface.

When the distance increases to infinity, the pressure goes to a finite value of the order of 10^{-7} Pa. Such a value is higher than the pressure in the interstellar pressure (10^{-13} Pa) by 6 orders of magnitude! One can conclude that the corona cannot be static: as will be shown in Sect. 2.3, material is flowing outwards.

2.3 The Sun and Its Permanent Loss Rate: The Solar Winds

So a wind blows (at least one), a fact which was actually predicted by Biermann as early as 1951 and confirmed by Parker [23] and finally detected/measured with

Fig. 2.9 An image of the
Hale-Bopp comet where one
clearly identifies two tails: the
curved one made of dust and
the (*blue*) straight one which
is aligned with the comet
coma along the solar
direction. Its shape and nature
(ions) are determined by the
blowing solar wind

Mariner 2 in 1962. It was identified as the force acting on comet tails (Fig. 2.9) by
Biermann, who wrote of a "solar corpuscular radiation" [6]. It corresponds to an
overall (average) mass loss of 10^9 kg s^{-1}.

2.3.1 The Two Kinds of Solar Wind

The two categories have been detected in situ at one AU or farther as in Fig. 2.10.
Above most solar regions, one finds a slow wind with a speed of about 400 km s^{-1}
while, mainly above the poles, one finds a fast solar wind of about 800 km s^{-1} (up to
1200 or even exceptionally more). This latter result has been beautifully confirmed
by the Ulysses probe which went above the poles (or quite) at a distance close to
2 UA. The results are shown in Fig. 2.10, where the wind speed is shown as a vector
in the plane perpendicular to the ecliptic. It can be noted that the fast wind is rather
steady while the slow wind is variable in time and latitude. It is now well established
that the fast wind comes from the above-mentioned low-density regions, the coronal
holes. Immediately, a question arises about the high speed and the temperature of
these regions.

2.3.2 A First Approach to the Wind Velocity

As shown by Meyer-Vernet [35, Sect. 5.1.1], in the simple adiabatic case, no wind
can be produced. On the contrary, in the isothermal case ($\gamma = 1$), it is easily shown
that at large distances, the material speed varies as

$$2c_s \sqrt{\mathrm{mod}(\ln r)}, \tag{2.8}$$

where c_s is the isothermal sound speed. Of course, this is a very crude approximation, but one can immediately conclude that since the sound speed is proportional to
$\sqrt{T}$, a faster flow should be emitted by a higher-temperature coronal plasma.

Fig. 2.10 Plot of the radial velocity of the solar wind in a plane perpendicular to the ecliptic, superimposed on a composite made of a UV image from EIT/SOHO, a white-light image of the inner solar corona from Mauna Loa, and a white-light image of the corona from C2/LASCO/SOHO. The *blue* (*red*) *regions* correspond to inward (outward) magnetic field, respectively. The velocity has been measured in situ with the SWOOPS instrument aboard the Ulysses probe which was at a distance of 2 AU when above the poles of the Sun. From [20]

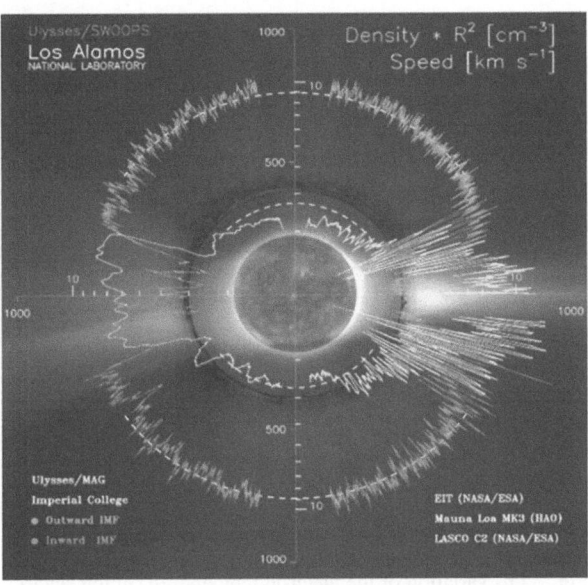

Only with the advent of SOHO and its two EUV-UV spectrometers, called SUMER and CDS, it has been possible to validate (or not) this result. The technique called for sophisticated atomic physics (the so-called Doppler dimming technique described in Sect. 2.3.3) and rather acrobatic measurements since it used two different spectrometers SUMER and CDS on SOHO and necessitated a roll of the spacecraft. The results, shown in Fig. 2.11, are nonequivoqual [8]: the electron temperature in the lower part of polar coronal holes is lower than in the "quiet corona" (i.e., all other closed magnetic regions at the same altitude)! This means that the simple isothermal model mentioned above is invalid and that, in order to build a valid one, one must start by identifying the solar structures and the altitude where the wind comes from.

2.3.3 The "Sources" of the Fast Wind

As far as the fast wind is concerned, the source has been identified as coronal holes, but this information is not sufficient for pinpointing the mechanism at work. The detection of velocities is usually made through the Doppler effect (spectral line shift), but when observing polar coronal holes, the projected velocity along the line-of-sight (LOS) is so small that the line shift is difficult to detect.

Another technique (the "Doppler dimming") is possible but can be worked out only at relatively high altitudes (see below). However the direct Doppler technique has been successfully applied to the lower boundaries of polar coronal holes where projection effects are minimized.

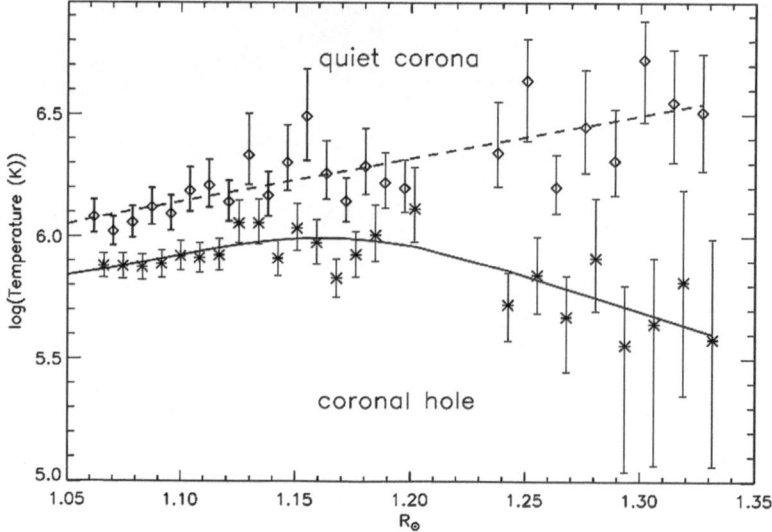

Fig. 2.11 Variation of the electron temperature (in log scale) with the altitude normalized to the solar radius [8]

Fig. 2.12 The green image was obtained in the Fe XII line at 19.5 nm with EIT/SOHO. It shows the coronal hole border where the SUMER profiles of Ne VIII (77 nm) have been obtained in and out the coronal hole. The blueshifts (coded in *blue*) correspond to outward flows, while the redshifts (coded in *red*) correspond to inward flows. Note that the strongest outflows in the coronal hole correspond to areas where contiguous chromospheric networks (delineated in *black*) converge [13]

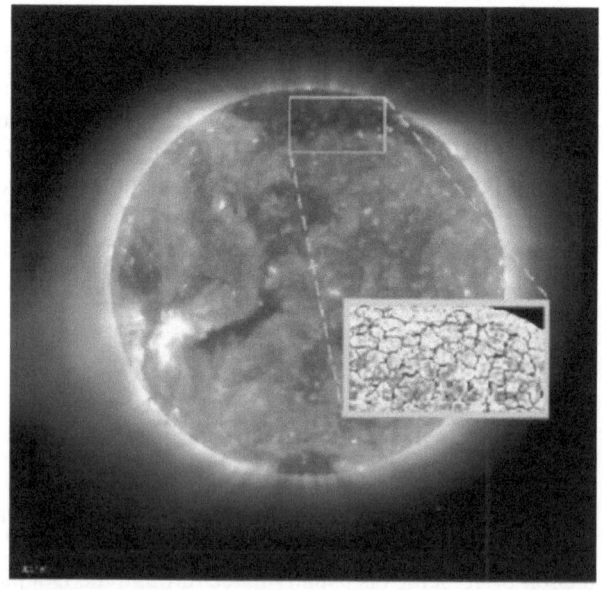

Using SUMER profiles of the UV Ne VIII line at 77 nm and formed at 6×10^5 K, Hassler et al. [13] were able to detect blueshifts of the order of a few km s^{-1} above the contours of the chromospheric network where operates what we called above the supergranulation (Fig. 2.12). So, at the 6×10^5 K level (high in the CCTR), we already have an outflow, but in the coronal hole, this ouflow is stronger at the

Fig. 2.13 EIT and eclipse images of the South Pole region where the spectroscopic SUMER observations took place. *Top*: EIT image in Fe IX/X at 17.1 nm, where polar plumes are well evidenced. *Bottom*: White-light (eclipse) image (courtesy J.-R. Gabryl, taken on 26/02 1998 at 18:33 UT). The SUMER slit is shown as a thin rectangle on both images: since it cuts the solar limb, the altitude is rigorously defined. From [24]

frontiers of the supergranulation pattern. It should be noted that lower values have been finally obtained by Wilhelm et al. [31].

Higher in the atmosphere, when observing out of the limb, one definitely loses any possibility of measuring radial flows with the Doppler method. Then, one can rely on the technique dubbed Doppler dimming initially developed by Noci [22] in the case where a chromospheric or CCTR line is resonantly scattered by remaining ions or atoms in the corona. It is then easy to understand that the coronal atoms and ions only fully absorb the incident radiation if it is not shifted by Doppler effect. When relatively strong radial flows are at work, these atoms or ions no longer see the incident line (or only the wings), and the scattered radiation decreases: this is why this effect is called "Doppler dimming."

This technique has been extensively used by the UVCS/SOHO Team and was developed in order to take into account the collisional contribution to the radiation, thereby allowing its use in the very low corona. However, it should be kept in mind that it requires, amongst others, the knowledge of the electron density. In a rather unique combined eclipse-SOHO observation (Fig. 2.13), Patsourakos and Vial [24] were able to derive the density from the eclipse white-light data, and from the ratio of the O VI doublet at 103.2 and 103.7 nm from SUMER they derived the radial velocity (Fig. 2.14). At 0.05 solar radius above the limb (or 35 Mm), they found 67 km s^{-1}, a rather important figure at such a low altitude. Moreover, their spectroscopic measurement took place in a "void" region between plumes, plumes being the radial regions of higher density found in coronal holes (Fig. 2.13). The authors could then claim that the fast wind originates from the "interplumes." The authors could also derive: $T_e = T_i = 0.9$ MK and $n_e = 1.8 \times 10^7$ cm^{-3} in this interplume region.

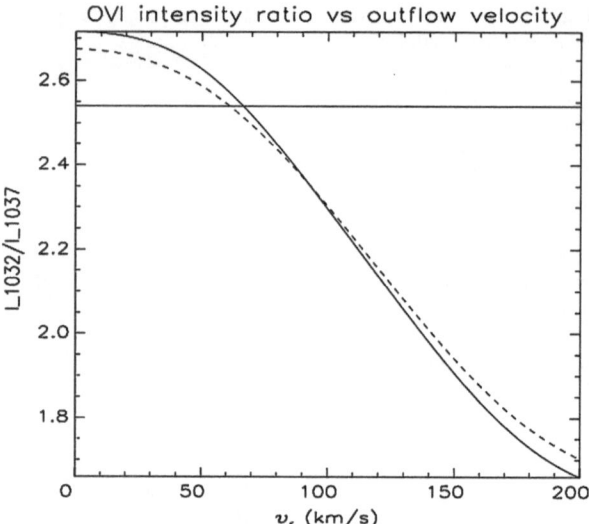

Fig. 2.14 Variation of the OVI 103.2 to 103.7 intensity ratio with the radial outflow velocity. The *horizontal line* corresponds to the measured value. From [24]

The issue of the plumes vs. interplumes origin of the fast wind is the subject of an ongoing hot debate (see [10]). Actually, the diagnostic is made difficult by line-of-sight (LOS) effects as discussed in [10], who suggest that some plumes could result of the superposition of network boundaries when seen out-of-limb. In view of the Hassler et al. results (outflows at network boundaries), this could explain why fast wind is also detected in plumes (for the most recent review, see [32]). Other sources are proposed such as small-scale ejecta and spicules of type II, now suspected to be the source of coronal heating [9]. Expanding "Funnels" have also been proposed by Tu et al. [28]: in these structures, high-frequency Alfvén waves (<10 kHz) would start in the chromosphere. A nice feature of this mechanism is that it could explain the "FIP" effect (overabundance of elements with First Ionisation Potential (or FIP) <10 eV) in the solar wind.

2.3.4 The "Sources" of the Slow Wind

As shown in Figs. 2.10 and 2.15, at the solar minimum of activity, the slow wind is confined in low-latitude regions with multipolar (consequently closed) magnetic field. So it has been suggested that the slow wind would be initiated from the boundary between the dominant coronal hole and the current sheet(s) resulting from the complex magnetic field [5, 26]. It is true that strong outflows have been found at the boundary between coronal hole and active region [4, 25], but "open" field lines could actually be long-range closed lines (as shown in Fig. 2.16 and in [7]). See also [12, 14]. Anyway, this is still an open question.

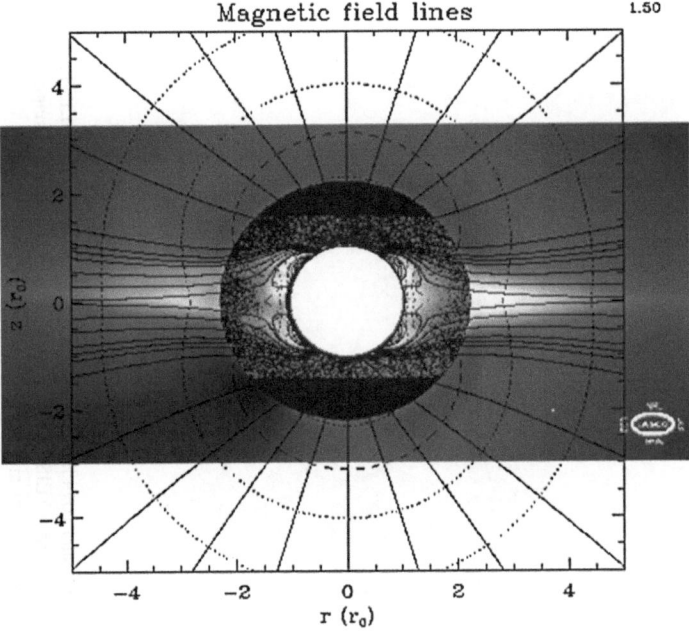

Fig. 2.15 On a montage of C1/C2 white-light images from LASCO/SOHO, magnetic field lines are superposed. The slow wind emanates from regions which are supposed to be closed fields [5]

2.4 Fluid Models of the Wind

These models are based upon a set of assumptions:

1. The thermal (electron) conductivity is given by the expression

$$\kappa \sim \frac{3}{2} n k_B v_{\text{th}} l_f, \tag{2.9}$$

where the symbols have the usual meaning (n density, k_B Boltzmann constant, v_{th} the electron speed, and l_f the mean free path). Equation (2.9) becomes

$$\kappa \sim 10^{-11} \times T^{5/2} \, \text{W} \, \text{m}^{-1} \, \text{K}^{-1}. \tag{2.10}$$

2. The mean free path (l_f) is much smaller than the plasma variation scale. This is obviously verified in the very low corona where $l_f \sim$ a few 100 km, but at 1 A.U. ($n \sim 5 \times 10^6$ m^{-3}; $T \sim 10^5$ K), $l_f \sim 1$ A.U.! Moreover, l_f varies as v^4, so $l_f(3v) \sim 100 l_f(v)$, which means that the assumption is no longer valid for velocities in the high wing of the distribution.

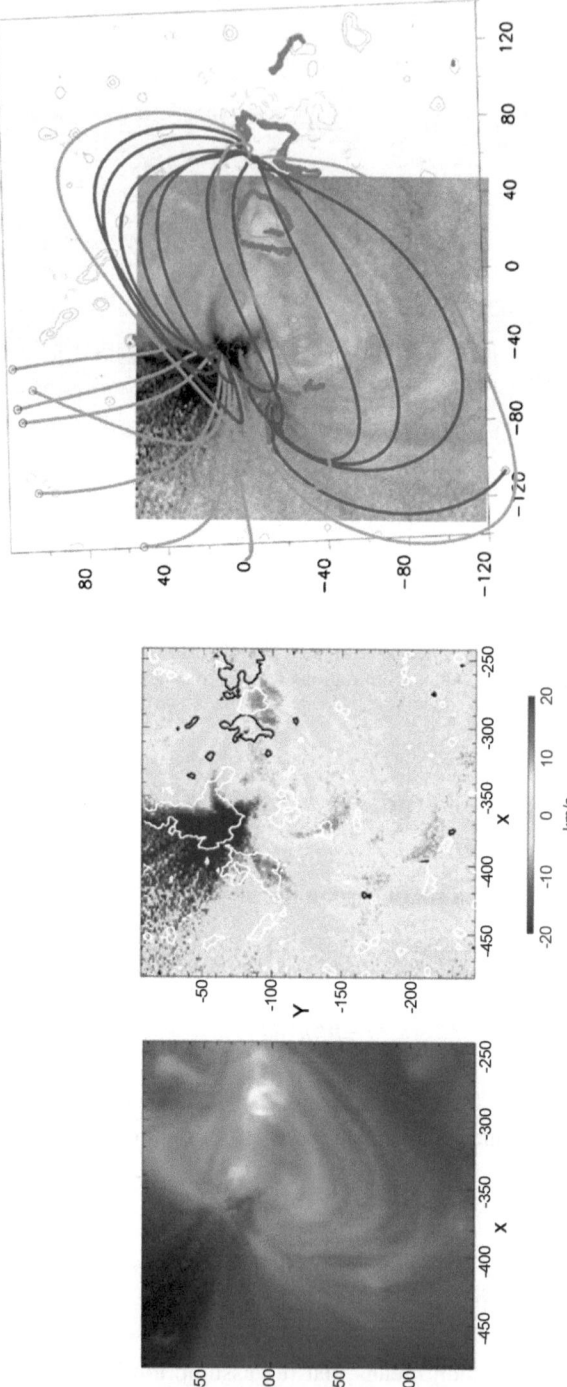

Fig. 2.16 Plasma flows and magnetic field lines in and between the two active regions AR10942 and 10943. *Left image*: FeXII 19.5 nm intensity (EIS/Hinode). *Middle image*: Velocities derived from FeXII Doppler shifts in the usual blue and red color convention (EIS/Hinode). *Right image*: extrapolated magnetic field over velocities field (FeXII Doppler shift with a black and white convention) [4]. All axis of these figures are in arcsecs

Fig. 2.17 *Top*: variation of the wind speed with altitude (in 10^3 Mm up to the Earth orbit) for different values of the temperature (from [23]). *Bottom*: wind velocity as measured with the so-called "blob" technique as a function of altitude (up to 25×10^3 Mm). (The technique consists in identifying and following pieces of plasma (or blobs) during their expansion) (from [27])

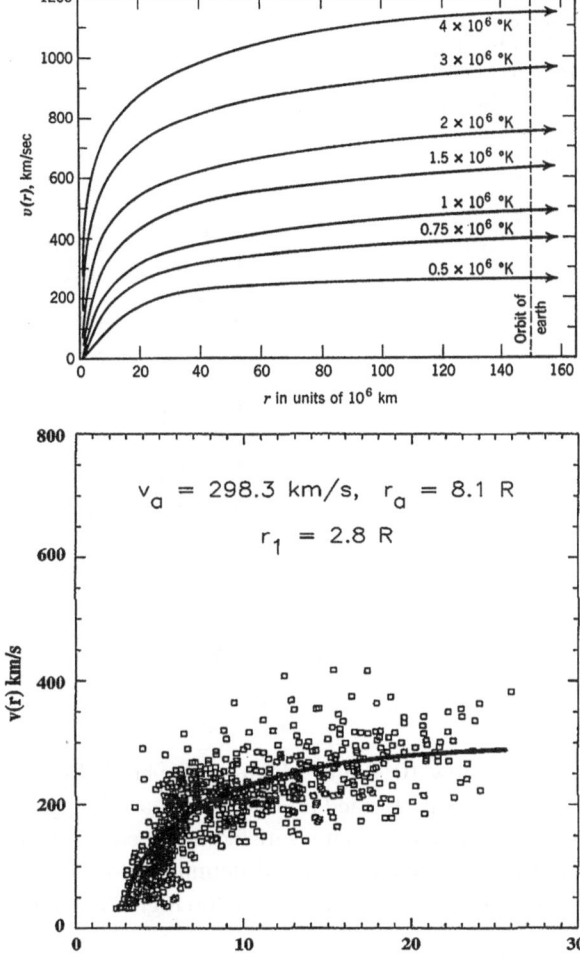

2.4.1 The Isothermal Parker Model

From the laws of conservation of mass and momentum one can derive the equation

$$(V/c_s)^2 - \ln(V/c_s)^2 = 4(\ln(r/r_c) + r_c/r) + \text{Constant},\qquad(2.11)$$

where c_s is the sound speed, and r_c the critical distance defined by $r_c = GM/(2c_s^2)$.

A positive velocity gradient $dV/dr > 0$ implies that for $r < r_c$, the flow is subsonic, while for $r > r_c$, it is supersonic. For the Sun, $c_s = 140$ km s^{-1}, $r_c = 4.5$ R, and the mass loss is found to be 1.6×10^9 kg s^{-1} (about the measured value).

The Parker solution can be compared to measurements made by Wang et al. [30] and Sheeley [27] from isolating plasma "blobs" in the slow wind and following their motions (Fig. 2.17). The overall agreement is rather good, at least in the observed

Fig. 2.18 The four sets of
solutions (I, II, III, IV) of the
(signed) Mach number as a
function of radial distance.
The stationary outflow
solution is marked in thick
line and is noted W

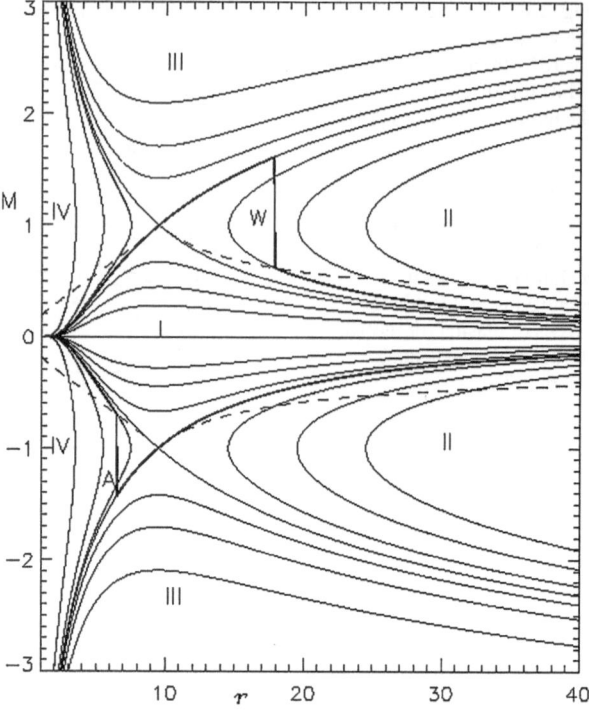

regions (below 10^3 Mm). Much farther but below one AU, the only (in situ) measurement was performed by Helios, and a velocity of about 300 km s^{-1} was found. However, the model is not satisfactory because of the low "constant" temperatures implied (less than 10^6 K) and definitely does not represent the fast wind, which would require, as we have seen, a too high temperature.

2.4.2 Solving the Bernoulli Equation [36]

The four sets of solutions (I, II, III, IV) in terms of the Mach number $M = V/c_s$ are shown in Fig. 2.18 as functions of radial distance normalized to the solar radius. Note that the graph also takes into account the accretion process (lower part of the figure where $V < 0$). If one adopts the right and unique pressure values at the lower boundary (the "surface") and at infinity, one has a stationary outflow solution (denoted W) with a subsonic breeze (which matches the observed velocity) until, above about 10 solar radii, the wind becomes supersonic. This supersonic flow leads to a shock below about 20 solar radii (which allows for a low terminal pressure). A thorough discussion why other solutions are discarded can be found in [35, Chap. 5].

Fig. 2.19 Distribution of the electron velocity. The parallel and perpendicular velocities are compared to the Maxwellian distribution. From [35] and courtesy I. Zouganelis

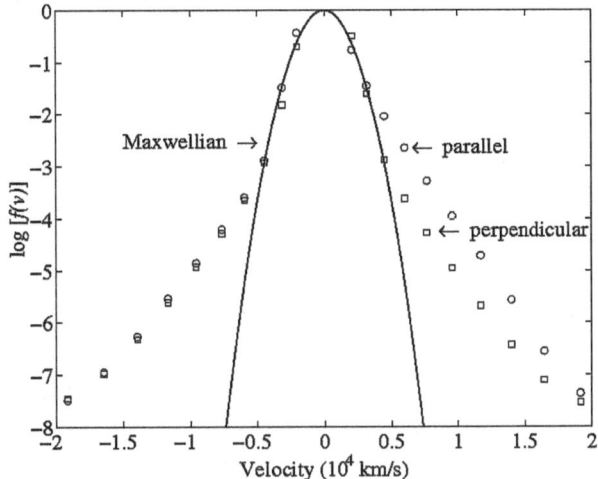

2.4.3 The Fast Wind and the Fluid Models

As we have seen, there is, for the fast wind, a temperature issue: in order to have a fast wind with a $T < 10^6$ K corona, one needs to deposit additional momentum in the flow. Several tricks have been proposed:

- Polytropic approximation with $\gamma < 5/3$, but why?
- Two-fluid model (in order to take into account the fact that $T_p > T_e$ as observed at increasing distances from remote-sensing UVCS/SOHO to in situ Ulysses measurements);
- Alfvén waves in a nonradial expansion geometry: with frequencies up to 10 kHz generated in the chromospheric network, it implies supersonic speeds in the very low corona.
- Ion gyroresonance (which couples incident high-frequency waves with ion gyration), derived from UVCS/SOHO line profiles measurements perpendicular to the open magnetic field and showing a strong anisotropy of line widths [16].
- And so on...

As said by N. Meyer-Vernet in her book (Basics in the Solar Wind), solar physicists have devised "more and more ingenious schemes reminiscent of the Ptolemaic system...". The main handicap of fluid models is that by definition they cannot take into account the fact that observed particles distributions are far from Maxwellian (or bi-Maxwellian), e.g., the electron distribution of Fig. 2.19 (courtesy I. Zouganelis) or the protons distributions found with the Helios mission in the fast wind at about 60 solar radii (Fig. 2.20). The proton distribution is not only non-Maxwellian but also highly anisotropic [18]. In view of these deficiencies, more and more physicists have turned to models which describe the velocity distributions and their various moments: density, mean velocity, etc. But can we speak of a Copernician revolution? See below.

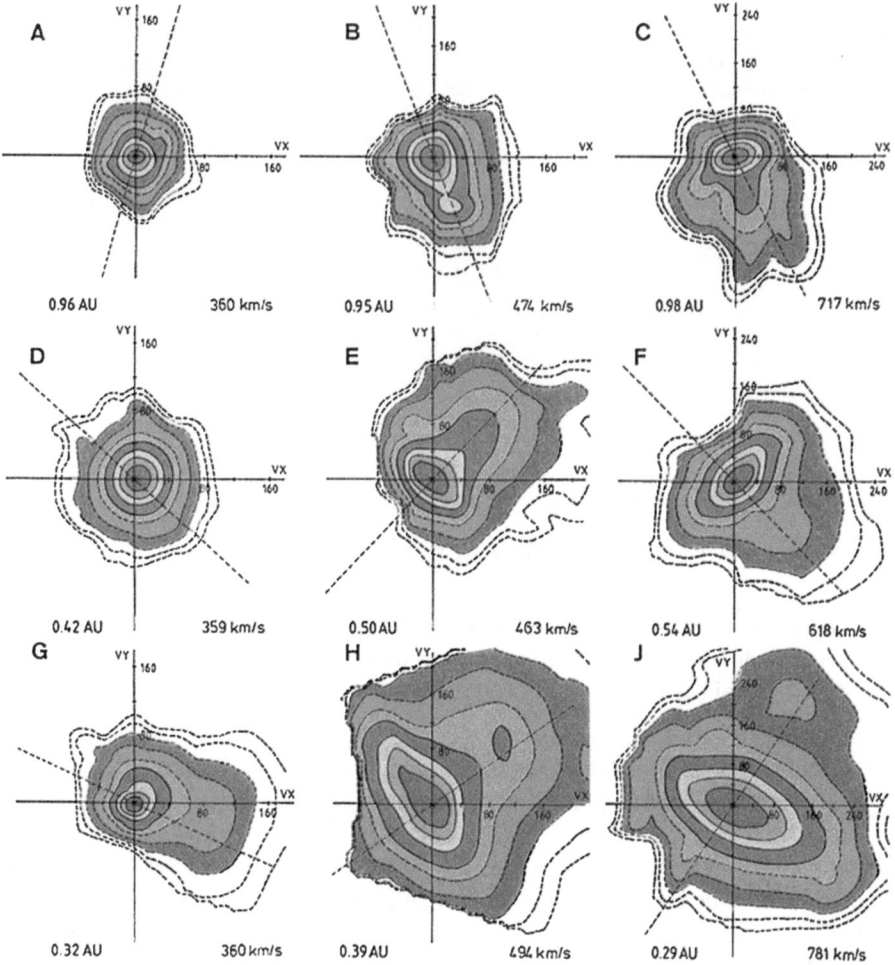

Fig. 2.20 *Top*: Angular distribution of protons as measured by Helios with variable distances from the Sun. The distance varies from 1 AU (*upper left-hand image* labeled **A**) to 0.29 AU (*lower right-hand image* labeled **J**). *Bottom*: Proton distribution at increasing distances from the Sun as measured by Helios. Note the strong asymmetry in the wings (from [18] and [35])

2.5 The Fast Wind and the Kinetic/Exospheric Models

2.5.1 The Kappa Distribution

The distribution is described by the equation

$$f_n(v) \propto \left[1 + \frac{v^2}{\kappa \, v_{\text{th}}^2} \right]^{-(n+1)} . \tag{2.12}$$

Fig. 2.20 (Continued)

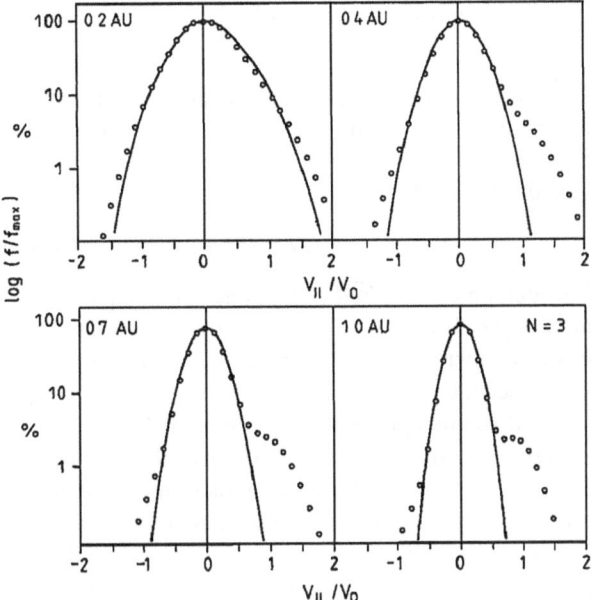

Fig. 2.21 The $\kappa = 3$ distribution as a function of v/v_{th} compared to Maxwellian and power law distributions. From [35]

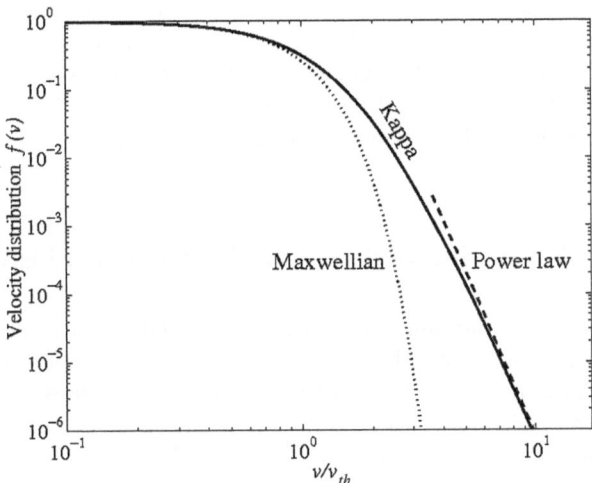

It is nearly Maxwellian at low speeds and decreases as a power law at high speed (Fig. 2.21). The power law part where energy accumulates at a rate proportional to itself reminds of the flare to nanoflare energy distribution. Could it be the same "nanoscale" phenomenon?

2.5.2 Exospheric Models

The electrons, being light, tend to concentrate in the outer coronal regions, contrary
to protons which, being heavy, drift into the inner regions. This leads to the formation of an electrostatic potential $\Phi_E(r)$. Expressing the total electron energy, one
gets the equation

$$M_e v^2/2 - e\Phi_E(r) = \text{Constant.} \tag{2.13}$$

Introducing $v_E = \sqrt{2e\Phi_E/m_e}$, one can compare v and v_E. When $v < v_E$, electrons
are trapped. If $v > v_E$, electrons move outward to infinity. But an excess of escaping
electrons increases the potential which rises the electric field which increases the
wind speed, etc.

So it seems appropriate to improve the model, e.g., introducing the invariants
of motion leads to a very small cone of escaping electrons. One can also include
collisions and wave–particle interactions. Actually, building a full model requires
putting the boundary conditions in the chromosphere where ... radiative losses are
important, the heat conductivity is not very well known, the magnetic field energy
dominates (plasma $\beta \ll 1$), but where the magnetic field is not measured, where the
geometry (including the introduction of a filling factor) is complex and the temporal
variations important. One can easily imagine that it is a formidable task for solar
physicists.

Finally, one can compare the situation of our Sun and solar-type stars where hot
coronae are associated to diluted, hot and fast winds to giant ($T = 10^5$ K and $v =$
200 km s^{-1}) and supergiant stars ($T = 10^4$ K and $v = 10$ km s^{-1}), where winds are
denser and cooler. Broadly speaking, the "Parker" law is satisfied. Radiation-driven
winds are of a completely different nature and out of the scope of this presentation.

2.6 The Sun and Its Activity-Related Plasma Losses

Not only the Sun loses mass in a continuous way but also episodically with three
types of events: flares, erupting prominences (EP), and Coronal Mass Ejections
(CME), see e.g. [38] and [41]. The three features are shown in Figs. 2.22, 2.23,
and 2.24, respectively. A flare (Fig. 2.22) can be defined as a strong brightness en-
hancement in all wavelengths, especially in the UV and X-rays, which means that
the plasma reaches very high temperatures (a few 10^7 K). An eruption of prominence is the lift-up of cool material (less than 10^4 K), which sometimes, but not always, leaves definitely the solar corona where it was initially embedded (Fig. 2.23).
A CME is the ejection of cool and hot material toward the interplanetary medium
and can only be observed with a coronagraph, very often through a technique of
subtracting two consecutive images (Fig. 2.24). Flares, EPs, and CMEs all imply
large energies (radiative, thermal, etc.), but what about their mass losses, which is
the today topic?

Fig. 2.22 Flare observed in the EUV by TRACE

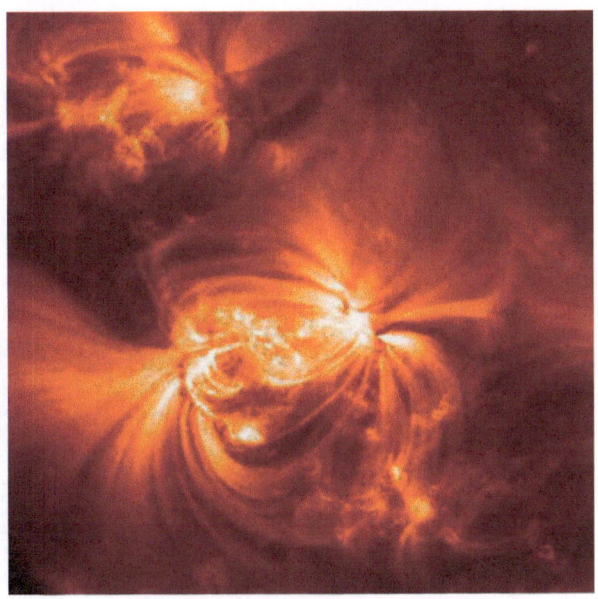

Fig. 2.23 Eruption of a prominence (EP) observed in the He II line 30.4 nm (formed at about 70 000 K) by EIT/SOHO

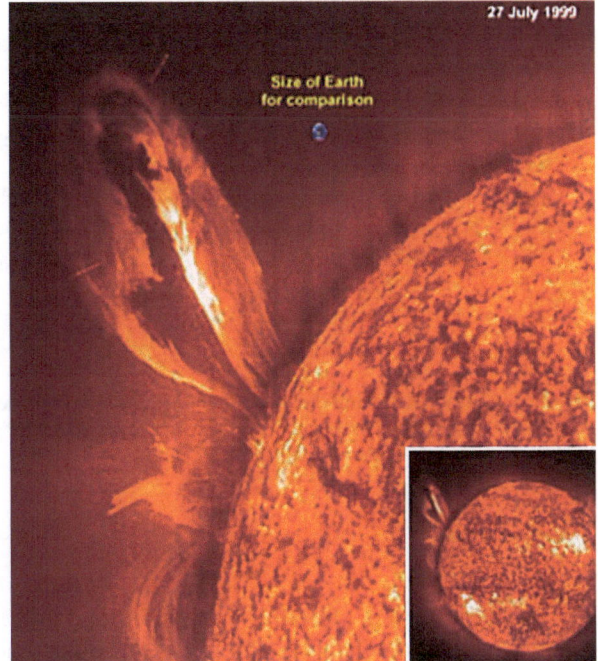

Fig. 2.24 Coronal Mass
Ejection (CME) observed by
the C3 coronagraph of SOHO
at about ten solar radii. Note
the "tennis racket" shape of
the CME

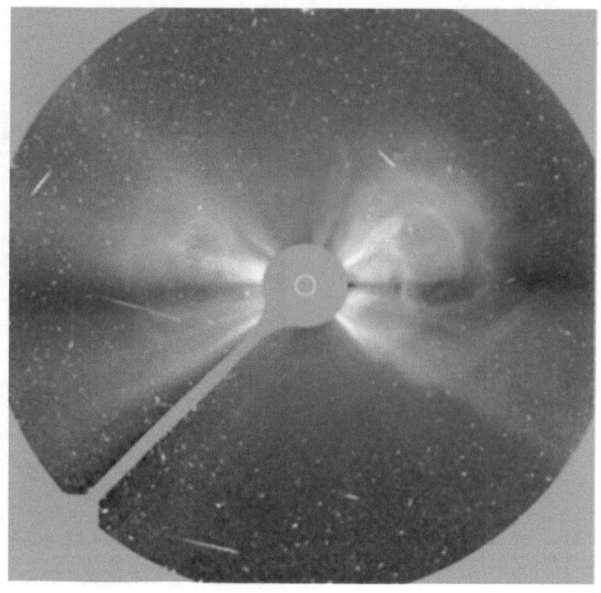

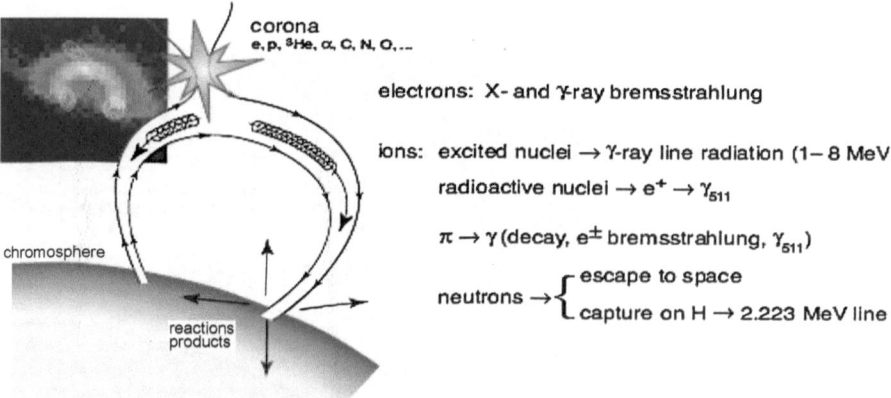

Fig. 2.25 Cartoon depicting the three main features of a flare: acceleration of particles, precipitation on the surface (and associated heating), and escape of SEP cartoon taken from Fig. 2 of the SMESE proposal [29] and adapted from [21] and [19]

2.6.1 Flares

The energy release is typically of the order of 10^{26} J, an energy which goes into heating, particle acceleration, the release of Solar Energetic Particles (SEP), and (sometimes) a CME. Let us discuss the cartoon of Fig. 2.25. The main ingredients of the common flare scenario can be found: some disruption (magnetic reconnection?) occurs in a loop-like magnetic structure, and particles are accelerated

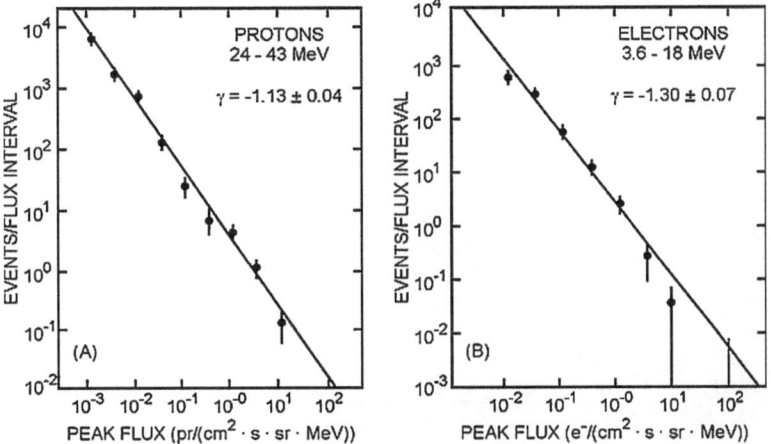

Fig. 2.26 Distributions of (IMP-8) SEP peak intensities of $24 < E < 43$ MeV proton events (*left*) and $E > 3$ MeV electron events (*right*). At higher energies (not shown), the slope of the power law increases in modulus. Taken from Fig. 19 of [15]

(magnetic reconnection?) at high energies. Part of them "fall" into the solar surface (emitting bremsstrahlung radiation) and impinging on the dense chromosphere and photosphere, they heat, through Coulomb collisions, the local plasma which intensely radiates (it "flares"). The other part, which is of interest here, leaves the Sun and propagates in the interplanetary medium. These SEPs have a typical spectrum of Fig. 2.25. They carry a mass less than 10^4 kg s^{-1} for an event duration of less than a few hours. This is orders of magnitude smaller than the solar wind in one second. But, as shown by Fig. 2.26, they carry energies per nucleon higher than one MeV. Finally, since the high chromosphere and low corona are severely perturbed, the new magnetic configuration can lead to plasma ejections such as CME. Actually SEPs are often associated with CME. Of course, their propagation times are widely separated: about an hour for protons with energies of 20 MeV and about two days for CME propagating with a speed of about 800 km s^{-1}.

2.6.2 Eruptive Prominences

A prominence consists in cool ($T \sim 10^4$ K) and dense (10^9–10^{11} cm^{-3}) material suspended and confined in the corona. It lies upon a magnetic inversion line in such a way that the material is prevented to fall by the (horizontal) magnetic tension. Its mass $M \sim 5 \times 10^{12}$ kg is known within a factor 10 (at least) because it depends on the volume (i.e., the morphology of the filament/prominence). For a complete overview, see [39] and [40]. Since $M_{prom} \sim M_{cor}$ or less, prominence eruptions are not frequent enough to "feed" the corona. How much material actually leaves the Sun? The set of movies obtained with SDO and STEREO easily shows that not all

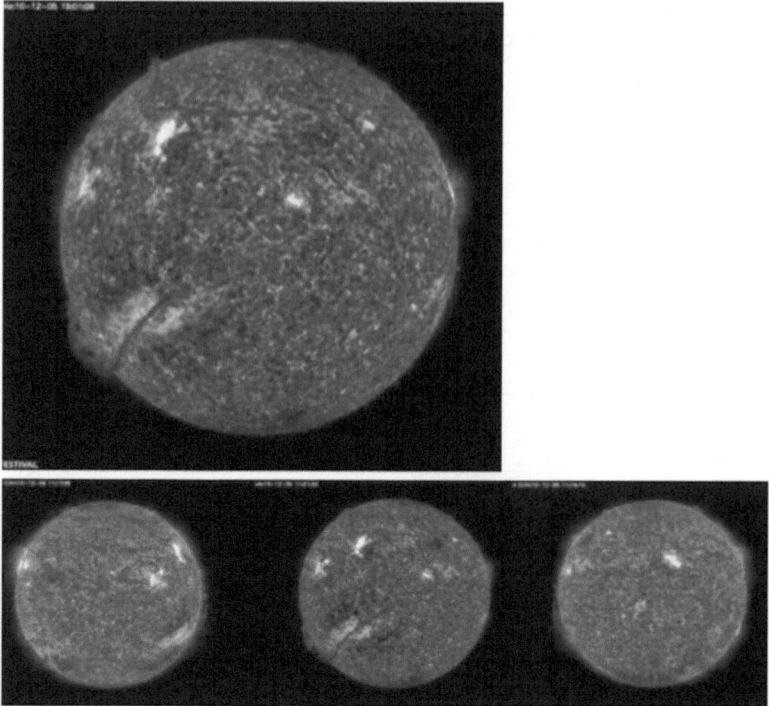

Fig. 2.27 Quasi-simultaneous movies obtained in the He II 30.4 nm line by the EUVI imagers on the STEREO A and B probes and the AIA imager on SDO on 6 Dec. 2010. *Top*: The large images, taken at a cadence of about 12 s, show part of the EP at the South-East (*lower and left*) part of the solar image. *Bottom*: The set of three movies come respectively from STEREO B (*left*), SDO (*center*), and STEREO A (*right*). The STEREO B images (at South-West) allow one to observe the leg of the EP invisible with SDO. (Movies available to the public, courtesy STEREO and SDO Teams)

material is carried out and that some material flows back toward the feet of the EP (Fig. 2.27). These are really unique observations made when the EUV imagers of the STEREO B (and A) and SDO were separated by nearly 90°. The evaluation of the respective parts of the CME leaving the Sun and returning to the Sun is not straightforward because there is no simple law relating the emissivity of the He II (30.4 nm) line and the density of the material (a usual problem in remote-sensing measurements).

2.6.3 Coronal Mass Ejections

The mass involved in a CME, which comes essentially from the EP material, is in the range 10^{12}–10^{13} kg. Of course, it is important to take into account the frequency of such events. At minimum of activity, there are two events per week on average: with

CME   Erupting Prominence

CME/Flare Ribbons   CME/Flare Loops

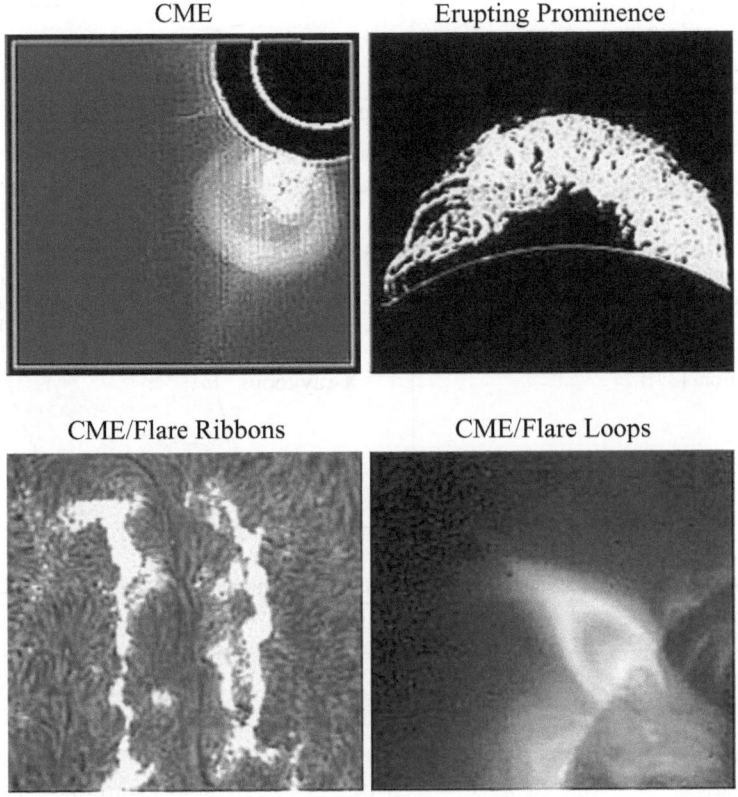

Fig. 2.28 The four images display a CME and an EP (*top row*) and two flare structures (flare ribbons and flare loops) associated with the CME. The different features are not as disparate as taken at face value. They can be unified with the cartoon of Fig. 2.29 (from [37])

a rate of 3×10^6–3×10^7 kg s^{-1}, this is much less than the Solar wind. At maximum of activity, with two events per day, the rate becomes 3×10^7–3×10^8 kg s^{-1}, still orders of magnitude smaller than the Solar wind. As for the kinetic energy, with a velocity ~ 1000 km s^{-1}, one gets 0.5×10^{24}–0.5×10^{25} J, smaller but of the order of flare energy (10^{26} J).

2.6.4 Flares, EPs, and CMEs: Closely Related Phenomena

The different aspects of the three phenomena are well evidenced in Fig. 2.28. The CME and EP (top row) have distinct shapes (sizes, altitudes, etc.). The CME is associated with a flare with its two main features (low row): loops (right-hand side) and ribbons (actually feet of the loops, left-hand side). Is it possible to catch these distinct features within a single process? and who starts first? Figure 2.29 summarizes

Fig. 2.29 The four images displayed in Fig. 2.28 are summarized in a cartoon where the lift-off of the magnetic structure leads to the following time sequence described from right to left: a shock bow in front of a plasma pile-up, followed by a coronal cavity, overlying the cool prominence no longer maintained by magnetic field filling X-ray loops whose activated feet delineate flare ribbons (from [37])

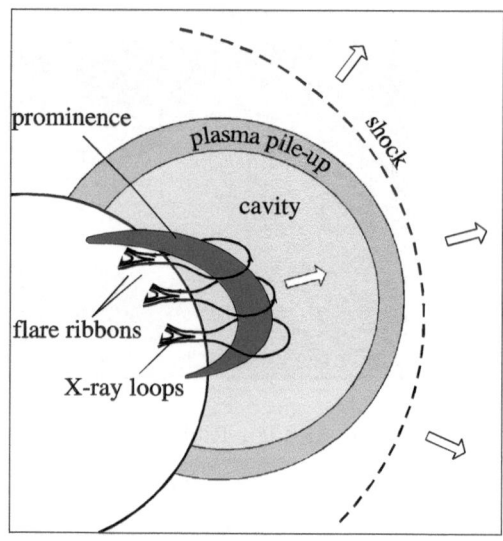

Table 2.4 Typical values of magnetic, thermal, kinetic and gravitational energies in the corona

Energy type	Formula	Value (J/m^3)	Parameter values
Magnetic	$\frac{B^2}{2\mu}$	40	$B = 100$ Gauss
Thermal	nkT	0.01	$n = 10^{15}$ m^{-3}, $T = 10^6$ K
Bulk kinetic	$\frac{m_p n v^2}{2}$	10^{-6}	$n = 10^{15}$ m^{-3}, $v = 1$ km/s
Gravitational	$m_p\, n\, g\, h$	0.04	$n = 10^{15}$ m^{-3}, $h = 10^8$ m

the eruption (EP, flare, and CME) in a simple cartoon, without providing the answer but implies a disruption of the magnetic field which is treated in Sect. 2.6.5. This disruption is made possible through the realization that a single energetic process is at work, namely the release of magnetic energy stored in the coronal magnetic field. Table 2.4 [37] shows how the magnetic energy in the lower coronal layers dominates all other energies and can be the energy reservoir, at least in active regions since the adopted value for the magnetic field, 100 G (or 0.01 T) is not the average coronal field! (Also note that the table does not take into account the solar wind kinetic energy.)

2.6.5 Models of Formation of EP and CME

This is a very active field where models flourish for the main reason that the triggering mechanism is still unknown. We focus here on two popular mechanisms and scenario, the flux rope and the break out models. The flux rope model is illustrated by Fig. 2.30 [1]. The left-hand drawing of the field lines shows how a magnetic flux

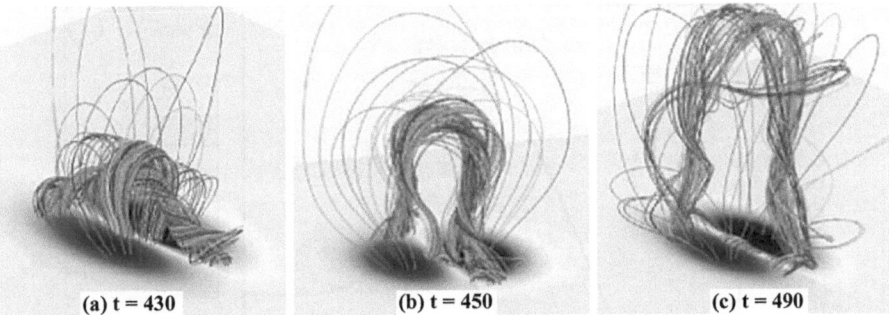

(a) t = 430   **(b) t = 450**   **(c) t = 490**

Fig. 2.30 A three-step model of an Eruptive Prominence. *On the left*, one sees a magnetic flux rope (mfr), highly twisted and elongated along the magnetic neutral line, which is lifting up, pushing the magnetic field of the coronal arcades above. The mfr expands (*middle drawing*) while becoming more perpendicular to the neutral line. Finally (*right drawing*), it gets free with a decrease of the twist of the field lines (from [1])

rope (mfr), highly twisted and elongated along the magnetic neutral line, lifts up, pushing the magnetic field of the coronal arcades above. The mfr expands (middle drawing) while its feet are getting closer on both sides of the neutral line. Finally (right drawing), it gets free with a decrease of the twist of the field lines. The timing is not clear in the sense that the mfr could emerge already twisted from the convection zone or emerge far in advance and become more and more twisted because, e.g., of photospheric motions of the footpoints. The break out model (e.g., [2]) is explained in Fig. 2.31 (from [37]). It starts with a quadrupolar magnetic structure (right hand drawings) where the region between the two dipoles is prone to flux rise (top drawing 1 and middle drawing 2). The end of phase 2 sees the appearance of an X-point and magnetic reconnection. Finally, the magnetic configuration (lower drawing 3) includes a plasmoid at the top, a current sheet below, and loops close to the surface. On the left-hand side, the free magnetic energy is plotted vs. time and the three different phases are shown: the free energy first strongly increases and then decreases when the reconnection takes place. Note that the Alfvén time is very large because of the large size of the structures located in the corona.

2.6.6 Interplanetary Coronal Mass Ejections (ICME)

Farther in the corona and closer to the Earth and other planets, the CME develop into an ICME, where the magnetic field, contained in a magnetic cloud, plays such an important role when it impacts the magnetospheric field. When its sign is opposite to the sign of the terrestrial field, it leads to a large range of phenomena (radio bursts, precipitation of particles in the polar cusps, etc.) pertaining to the Space Weather. The ICME propagation is depicted in the cartoon of Fig. 2.32 from [33], which combines magnetic field, plasma, and solar wind suprathermal electron flows. The upstream shock contributes to accelerate particles. The shape of the plasma cloud

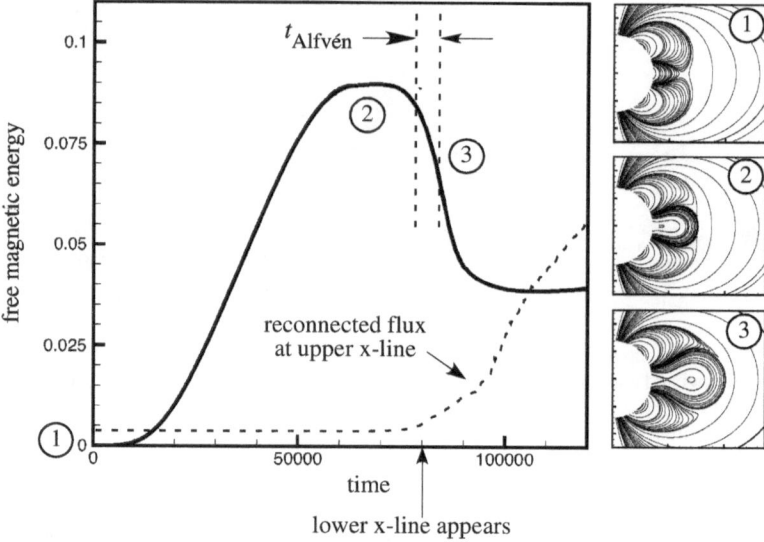

Fig. 2.31 Break out model. Right hand drawings display three different times of the eruption. In the quadrupolar magnetic structure (*top drawing 1*), the region between the two dipoles is prone to flux rise which takes place in the *middle drawing 2*. At the end of phase 2, an X-point appears, and magnetic reconnection takes place. Finally, the magnetic configuration (*lower drawing 3*) includes a plasmoid at the top, a current sheet below, and loops close to the surface. On the left-hand side, the free magnetic energy is plotted vs. time and the three different phases are shown: the free energy first strongly increases and then decreases when the reconnection takes place. Note that the Alfvén time is very large because of the large size of the structures located in the corona (from [37])

can be observed when it escapes in the plane of the sky of the instruments with the help of Heliospheric Imagers, as is the case for STEREO (Fig. 2.33). The quantity of material, and consequently the emission, is so faint that this requires very sensitive and clean instruments and a technique of image subtraction which allows one to visualize propagating plasmas. (It should be realized that the signal is smaller than that from a 12th magnitude star!)

2.7 The Heliosphere

It is an elongated bubble (Fig. 2.34) with a "radius" of about 100 AU, a distance which can be deduced from the equality of solar wind and interstellar medium pressures. It is bound by the interstellar medium of the Local cloud, "the low-pressure exit of the SW nozzle" according to [35]. Typical values there are: $n_H \sim 0.2$ cm^{-3}, $V = 26$ km s^{-1}, and $B = 2$–3 μG. With such figures, as shown by [35], the energy of the bulk plasma motion and the thermal and magnetic energies are about equal (a few 10^{-14} J m^{-3}). This energy density is of the order of the energy required

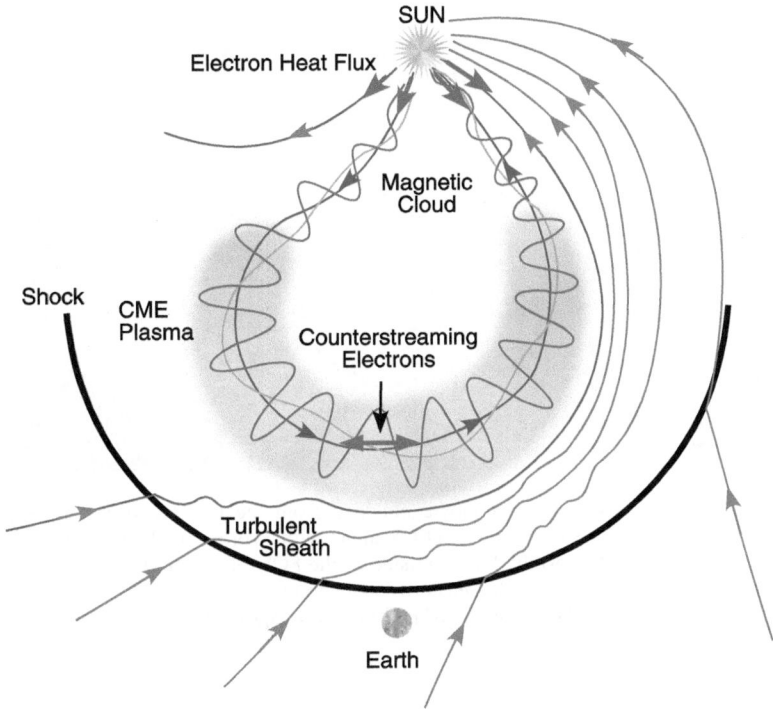

Fig. 2.32 Cartoon depicting magnetic field, plasma, and suprathermal flows in an ICME. Note the shock, already mentioned in Fig. 2.29 which also allows further particles acceleration (from [33])

by cosmic rays. Moreover, in this interface region, the importance of charge exchange between the fully ionized solar wind and the neutral interstellar medium is evidenced by "hydrogen walls," the generation of an important X-ray emission, the so-called anomalous cosmic rays. The interaction between heliosphere and interstellar medium is schematically depicted in Fig. 2.34, where one sees the backward return of the solar wind when it meets the interstellar medium at the level of a shock. What is fascinating is the fact that such bow shocks have been visualized close to their respective young stars in the Orion Nebula (Fig. 2.35). Their sizes are about 10^4 times the size of the heliosphere, and their mass flux is about 10^8 the mass flux in our heliosphere.

2.8 Conclusions and Prospects

As far as the solar wind is concerned, the need to fill the gap between measurements at a few solar radii and at 1 AU is evident since the remote-sensing spectroscopic measurements performed with UVCS/SOHO did not go higher than 5 solar radii on one hand, and the in situ measurements are mostly performed at 1 AU (apart

Fig. 2.33 This image is actually a movie obtained from the two Heliospheric Imagers on STEREO. On the right, the Sun is shown in EUV along with a coronagraph picture (in *blue*) from STEREO. The Earth is close to the left-hand side of the FOV of the most external heliospheric imager (the rocket-nose shaped black area corresponds to the Earth occulter). The movie catches the propagating CME with the technique of two-by-two images subtraction (courtesy STEREO Teams)

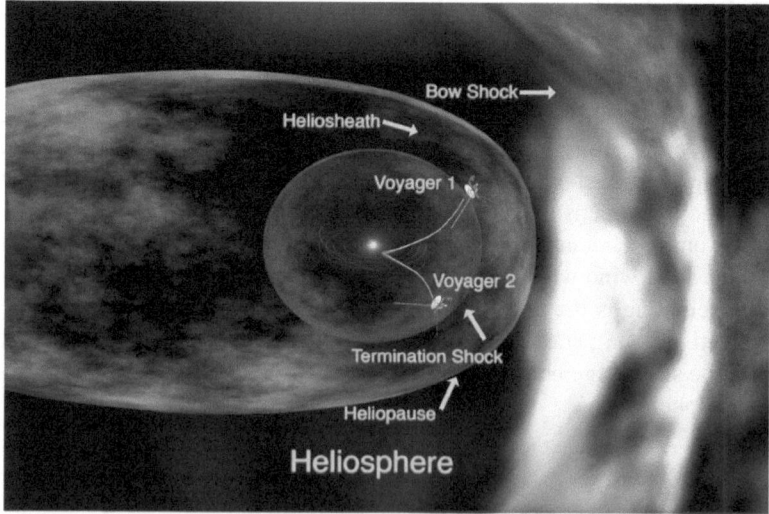

Fig. 2.34 Cartoon of the interface between the heliosphere and the interstellar medium. The Sun is drawn as a *yellow dot* with circling planet orbits. The *bright blue* sphere contains the interplanetary medium where the solar wind is supersonic. The termination shock marks the distance from where the solar wind becomes subsonic (heliosheath in *dark blue*). Further out, the heliopause marks the frontier with the interstellar medium. Note the bow shock ahead. Note the trajectories of Voyager 1 and 2 (courtesy Ed. Stone, from NASA/Walt Feimer)

Fig. 2.35 This is not a cartoon or an artist's rendering. These images of the Orion Nebula show, at the center and at the upper right, two bow shocks located very close to their respective young stars (courtesy Ed. Stone, from NASA/STScI/AURA)

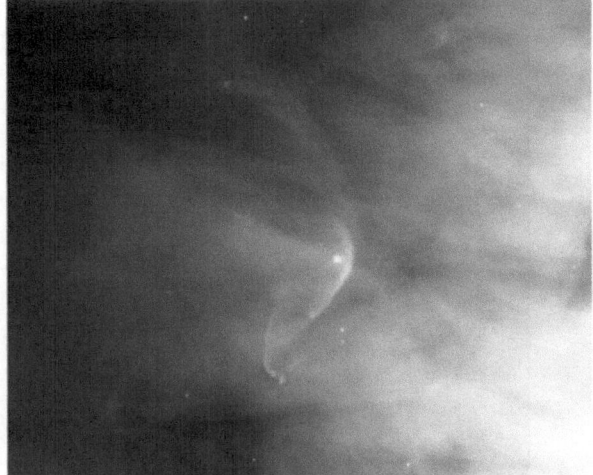

Fig. 2.36 Electron density measurements performed in a current sheet (*top curve*) and in a coronal hole (*lower curve*). The measurements gap between 5 and 215 solar radii is well obvious (from [11])

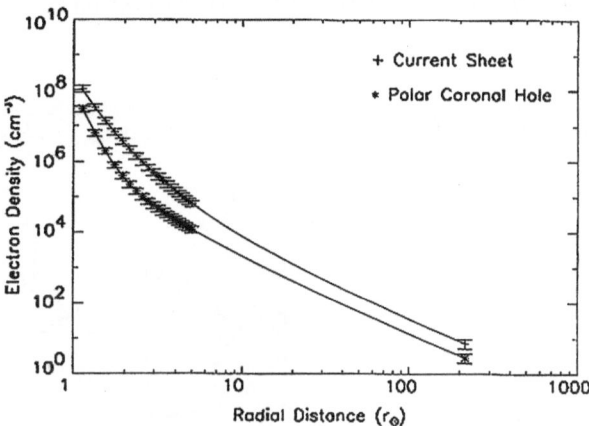

from the exceptional measurements of Helios which got closer than about 60 solar radii) on the other hand (Fig. 2.36). An answer to this gap consists in going closer to the Sun: the Solar Orbiter, approved on October 2011 by ESA, will sound the solar corona and the solar wind closer than 0.3 AU with a package of remote-sensing and in situ instruments far superior to SOHO instrumentation (Fig. 2.37). A peculiarity of the mission consists in reaching a 30 degrees latitudes allowing one to have a view on the polar coronal holes and the fast wind which originates from. The measurements will tackle the sources of the continuous and episodic flows of material (including energetic particles) and will record their transformations when they reach the S/C. The NASA Solar Probe+ will go closer to the Sun (about 10 solar radii), and the set of in situ instruments will provide the "ground truth" for diagnosing the encountered plasma and magnetic field (Fig. 2.38). Scientists also dream of the combined information obtained with the two missions. Another major step consists to measure the coronal magnetic field in the lower corona (lower than the 10 solar

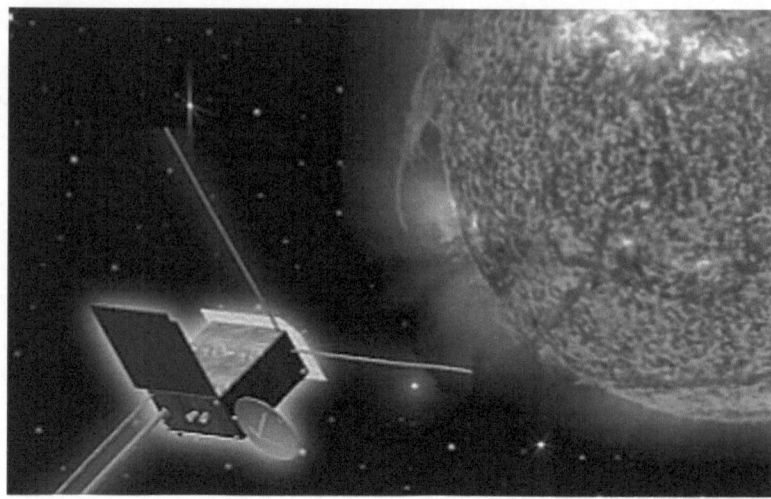

Fig. 2.37 Artist view of the ESA Solar Orbiter platform

Fig. 2.38 Artist view of the
NASA Solar Probe$^+$ platform

radii of Solar Probe$^+$): this is the aim of some Cosmic Vision proposals to ESA
(SOLMEX, LEMUR, etc.). Finally, the most challenging step is to understand better the sources of the various events (including the wind) in the low corona, the
CCTR, and especially in the complex chromosphere. This will require better diagnostic tools and observations of the magnetic field, non-Maxwellian distributions,
ionization degrees, flows, etc., close to the (unknown) spatial scales where most
physical processes take place!

Acknowledgements   I deeply thank the two organizers, Coralie Neiner and Jean-Pierre Rozelot,
for inviting me to give this talk and for transforming the present document from a mere draft into a
(I hope!) readable paper. I also thank the colleagues who kindly provided me their original figures.

References

Main References

1. Amari, T., Luciani, J.F., Aly, J.J., Mikic, Z., Linker, J.: Coronal mass ejection: initiation, magnetic helicity, and flux ropes, II: turbulent diffusion-driven evolution. Astrophys. J. **595**, 1231–1250 (2003)
2. Antiochos, S.K., DeVore, C.R., Klimchuk, J.A.: A model for solar coronal mass ejections. Astrophys. J. **510**, 485–493 (1999)
3. Aschwanden, M.J., Poland, A.I., Rabin, D.M.: The new solar corona. Annu. Rev. Astron. Astrophys. **39**, 175 (2001)
4. Baker, D., van Driel-Gesztelyi, L., Mandrini, C.H., Démoulin, P., Murray, M.J.: Magnetic reconnection along quasi-separatrix layers as a driver of ubiquitous active region outflows. Astrophys. J. **705**, 926–935 (2009)
5. Banaszkiewicz, M., Axford, W.I., McKenzie, J.F.: An analytic solar magnetic field model. Astron. Astrophys. **337**, 940–944 (1998)
6. Biermann, L.: Solar corpuscular radiation and the interplanetary gas. Observatory **77**, 109 (1957)
7. Boutry, C., Buchlin, E., Vial, J.-C., Régnier, S.: Flows at the edge of an active region: observation and interpretation. Astrophys. J. **752**, 13 (2012)
8. David, C., Gabriel, A.H., Bely-Dubau, F., Fludra, A., Lemaire, P., Wilhelm, K.: Measurement of the electron temperature gradient in a solar coronal hole. Astron. Astrophys. **336**, L90 (1998)
9. De Pontieu, B., McIntosh, S.W., Carlsson, M., Hansteen, V.H., Tarbell, T.D., Boerner, P., Martinez-Sykora, J., Schrijver, C.J., Title, A.M.: The origins of hot plasma in the solar corona. Science **331**, 55 (2011)
10. Gabriel, A., Bely-Dubau, F., Tison, E., Wilhelm, K.: The structure and origin of solar plumes: network plumes. Astrophys. J. **700**, 551–558 (2009)
11. Guhathakurta, M., Sittler, E.: Importance of global magnetic field geometry and density distribution in solar wind modeling. In: Habbal, S.R., Esser, R., Hollweg, J.V., Isenberg, P.A. (eds.) American Institute of Physics Conference Series, vol. 471, pp. 79–82 (1999)
12. Harra, L.K., Sakao, T., Mandrini, C.H., Hara, H., Imada, S., Young, P.R., van Driel-Gesztelyi, L., Baker, D.: Outflows at the edges of active regions: contribution to solar wind formation? Astrophys. J. Lett. **676**, L147–L150 (2008)
13. Hassler, D.M., Dammasch, I.E., Lemaire, P., Brekke, P., Curdt, W., Mason, H.E., Vial, J.-C., Wilhelm, K.: Solar wind outflow and the chromospheric magnetic network. Science **283**, 810 (1999)
14. He, J.-S., Marsch, E., Tu, C.-Y., Guo, L.-J., Tian, H.: Intermittent outflows at the edge of an active region—a possible source of the solar wind? Astron. Astrophys. **516**, A14 (2010)
15. Kahler, S.W.: Solar sources of heliospheric energetic electron events: shocks or flares? Space Sci. Rev. **129**, 359 (2007)
16. Kohl, J.L., Noci, G., Antonucci, E., Tondello, G., Huber, M.C.E., Cranmer, S.R., Strachan, L., Panasyuk, A.V., Gardner, L.D., Romoli, M., Fineschi, S., Dobrzycka, D., Raymond, J.C., Nicolosi, P., Siegmund, O.H.W., Spadaro, D., Benna, C., Ciaravella, A., Giordano, S., Habbal, S.R., Karovska, M., Li, X., Martin, R., Michels, J.G., Modigliani, A., Naletto, G., O'Neal, R.H., Pernechele, C., Poletto, G., Smith, P.L., Suleiman, R.M.: UVCS/SOHO empirical determinations of anisotropic velocity distributions in the solar corona. Astrophys. J. Lett. **501**, L127 (1998)
17. Le Chat, G., Issautier, K., Meyer-Vernet, N., Hoang, S.: Large-scale variation of solar wind electron properties from quasi-thermal noise spectroscopy: Ulysses measurements. Sol. Phys. **271**, 141 (2011)

18. Marsch, E., Schwenn, R., Rosenbauer, H., Muehlhaeuser, K.-H., Pilipp, W., Neubauer, F.M.: Solar wind protons—three-dimensional velocity distributions and derived plasma parameters measured between 0.3 and 1 AU. J. Geophys. Res. **87**, 52–72 (1982)
19. Masuda, S., Kosugi, T., Hara, H., Tsuneta, S., Ogawara, Y.: A loop-top hard X-ray source in a compact solar flare as evidence for magnetic reconnection. Nature **371**, 495–497 (1994)
20. McComas, D.J., Bame, S.J., Barraclough, B.L., Feldman, W.C., Funsten, H.O., Gosling, J.T., Riley, P., Skoug, R., Balogh, A., Forsyth, R., Goldstein, B.E., Neugebauer, M.: Ulysses' return to the slow solar wind. Geophys. Res. L ett. **25**, 1–4 (1998)
21. Murphy, R.J., Share, G.H.: What gamma-ray deexcitation lines reveal about solarflares. Adv. Space Res. **35**, 1825–1832 (2005)
22. Noci, G.: Diagnostic methods for the extended corona. Adv. Space Res. **11**, 263 (1991)
23. Parker, E.N.: Dynamics of the interplanetary gas and magnetic fields. Astrophys. J. **128**, 664 (1958)
24. Patsourakos, S., Vial, J.-C.: Outflow velocity of interplume regions at the base of polar coronal holes. Astron. Astrophys. **359**, L1–L4 (2000)
25. Sakao, T., Kano, R., Narukage, N., Kotoku, J., Bando, T., DeLuca, E.E., Lundquist, L.L., Tsuneta, S., Harra, L.K., Katsukawa, Y., Kubo, M., Hara, H., Matsuzaki, K., Shimojo, M., Bookbinder, J.A., Golub, L., Korreck, K.E., Su, Y., Shibasaki, K., Shimizu, T., Nakatani, I.: Continuous plasma outflows from the edge of a solar active region as a possible source of solar wind. Science **318**, 1585 (2007)
26. Schwenn, R., Inhester, B., Plunkett, S.P., Epple, A., Podlipnik, B., Bedford, D.K., Eyles, C.J., Simnett, G.M., Tappin, S.J., Bout, M.V., Lamy, P.L., Llebaria, A., Brueckner, G.E., Dere, K.P., Howard, R.A., Koomen, M.J., Korendyke, C.M., Michels, D.J., Moses, J.D., Moulton, N.E., Paswaters, S.E., Socker, D.G., St. Cyr, O.C., Wang, D.: First view of the extended green-line emission corona at solar activity minimum using the Lasco-C1 coronagraph on SOHO. Sol. Phys. **175**, 667–684 (1997)
27. Sheeley, N.R. Jr., Wang, Y.-M., Hawley, S.H., Brueckner, G.E., Dere, K.P., Howard, R.A., Koomen, M.J., Korendyke, C.M., Michels, D.J., Paswaters, S.E., Socker, D.G., St. Cyr, O.C., Wang, D., Lamy, P.L., Llebaria, A., Schwenn, R., Simnett, G.M., Plunkett, S., Biesecker, D.A.: Measurements of flow speeds in the corona between 2 and 30 R sub sun. Astrophys. J. **484**, 472 (1997)
28. Tu, C.-Y., Zhou, C., Marsch, E., Xia, L.-D., Zhao, L., Wang, J.-X., Wilhelm, K.: Solar wind origin in coronal funnels. Science **308**, 519–523 (2005)
29. Vial, J.-C., Auchère, F., Chang, J., Fang, C., Gan, W.Q., Klein, K.-L., Prado, J.-Y., Rouesnel, F., Sémery, A., Trottet, G., Wang, C.: SMESE (SMall Explorer for Solar Eruptions): a microsatellite mission with combined solar payload. Adv. Space Res. **41**, 183–189 (2008)
30. Wang, Y.-M., Sheeley, N.R. Jr., Walters, J.H., Brueckner, G.E., Howard, R.A., Michels, D.J., Lamy, P.L., Schwenn, R., Simnett, G.M.: Origin of streamer material in the outer corona. Astrophys. J. Lett. **498**, L165 (1998)
31. Wilhelm, K., Dammasch, I.E., Marsch, E., Hassler, D.M.: On the source regions of the fast solar wind in polar coronal holes. Astron. Astrophys. **353**, 749–756 (2000)
32. Wilhelm, K., Abbo, L., Auchère, F., Barbey, N., Feng, L., Gabriel, A.H., Giordano, S., Imada, S., Llebaria, A., Matthaeus, W.H., Poletto, G., Raouafi, N.-E., Suess, S.T., Teriaca, L., Wang, Y.-M.: Morphology, dynamics and plasma parameters of plumes and inter-plume regions in solar coronal holes. Astron. Astrophys. Rev. **19**, 35 (2010)
33. Zurbuchen, T.H., Richardson, I.G.: In-situ solar wind and magnetic field signatures of interplanetary coronal mass ejections. Space Sci. Rev. **123**, 31–43 (2006)

General References About Solar Wind

34. Marsch, E.: In: Space Solar Physics. Lecture Notes in Physics, vol. 507, p. 107 (1998)
35. Meyer-Vernet, N.: Basics of the Solar Wind. Cambridge University Press, Cambridge (2007)
36. Velli, M.: In: Space Solar Physics. Lecture Notes in Physics, vol. 507, p. 217 (1998)

General References About Flares, CMEs, Erupting Prominences

37. Forbes, T., et al.: CME theory and models. Space Sci. Rev. **123**, 251–302 (2006)
38. Schrijver, C.J.: Driving major solar flares and eruptions: a review. Adv. Space Res. **43**(5), 739–755 (2009)

General References About Prominences

39. Labrosse, N., Heinzel, P., Vial, J.-C., et al.: Space Sci. Rev. **151**, 243 (2010)
40. MacKay, M., Karpen, J., Ballester, J.L., et al.: Space Sci. Rev. **151**, 1 (2010)
41. Schrijver, C.J.: Solar energetic events, the solar-stellar connection, and statistics of extreme space weather. In: Proceedings of the 16th Workshop on Cool Stars, Stellar Systems, and the Sun. PASP Conference Series (2010)

Chapter 3
Space Weather and Space Climate— What the Look from the Earth Tells Us About the Sun

Katya Georgieva

Abstract Data for geomagnetic activity can be used to derive long-term variations in solar dynamo parameters and to estimate the regime of operation of the dynamo. It is found that the dynamo operates in moderately advection-dominated regime in the upper part of the convection zone: a part of the flux diffuses directly to the tachocline, "short-circuiting" the meridional circulation; another part makes a full circle to the poles, down to the base of the convection zone and equatorward to sunspot latitudes. This provides an explanation of the two peaks in the sunspot cycle due to the two surges of the toroidal field. In the lower part of the convection zone, diffusion is more important than advection, so the sunspot cycle amplitude increases with the speed of the deep meridional circulation. The well-known Waldmeier rule relating the rise time of a cycle with its amplitude is a direct consequence of this dependence. It is found that periods of grand minima are characterized by a change in the regime of operation of the dynamo in the lower part of the convection zone to advection-dominated. The correlations between the dynamo parameters in the solar cycles allow us to conclude that the factor determining the cycle amplitude is the speed of the surface meridional circulation and to find its dependence on planetary tidal forces. From the changing correlation between sunspot and geomagnetic activity the long-term variations in the toroidal and poloidal components of the solar magnetic field can be estimated.

3.1 Introduction

The term "space weather" was probably first mentioned by Joe Allen in 1985 at the conference on "Space environment effects on satellites" in Albuquerque, USA. By that time, convincing evidence was produced that the many anomalous events on satellites were a product of susceptibility to galactic cosmic rays and to solar energetic protons and heavier ions. As historically the term was first used in relation

K. Georgieva (✉)
Space Research and Technologies Institute, Bulgarian Academy of Sciences, Bl.1 "Acad.Georgi Bonchev" str., 1113 Sofia, Bulgaria
e-mail: KGeorgieva@space.bas.bg

J.-P. Rozelot, C. Neiner (eds.), *The Environments of the Sun and the Stars*,
Lecture Notes in Physics 857,
DOI 10.1007/978-3-642-30648-8_3, © Springer-Verlag Berlin Heidelberg 2013

to the effects of space agents on satellites in particular and on technological systems in general, the technological aspect of space weather was emphasized in the original definition of the term. According to the US "National Space Weather Program Strategic Plan" (Washington DC, 1995), "Space weather refers to conditions on the Sun and in the solar wind, magnetosphere, ionosphere, and thermosphere that can influence the performance and reliability of space-borne and ground-based technological systems and can endanger human life or health. Adverse conditions in the space environment can cause disruption of satellite operations, communications, navigation, and electric power distribution grids, leading to a variety of socioeconomic losses."

In the years to follow, the space weather concept evolved to much broader than just effects on the performance and reliability of space-borne and ground-based technological systems, and now the definition, for example, in the European Space Weather Portal (http://www.spaceweather.eu/en/glossary) reads: "Space weather is the physical and phenomenological state of natural space environments. The associated discipline aims, through observation, monitoring, analysis and modelling, at understanding and predicting the state of the Sun, the interplanetary and planetary environments, and the solar and non-solar driven perturbations that affect them; and also at forecasting and nowcasting the possible impacts on biological and technological systems."

Space climate, on the other hand, "denotes long-term variations in solar activity, as well as the related long-term changes in the heliosphere, the solar wind and the heliospheric magnetic field, and their effects in the near-Earth environment, including the magnetosphere and ionosphere, the upper and lower atmosphere, climate and other related systems" [74, 95]. Therefore, the difference between space weather and space climate is analogous to the difference between meteorological weather and climate. While space weather describes short-term variations in the different forms of solar activity and their terrestrial effects, space climate deals with time scales longer than a few solar rotations, going to decades, centuries, and millennia.

By the term "solar activity" any variation in the appearance and energy output from the Sun is denoted. The solar activity is a direct consequence of the fact that the Sun has a magnetic field maintained by a magnetohydrodynamic dynamo whose operation is responsible for both short-term manifestations of solar activity, like sunspots, solar flares, coronal mass ejections, and coronal holes, and the long-term evolution of solar activity, including periods of several decades with increased activity like the one observed in the 20th century, and episodes of deep prolonged minima such as the Maunder minimum (1645–1715 AD), when almost no sunspots were observed. Many aspects of the solar dynamo are still unknown, especially its long-term variations.

The solar wind carries solar plasma and magnetic fields to the Earth. Their interactions with the terrestrial magnetosphere result in disturbances of the geomagnetic field which provide information about the solar drivers which have caused them [96]. Therefore, the Earth is sort of a probe registering the variations in solar activity, and records of geomagnetic activity can be used to better understand the operation of the solar dynamo.

3.2 The Sun—Basic Characteristics

Sun is a typical G-type star in the middle of its lifetime on the main sequence. It has an inner core spanning up to ~0.3 solar radii where thermonuclear reactions of converting hydrogen into helium through the p–p (proton–proton) chain take place. The energy produced in these reactions can be transmitted outward by means of two processes, radiation and convection. Above the core, in the so-called radiative zone between ~0.3 and ~0.7 solar radii, matter is hot and dense enough for the energy to be transmitted radiatively. Above it, separated by a thin layer called "tachocline," there is the solar convection zone in which energy is transported to the surface by turbulent convection.

3.2.1 Importance of the Convective Envelope

In a star with a convective envelope, the interaction between rotation and convection modifies rotation making it "differential," with different rotation rates at different latitudes. In the Sun and almost all other known stars, the rotation is "Sun-like," that is, fastest at the equator and slowing down toward the poles. In only about ten stars an "anti-Sun-like" rotation has been observed so far, with the rotation rate increasing from the equator to the poles. The differential rotation in the convection zone leads to the establishing of a large-scale meridional circulation between the equator and the poles. In the Sun and in stars with Sun-like differential rotation, the meridional circulation is poleward in the surface layers and equatorward at the base of the convection zone. In stars with anti-Sun-like differential rotation, the direction of the meridional circulation is the opposite: equatorward at the surface and poleward at the base of the convection zone [60].

The magnetic activity of the Sun and Sun-like stars is due to the presence of a convective envelope. The convection of conducting plasma leads to the generation of a dipolar magnetic field, and the observed cyclic variations in the magnetic field are a result of the action of a magnetohydrodynamic dynamo transforming the magnetic field from dipolar (poloidal) into toroidal and back.

3.2.2 Sunspots and Sunspot Cycles

The most obvious manifestations of the solar magnetic activity, and with the longest data record, are sunspots. Very big sunspots can be seen with naked eye, and observations of sunspots have been reported in old chronicles. The earliest records of sunspots observed by Chinese astronomers are from 28 BC, and there is evidence that the Greeks knew of them at least by the 4th century BC. Systematic observations of sunspots began only after the telescope was invented, early in the 17th century. Probably the first one who observed telescopically sunspots was the English scholar

Thomas Harriot (1560–1621), who recorded one such observation in a note dated 8 December 1610. The sunspot cycle was discovered by the German pharmacist and amateur astronomer Heinrich Schwabe, who was actually looking for a planet inside the orbit of Mercury and tried to detect it as a spot moving across the solar disc. Schwabe did not find the planet, but in his daily observations from 1826 to 1843 he noticed the regular variation in the number of sunspots with a period of approximately 11 years [102]. Hale [41], who found that sunspots are associated with strong magnetic fields, was the first one to associate the sunspot cycle with the magnetic activity of the Sun.

As the sunspots are manifestations of the magnetic activity of the Sun, any theory of the solar dynamo must be able to explain the observed basic features of the sunspot cycle:

- The number and area of sunspots increase and decrease cyclically every about 11 years;
- Sunspot cycles vary in length and amplitude;
- Long periods occur of few or no spots;
- Some sunspot cycles are double peaked;
- Sunspots tend to occur in pairs. On one hemisphere, the leading (with respect to the direction of the solar rotation) spots in all pairs have the same polarity, and the trailing spots in the pairs have the opposite polarity, while on the other hemisphere, the polarities are reverse. In each hemisphere, the polarity of the leading spots is the polarity of the respective pole.
- In subsequent 11-year sunspot cycles the polarities in the two hemispheres reverse [42].
- When the new solar cycle begins, sunspots first appear at higher heliolatitudes, and as the cycle approaches minimum, their emergence zone moves equatorward, a rule known as Spörer's law but actually discovered by Spörer's contemporary Carrington [13].
- The leading sunspot in a sunspot pair appears at lower heliolatitudes than the trailing sunspot, and the inclination of the sunspot pair relative to the equator increases with increasing latitude, the so-called Joy's law [43].

3.3 The Solar Dynamo Theory

According to the solar dynamo theory developed by Parker [82], the sunspot cycle is produced by an oscillation between toroidal and poloidal components, similar to the oscillation between kinetic and potential energies in a simple harmonic oscillator. Because the magnetic Reynolds number in the Sun is high, magnetic fields are frozen in the plasma and are carried by the velocity fields of the plasma. The differential rotation stretches the north–south field lines of the poloidal magnetic field predominant in sunspot minimum in east–west direction, thus creating the toroidal component of the field. This mainly occurs where the gradient of the angular velocity is greatest; it is the tachocline, the thin boundary between the convection zone, where the rotation is differential, and the radiative zone, which is characterized by rigid rotation. The process of transforming the poloidal field into toroidal field is generally known as "Ω-effect." Due to the interaction with convection, it is supposed that the magnetic field concentrates in bundles of field lines, magnetic field tubes. The pressure inside the magnetic field tube is a sum of the gas pressure and the magnetic field pressure $B^2/2\mu$, and this pressure is balanced by the pressure outside the field tube, which is only due to the gas pressure:

$$p_{\text{gas out}} = p_{\text{gas in}} + p_{\text{mag in}}.$$

Therefore, the gas pressure inside a field tube $p_{\text{gas in}} = p_{\text{gas out}} - p_{\text{mag in}}$ is lower than the outside pressure $p_{\text{gas out}}$, and when the magnetic field is strong enough, the field tube becomes buoyant and emerges, piercing the solar surface in two spots with opposite magnetic polarities, sunspots.

This theory explains why sunspots appear in pairs, why the leading sunspots have the polarity of their respective poles, and why the polarities in the two hemispheres are opposite. To complete the cycle, the toroidal field must be next transformed back into poloidal field with the opposite magnetic polarity.

In the original Parker's theory, helical turbulent motions in the convection zone due to the Coriolis force twist the toroidal field to produce the poloidal field of the next cycle. However, this process, known as "α-effect," can only exist if the maximum toroidal field does not exceed 10^4 G, so that the magnetic energy density does not exceed the kinetic energy of turbulence, while flux tube simulations for modeling sunspot formation suggest that the toroidal field is $\sim 10^5$ G, that is, one order of magnitude stronger. With such values of the toroidal magnetic field, the α-effect cannot work (see, e.g., [17], and the references therein).

Recently the most promising mechanism for poloidal field regeneration seems to be the one based on the idea first proposed by Babcock [5] and mathematically developed by Leighton [66]: Due to the Coriolis force acting on the emerging field tube, the bipolar pair of spots is tilted with respect to the meridional plane with the leading (in the direction of solar rotation) spot at lower heliolatitude than the trailing spot (Joy's law). As the sunspot cycle is progressing, sunspot pairs appear closer and closer to the equator. Late in the sunspot cycle, the leading spots diffuse across the equator where their flux is canceled by the opposite polarity flux of the leading spots in the other hemisphere. The flux of the trailing spots and of the remaining sunspot pairs is carried toward the poles. Opposite leading and trailing polarity flux cancels on the way to the poles, but as there is excess trailing polarity flux, the net flux reaching the poles has the polarity of the trailing sunspots. This trailing polarity flux first cancels the polar field of the previous solar cycle and then accumulates to form the poloidal field of the next cycle with polarity opposite to the one in the preceding cycle.

According to the original Babcock–Leighton mechanism, the flux is carried to the poles by random-walk diffusion-like process caused by supergranulation convection currents in the solar outer layers [65]. Wang et al. [113] suggested that the main flux transport factor is a large-scale meridional circulation with a surface flow toward the poles where the poloidal flux accumulates, sinks to the base of the solar convection zone, and the counterflow there carries it like a conveyor belt back to low latitudes to be transformed during the journey into toroidal flux and to emerge as the sunspots of the nest solar cycle.

This so-called flux-transport dynamo mechanism explains additionally the reversal of the polar fields and sunspot polarities in subsequent sunspot cycles, as well as the equatorward movement of the sunspot occurrence zone. Moreover, it is directly observed, both the forming of unipolar magnetic regions from decaying sunspot pairs and the meridional circulation carrying them to the poles. The near-surface poleward flux has been measured by different methods: helioseismology ([45], and the references therein; [36, 67, 117], magnetic butterfly diagram [54, 108], latitudinal drift of sunspots [57], Doppler shift of gas velocity [111]); it is found to be of order 10–20 m/s, varying with latitude, with the sunspot cycle phase [7, 47, 55, 57, 111], and from cycle to cycle [20, 55, 111].

The deep counterflow has not been yet observed but has been estimated from helioseismology to be 2–5 m/s, dependent on latitude and asymmetric about the equator, with the depth where the circulation reverses from poleward at the surface to equatorward in the lower part of the convection zone at about 0.8 solar radii [32]. An indirect estimation of the deep equatorward circulation was made based on the equatorward drift of the sunspot occurrence latitudes, and it was also found to depend on the phase in the sunspot cycle, decreasing from maximum to minimum, and to depend on latitude, increasing from 0 m/s at the equator to 2 m/s at 30° [50, 51].

3.4 Long-Term Variations in Solar Dynamo

The comparison of the variations in the speed of the deep and surface meridional circulation with the amplitude of the sunspot cycle is vital for verifying the solar dynamo theory and for identifying the regime in which the dynamo operates. However, direct measurements of the surface meridional circulation are only available for less than three sunspot cycles, and direct measurements of the deep meridional circulation are not available at all.

Even though we have no direct long-term observations of various solar processes, we do have means to reconstruct them from geomagnetic data [26–28]. An index of the geomagnetic activity (aa-index) based on measurements in two antipodal stations in the northern and southern hemispheres has been computed since 1868 [70], and the dataset has been extended back for two more sunspot cycles, since 1844, with data from only one station, Helsinki [79]. This extended ak-index covers a total of 15 sunspot cycles.

3.4.1 Geomagnetic Activity and Its Solar Drivers

The Earth's intrinsic magnetic field is basically dipolar, resembling the field of a bar magnet. Its magnitude is of order ∼0.3 G (30,000 nT) at the Earth's surface

Fig. 3.1 Geomagnetic *aa*-index (*black*) and sunspot number (*grey*) since the beginning of the *aa*-index record

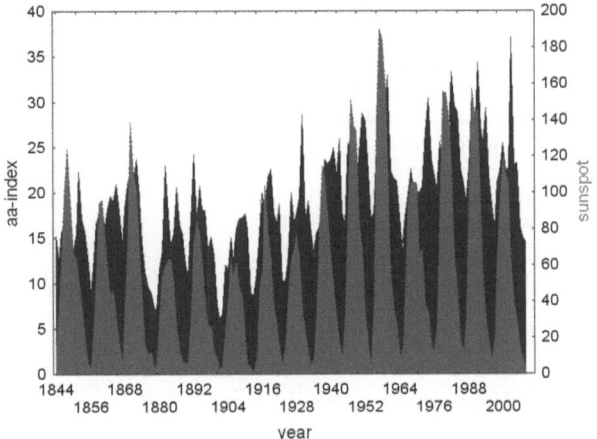

near the magnetic equator and twice that at the poles. At times of enhanced energy input, as a result of the action of solar activity agents, the magnetic field is disturbed, and its magnitude varies by about a percent of the main field due to currents in the ionosphere and magnetosphere.

Two types of solar activity agents are responsible for these geomagnetic disturbances, corresponding to the two faces of the sun's magnetism, the toroidal and poloidal components of the solar magnetic field: the coronal mass ejections (CMEs), huge bubbles of plasma with embedded magnetic fields ejected from the solar corona, which like the sunspots are manifestation of the solar toroidal field, and the high-speed solar wind streams, emanating from solar coronal holes, manifestation of the solar poloidal field. At any phase of the solar cycle, solar agents of both types affect the Earth; however, in the course of the solar cycle their relative importance changes. Richardson et al. [91–93] studied the sources of geomagnetic storms and their solar cycle distribution, and found that the most intense storms at both sunspot minimum and sunspot maximum are almost all generated by CMEs, while weaker storms are preferentially associated with high-speed solar wind streams from coronal holes at solar minimum, and with CMEs at solar maximum. In spite of the geoeffectiveness of the individual solar events, the contribution of the different types of solar drivers to the overall (e.g., yearly averaged) level of the geomagnetic activity depends on their abundance, and geomagnetic activity is mainly due to CMEs (that is, toroidal field-related solar activity) in sunspot maximum, and to high-speed solar wind (poloidal field-related solar activity) during the sunspot declining and minimum phase [29], in accordance with the relative intensity of the solar toroidal and poloidal magnetic fields. Consequently, geomagnetic activity has two peaks in the 11-year solar cycle (Fig. 3.1).

The first peak is due to solar coronal mass ejections which have maximum in number and intensity at sunspot maximum, and hence it coincides with the sunspot maximum. Often this peak is double, as is the peak in sunspot activity itself; we will come back to this issue later. The second peak is caused by high-speed solar wind streams from solar coronal holes which have maximum on the sunspot declining

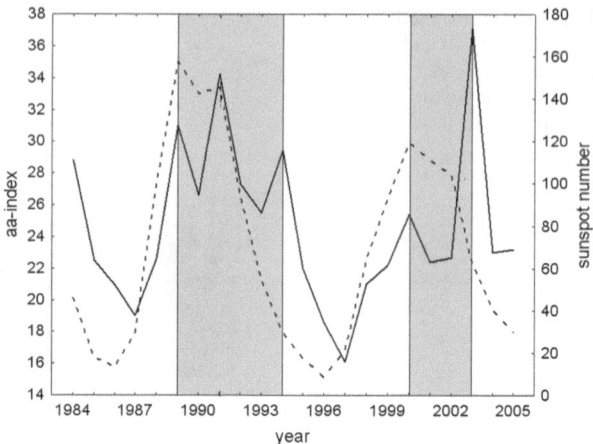

Fig. 3.2 Two geomagnetic activity maxima in the sunspot cycle used for the calculation of the speed of the surface poleward circulation: from the time between sunspot max and the following *aa* max (*grey*), and of the deep equatorward circulation from the time between *aa* max and the following sunspot max (*white*)

phase [37, 38]. Figure 3.2 demonstrates the two peaks in geomagnetic activity in cycles 22 and 23. Cycle 22 is an example of a cycle with a clearly visible double peak in the number of sunspots reflected in geomagnetic activity (1989 and 1991), and another peak in geomagnetic activity is on the declining phase of the sunspot cycle (1994).

The maximum geomagnetic activity during the sunspot declining phase is due to the interaction of the Earth's magnetosphere with high-speed solar wind from big long-lived low latitude coronal holes and equatorward extensions of polar coronal holes. The coronal holes are formed from the remnants of the sunspot pairs. At and just after sunspot maximum, these low latitude holes are small, narrow, and short-lived, so the Earth is only embedded for a short time in the high-speed solar wind emanating from them. When the trailing polarity flux reaches the poles and forms the polar coronal holes of the new cycle, the low-latitude holes begin attaching themselves to the equatorward extensions of the polar holes and start growing, so the Earth is repeatedly embedded in wide, long-lasting, long-lived and hence recurrent streams of fast solar wind [112] leading to peak geomagnetic activities. Finally, in sunspot minimum there are almost no low-latitude holes, and the big polar coronal holes do not affect the Earth because the fast solar wind emanating from them does not reach the ecliptic.

Therefore, the geomagnetic activity maximum on the declining phase of the sunspot cycle appears when the flux from sunspot latitudes has reached the poles.

3.4.2 Calculation of the Speeds of the Surface and Deep Circulation

The intervals between the sunspot maximum and geomagnetic activity maximum during the sunspot declining phase, and between this geomagnetic activity maximum and the following sunspot maximum may reflect different processes, and therefore can be used to estimate different parameters, depending on the relative importance of diffusion and advection.

3.4.2.1 Diffusion Versus Advection in the Surface Layer of the Solar Convection Zone

Two processes provide the poleward transport of the flux on the surface of the Sun, diffusion [65] and meridional circulation [113]. The dependence of the poloidal field during the next sunspot minimum and the amplitude of the following sunspot cycle on the time for which the flux reaches the poles is an indication which of the two time-scales is shorter: the advection time-scale $\tau_{\text{adv\_surf}} = L_{\text{surf}}/V_{\text{surf}}$ or the diffusion time-scale $\tau_{\text{dif\_surf}} = L_{\text{surf}}^2/\eta_{\text{surf}}$, where L_{surf} is the distance from sunspot latitudes to the poles, η_{surf} is the diffusivity in the upper part of the solar convection zone, and V_{surf} is the speed of the surface poleward circulation, or, in other words, which of the two processes is more efficient in carrying the flux to the poles at the surface.

If the advection timescale is shorter than the diffusion timescale, the meridional circulation will carry the flux to the poles before it can reach there by means of random supergranular diffusive walk. The fast poleward flow in this case means less time for the leading polarity flux to diffuse across the equator and to cancel with the leading polarity flux in the opposite hemisphere, so a larger fraction of the leading polarity flux remains in its own hemisphere and cancels on the way to the poles with the trailing-polarity flux, and less uncanceled trailing-polarity flux reaches the pole to form the polar field of the next cycle. From the weaker polar field, a weaker toroidal field is generated in the base of the convection zone [114].

If the time scale of the surface diffusion is shorter than that of the advection, a significant part of the poloidal field radially diffuses down before it can converge at the pole under the action of the meridional circulation [52], and all of the toroidal field is generated from the flux which has short circuited the meridional circulation. Only a small part of the trailing polarity flux reaches high latitudes before being diffused and reverses the polar field there. In this case, a shorter time for transporting the flux to the poles means a shorter time for the flux diffusion, and therefore a stronger poloidal field.

Figure 3.3 presents the dependence of the polar field on the flux transport time to the poles. Though we have values of the polar field for only four sunspot cycles so far (Wilcox Solar Observatory data obtained via the web site http://wso.stanford.edu, courtesy of J.T. Hoeksema), there is clearly seen the strong ($r = 0.8$) positive correlation between the time for the poleward transport of the flux and the strength of the resulting polar field, which indicates that transport by the meridional circulation is more important than supergranular diffusion.

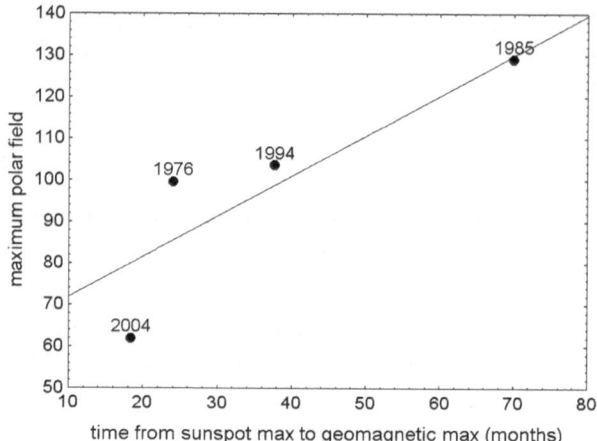

Fig. 3.3 Time between sunspot maximum and the following geomagnetic activity maximum on the sunspot declining phase (*dashed line*) and maximum magnitude of the polar field in sunspot minimum (*solid line*)

3.4.2.2 Calculation of the Speed of the Surface Meridional Circulation

If the advective transport is the main mechanism carrying the flux to the poles, the time between sunspot maximum and geomagnetic activity maximum on the sunspot declining phase is the time it takes the solar surface meridional circulation to carry the remnants of sunspot pairs from sunspot maximum latitudes ($\sim$15°) to the poles (Fig. 3.4a), so from this time we can calculate the average speed of the surface poleward circulation V_{surf}. With the solar radius equal to 6.96×10^8 m, these 75° on the solar surface are 9.11×10^8 m. If we divide this value by the time between the sunspot maximum and the geomagnetic activity maximum on the sunspot declining phase expressed in seconds (=months $\times$ 2.6352×10^6), we get directly the speed of the surface circulation V_{surf} in m/s. In the interval between sunspot cycle 10 and 23, V_{surf} calculated by this method varies between 5 and 20 m/s, averaged over latitude and over time between sunspot maximum and the moment the flux reaches the poles. In the last cycle 23, for which direct observations are available, the calculated speed is 16 m/s and agrees with measured velocities varying from 0 m/s at the equator to 20–25 m/s at midlatitudes to 0 m/s at the poles [111].

Figure 3.4b compares the calculated V_{surf} to the magnitude of the polar field of the next sunspot maximum, and Fig. 3.4c shows the dependence of the amplitude of the following sunspot maximum on the speed of the poleward surface circulation.

From Fig. 3.4 it is obvious that both the polar field B_{pol} and the maximum sunspot number of the following cycle R_{max} are negatively correlated with the preceding surface poleward meridional circulation, as it should be if the meridional circulation is the main transport mechanism in the upper part of the convection zone. The correlation coefficient $r(V_{surf}, R_{max})$ is -0.7 with $p = 0.03$. Though data for the polar field are available for only the last four cycles and the statistical significance cannot be evaluated, the negative correlation is very well expressed. The correlation coefficient $r(V_{surf}, B_{pol})$ estimated by the methods of nonparametric statistics (Statistica, Statsoft Inc. 1994–2001) is -0.8.

Fig. 3.4 (a) The distance traversed by the surface poleward circulation between sunspot maximum and the following geomagnetic activity maximum; (b) The calculated speed of the surface poleward meridional circulation V_{surf} (*solid line*) and the maximum polar field during the following sunspot minimum (*dashed line*). (c) The calculated speed of the surface poleward meridional circulation V_{surf} (*solid line*, note the reversed scale) and the amplitude of the following sunspot maximum (*dashed line*). From [26]

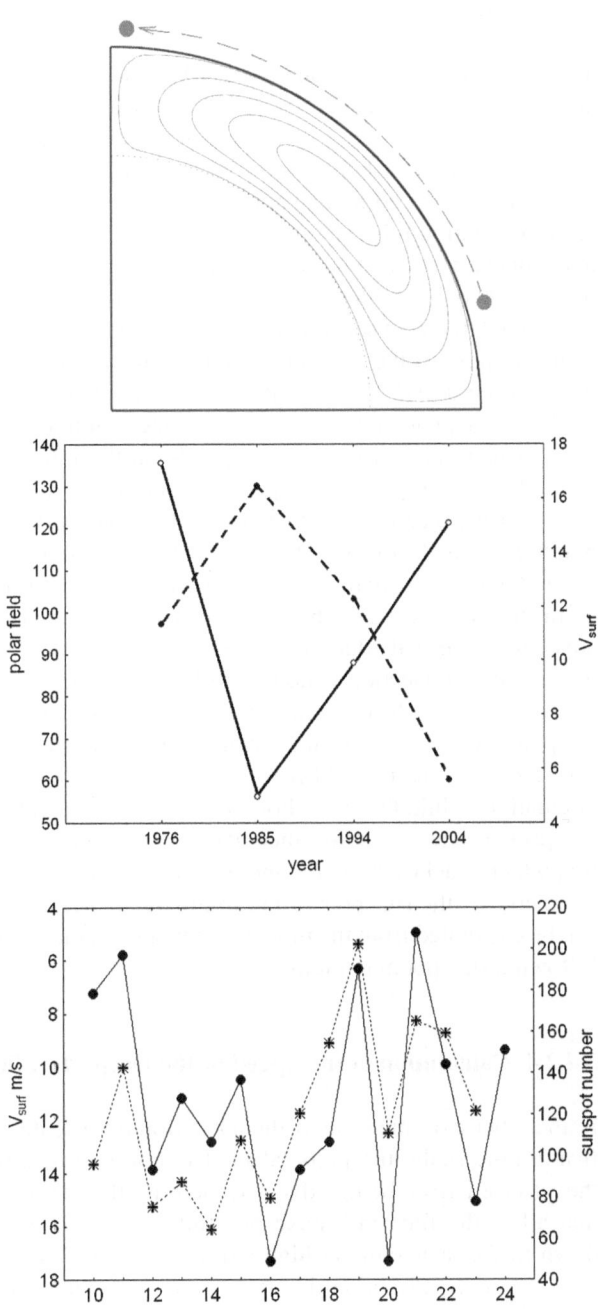

3.4.2.3 Diffusion and Advection in the Upper Part of the Solar Convection Zone

The fact that the correlation between V_{surf} and both the polar field of the next cycle and the following sunspot maximum is negative may not necessarily mean that the sunspot maximum is purely a product of the polar field. As pointed out by Jiang et al. [58], the polar field at sunspot minimum and the strength of the next sunspot maximum may be correlated not because the polar field is the direct cause of the next sunspot maximum by being advected from the pole to the tachocline. Rather, they may appear correlated because both of them arise from the poloidal field produced by the Babcock–Leighton process in the mid-latitudes.

Whether the next sunspot maximum is entirely created by the polar field advected from the pole to the tachocline and equatorward to sunspot latitudes, or whether the polar field *and* the toroidal field whose manifestation is the sunspot maximum are both independently produced from the poloidal field created from remnants of sunspot pairs in mid-latitudes, depends on the ratio of the time scales of the surface advection $\tau_{adv\_surf} = L_{surf}/V_{surf}$ and on the diffusion in the upper part of the solar convection zone (the part involved in poleward motion) $\tau_{dif\_surf} = L_{conv}^2/\eta_{surf}$, where L_{conv} is the distance between the surface and the base of the convection zone where the flux eventually reaches by diffusion to be transformed into toroidal field.

In the extreme case where the diffusion time scale is so much shorter than the advection time scale that the entire toroidal field is produced from the flux diffused through the convection zone before being able to reach the poles, the polar fields will be produced from a small part of the poloidal field which eventually reaches the poles. We are not dealing with this case because, as pointed out above, in this case the polar field would increase with increasing speed of the surface poleward circulation, while the data show the opposite dependence (Fig. 3.4b). In the other two possible cases, where either all or a part of the toroidal field is produced from the flux which reaches the poles and is carried down to the tachocline and equatorward to emerge as the sunspots of the following cycle, the speed of the deep circulation can be calculated from the time between the geomagnetic activity maximum and the following sunspot maximum.

3.4.2.4 Calculation of the Speed of the Deep Meridional Circulation

In the other extreme case, if the advection time scale is much shorter than the diffusion time scale, the poloidal field is not able to diffuse to the tachocline from the surface before being advected there by the meridional circulation. This means that all of the flux will make one full circle from sunspot latitudes to the poles, down to the tachocline at high latitudes and back to sunspot latitudes (Fig. 3.5a). This "strongly advection-dominated" regime can exist if the diffusivity is of order 10^6 m^2/s [58]. In this case, the time between the geomagnetic activity maximum during the sunspot declining phase and the next sunspot maximum is the time for the flux to sink to the base of the convection zone at polar latitudes, to be carried

Fig. 3.5 (**a**) The distance traversed by the circulation between geomagnetic activity maximum and the following sunspot maximum in the case of very low diffusivity; (**b**) Speed of the deep meridional circulation V_{deep} (*solid line*) and amplitude of the following sunspot cycle (*dashed line*)

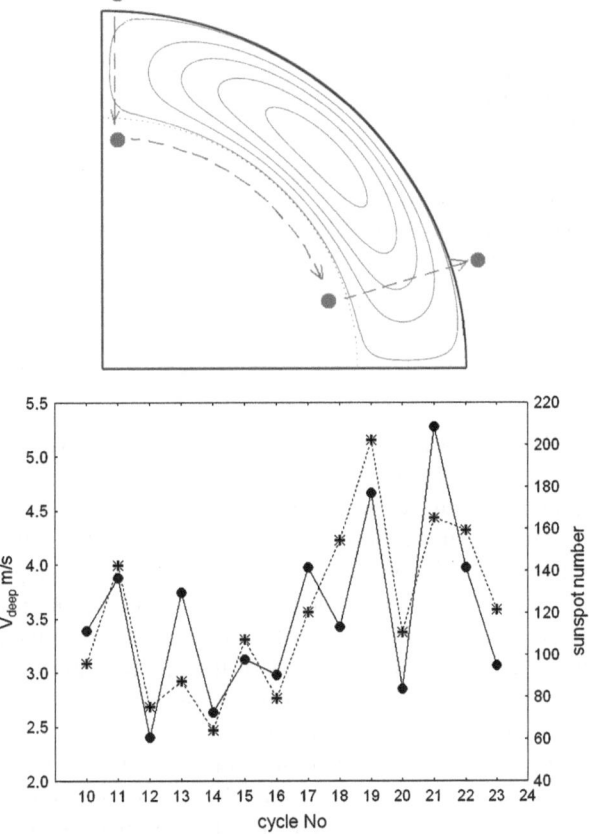

by the deep equatorward circulation to low latitudes, and to emerge as the sunspots of the new cycle. Assuming the speed of the downward transport of the flux equal to the speed of the deep equatorward circulation [58] and the time for the field tubes to emerge from the base of the convection zone to the surface equal to three months [24], we can calculate the speed of the deep meridional circulation V_{deep}. In Fig. 3.5b the calculated speed of the deep equatorward meridional circulation V_{deep} is presented, together with the amplitude of the following sunspot maximum. The values of V_{deep} vary between 2.5 and 5 m/s, in good agreement with estimations from helioseismology and latitudinal drift of the sunspot occurrence zone.

In a "moderfately advection-dominated regime, if the diffusivity is not so small but still less important than advection, $\eta \sim 1$–2×10^8 m^2/s [58], a part of the flux diffuses to the tachocline during the poleward transport before reaching the poles, short-circuiting the meridional circulation, and another part makes the full circle to the poles, down to the base of the convection zone and equatorward to sunspot latitudes (Fig. 3.6).

Fig. 3.6 Moderately
advection-dominated regime:
a part of the flux diffuses
through the convection zone,
"short-circuiting" the
meridional circulation,
another part makes a full
circle to the poles, down to
the base of the convection
zone and equatorward to
sunspot latitudes

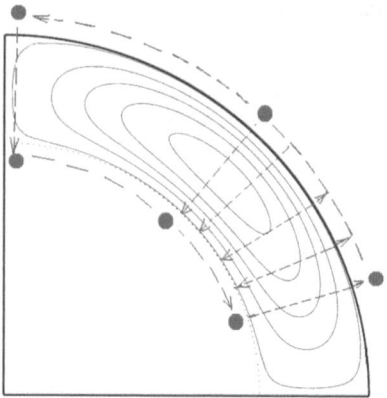

This, according to Jiang et al. [58], explains why the polar field at sunspot minimum and the strength of the next maximum are correlated: not because the toroidal
field of the sunspot cycle originates from the polar field during the preceding sunspot
minimum, but because they both arise from the poloidal field produced from the
remnants of sunspot pairs the mid-latitudes. A more probable explanation, however,
is that the polar field at the minimum and the strength of the next maximum are correlated *not only* because they both originate from the poloidal field produced from
the remnants of sunspot pairs of the previous cycle, but also because a part of the
toroidal field responsible for the sunspot maximum does originate from the polar
field during the preceding sunspot minimum. If this is the case, the sunspot maximum will be a superposition of the two surges of toroidal field, the one produced
from the part of the flux which has short-circuited the meridional circulation and
diffused directly to the tachocline to be transformed by the differential rotation into
toroidal field, and the other one produced from the part of the flux which has made
the full circle to the poles, down to the base of the convection zone and equatorward, again being transformed into toroidal field on the way to sunspot latitudes.
These two surges will follow closely one after the other because the time scales of
diffusion and advection are not much different in this regime, but will not exactly
coincide because the flux tubes from which they originate follow different paths, so
the result will be two maxima of toroidal field.

3.4.2.5 The Two Peaks in the Sunspot Cycle

The two maxima in solar activity were first found by Gnevyshev [35], who studied
the evolution of the intensity of the solar coronal line at 5303 Å in different latitudinal bands during the 19th sunspot cycle. While in globally and hemispherically
averaged data cycle 19 is single peaked, two peaks are seen in the separate latitudinal bands: the first peak, during which the coronal intensity increased and subsequently decreases simultaneously at all latitudes, appeared in 1957; the second
peak appeared in 1959–1960 and was only observed at low latitudes, but below 15°,

Fig. 3.7 Sunspot area in different latitudinal bands in sunspot cycle 16. The peak area at higher latitudes appears simultaneously in a wide latitudinal band (**a**) and at lower latitudes moves equatorward (**b**). (**c**) The white lines show the positions of the two maxima in time (from [26])

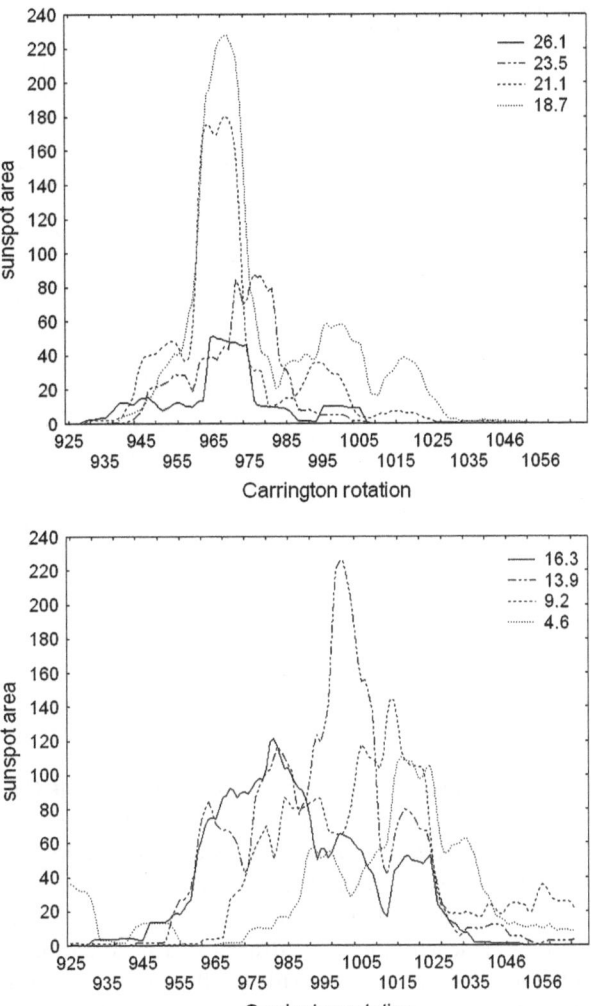

it was even higher than the first maximum. Antalova and Gnevyshev [4] checked whether this is a feature of only sunspot cycle 19 or of all cycles. They superposed the sunspot curves of all cycles from 1874 to 1962 and found that there are always two maxima in the sunspot cycle: the first one applies to all latitudes and appears simultaneously at all latitudes, and the second one occurs later and only at low latitudes. The relative amplitude of the two peaks and the time interval between them vary, so in some cycles they are seen as a single peak in latitudinal averages, while in other cycles the gap between them known as "Gnevyshev gap" is clearly seen (Fig. 3.1).

Figure 3.7 demonstrates two peaks in the sunspot area in cycle 16 (sunspot area data available online at http://solarscience.msfc.nasa.gov/greenwch.shtml, courtesy

Fig. 3.7 (Continued)

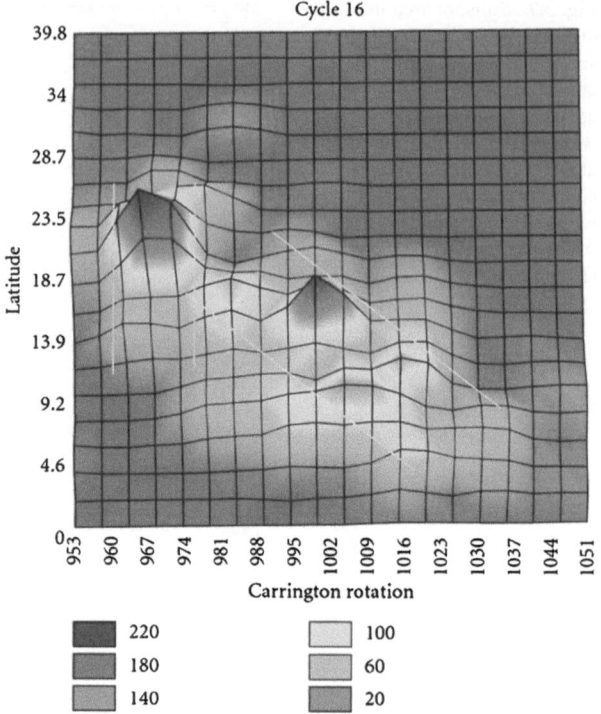

of David Hathaway). The first one is centered at Carrington rotation 965 and, as found by Gnevyshev [35] and Antalova and Gnevyshev [4], appears simultaneously in a wide latitudinal range, between 26.1 and 18.7° heliolatitude (Fig. 3.7a), the second one moves from 16.3° in Carrington rotation 981, to 13.9° in rotation 1003, to 9.2° in rotation 1018, and to 4.6° in rotation 1024 (Fig. 3.7b). We can identify the first peak with the flux diffused in a wide latitudinal area across the convection zone, and the second one with the flux advected all the way to the poles, down to the tachocline, and carried like on a conveyor belt by the deep meridional circulation equatorward to sunspot latitudes. In Fig. 3.7c the total sunspot area in cycle 16 is plotted as a function of time and latitude.

The diffusion-generated peak appears earlier and at higher heliolatitudes in all cycles from 15 to 19. The order is reversed in cycles 12–14 and 20–23: first, the advection-generated peak at higher latitudes and then the diffusion generated peak at lower latitudes. An example (cycle 21) is shown in Fig. 3.8. First, the advection generated peak appears in Carrington rotation 1650 and until rotation 1690 moves from about 20° to about 10° (Fig. 3.8a); the second peak, centered around Carrington rotation 1705, appears simultaneously at all latitudes below 15° (Fig. 3.8b).

Figure 3.9 presents the surface plots of the sunspot area in all cycles from 12 to 23 (averaged over the two hemispheres) as functions of time and latitude.

It seems that the order in which the diffusion-generated and the advection-generated peaks appear changes either in ascending and descending phases of the

Fig. 3.8 The same item as in Fig. 3.7, but for sunspot cycle 21. From [26]

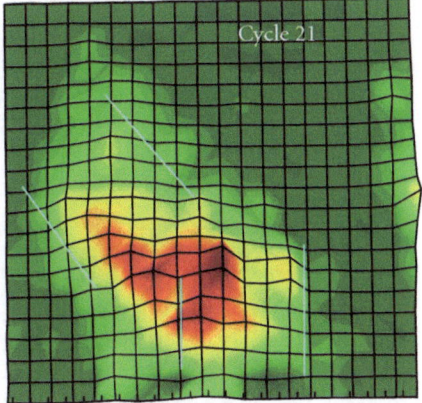

secular cycle or, if cycle 20 is indeed already on the ascending branch of the next secular cycle [62, 116], in consecutive secular cycles (Fig. 3.10). This obviously has a connection with the long-term variations in solar activity and can give additional information about solar dynamo.

3.4.2.6 Calculation of the Diffusivity in the Upper Part of the Solar Convection Zone

From the time between the geomagnetic activity peak on the sunspot declining phase and the diffusion-generated peak in sunspot activity, the average diffusivity in the upper part of the convection zone can be evaluated. This time is the time it takes the flux to reach by means of diffusion from the surface to the lower part of the convection zone involved in equatorward motion $t = L_{conv}^2 / \eta_{surf}$, plus the time for the flux to be carried by the deep meridional circulation from mid-latitudes to the sunspot occurrence zone, being transformed on the way into toroidal field, plus the time for the field tubes to emerge as the sunspots of the next cycle. The speed of the deep circulation, used to derive the diffusivity, can be calculated more accurately from the time between the geomagnetic activity peak on the sunspot declining phase and the advection-generated peak in sunspot activity. The calculated values of the diffusivity (Fig. 3.11) are in agreement with estimations based on the observed turbulent velocities and size of convection cells [66, 97] and with considerations about the correlation between the two solar hemispheres and between the strength of the polar field and the amplitude of the following sunspot maximum [58]. Moreover, they correspond to the estimated range of values for which moderately advection-dominated regime exists [58].

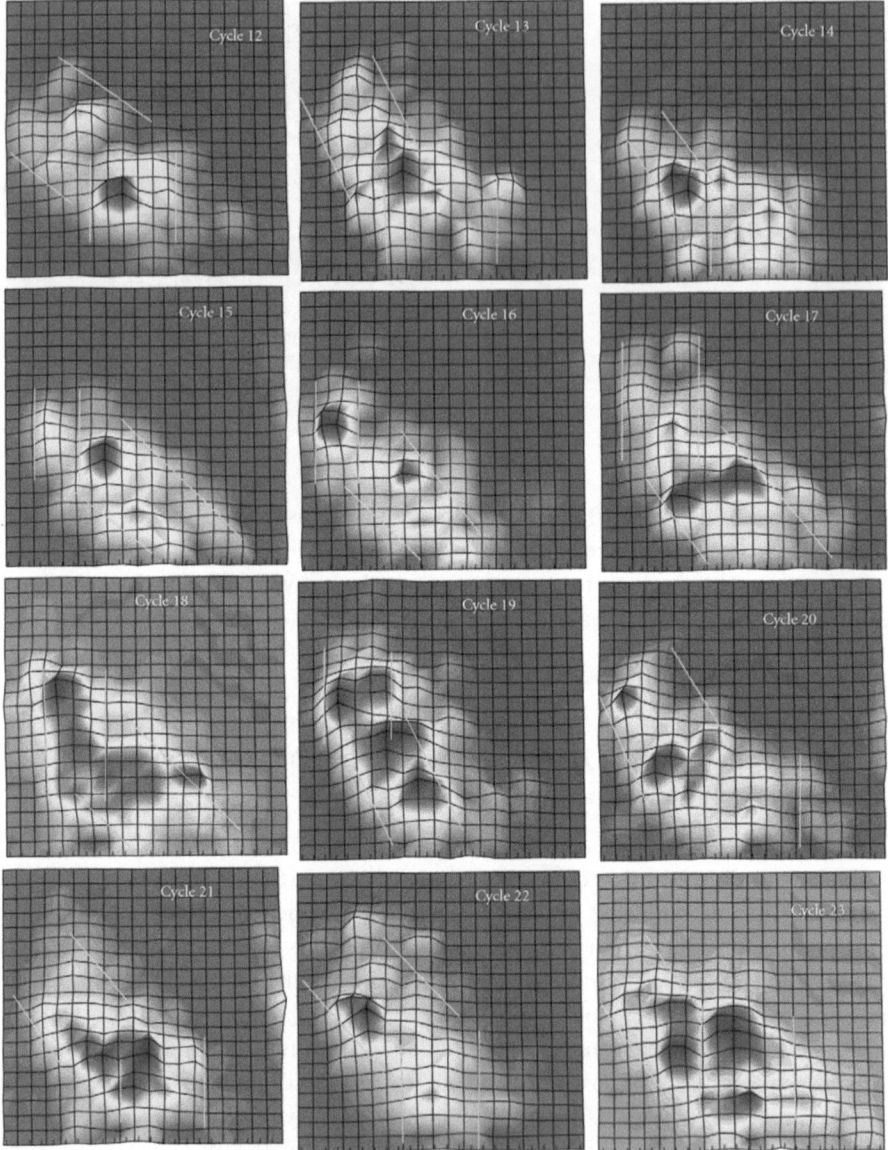

Fig. 3.9 The same item as in Fig. 3.7c, but for cycles from 12 to 23. From [26]

3.4.3 Regimes of Operation of the Solar Dynamo

According to the Babcock–Leighton dynamo mechanism, the solar poloidal and toroidal fields are generated in two separate domains, the poloidal field in the upper part of the convection zone and the toroidal field at the tachocline at the base of the convection zone. The physical characteristics of these two domains are quite

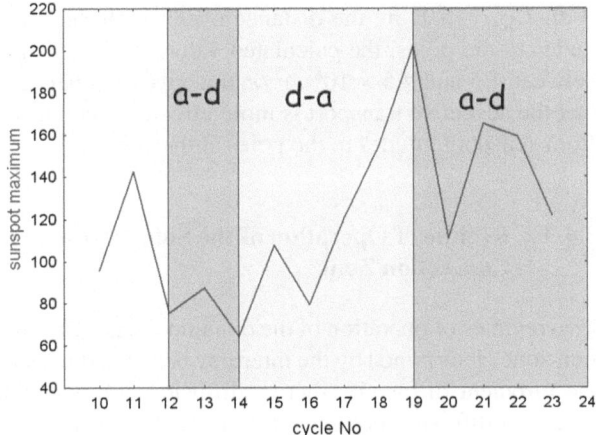

Fig. 3.10 Advection-dominated before diffusion-dominated peak ("a–d") and diffusion-dominated before advection-dominated peak ("d–a") in the secular solar cycle. From [26]

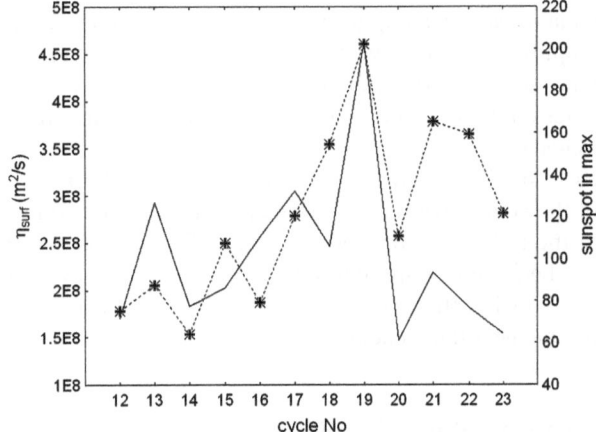

Fig. 3.11 Diffusion through the convection zone in the case of moderately advection-dominated regime in the upper part of the convection zone. From [26]

different; therefore the regimes of operation of the dynamo mechanism there may be also different.

3.4.3.1 Regime of Operation in the Upper Part of the Solar Convection Zone

The negative correlation found between the speed of the surface poleward meridional circulation and the magnitude of the polar field (Fig. 3.3) indicates that advection is more important than diffusion in the upper part of the solar convection zone. Now that we have both the speed of the surface circulation and the diffusivity in the upper part of the solar convection zone, we can calculate the magnetic Reynolds number, which is a measure of the relative importance of advection versus diffusion:

$$R_m = V_{surf} L_{surf} / \eta_{surf}.$$

With $L_{surf} \sim 10^9$ m, the distance over which the flux is carried from sunspot latitudes to the poles, the calculated values of V_{surf} between 5 and 20 m/s and η_{surf} between 1.5 and 4.5×10^8 m^2/s, this gives R_m between 10 and 60, which confirms that the advective transport is more effective than the diffusion for carrying the flux from sunspot latitudes to the poles at the surface.

3.4.3.2 Regime of Operation of the Solar Dynamo in the Lower Part of the Convection Zone

Two regimes of operation of the dynamo are possible in the lower part of the convection zone, determined by the interplay between the deep equatorward circulation and the turbulent diffusivity there, "diffusion-dominated" and "advection-dominated" [115]. In diffusion-dominated regime, if diffusion is more important than advection in carrying the flux at the base of the convection zone, a higher circulation speed means less time for diffusive decay of the poloidal field during its transport equatorward, leading to more generation of toroidal field and hence a higher cycle amplitude. In the advection-dominated regime, the diffusive decay is not so important, and a higher circulation speed leads to lower cycle amplitude because there is less time to generate toroidal field in the tachocline through which the magnetic fields are swept at a faster speed. As obvious in Fig. 3.5b, a high and highly statistically significant positive correlation ($r = 0.81$, $p < 0.01$) exists between the speed of the equatorward circulation at the base of the convection zone and the amplitude of the following sunspot maximum. This means that in the last 13 cycles diffusion has been more important than advection in the bottom part of the solar convection zone, or in other words, solar dynamo has been operating in diffusion-dominated regime near the tachocline.

3.4.3.3 Waldmeier's Rule

The dependence of the cycle amplitude on the speed of the preceding deep circulation in the diffusion-dominated mode explains why the rise time of a cycle anticorrelates with the amplitude of the cycle; this is the so-called Waldmeier rule (e.g., [12, 48, 49, 101]). The rise time is determined by the time it takes the deep meridional circulation to carry the flux from the poles to sunspot maximum latitudes, which is inversely proportional to the speed of the deep meridional circulation (Fig. 3.12), and in the diffusion-dominated regime in which the solar dynamo has been operating at least since the middle of the 19th century, the faster the flux is carried through the tachocline, the higher the following sunspot maximum.

3.4.3.4 Maunder Minimum and Other Grand Minima

For periods before the beginning of systematic solar and geophysical observations, we have to rely on proxy data. Cosmogenic radionuclides stored in natural archives

Fig. 3.12 Dependence of the sunspot cycle rise time on the speed of the deep equatorward circulation

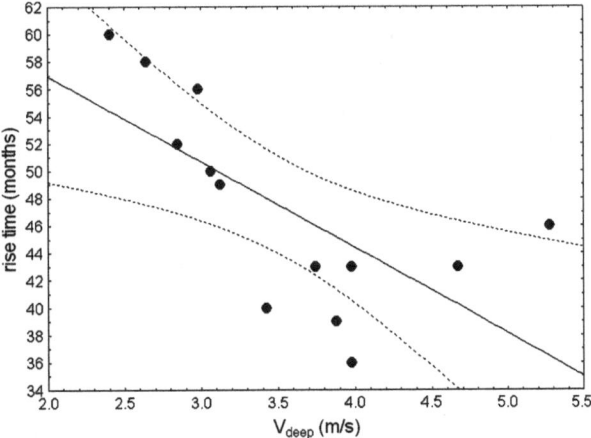

Fig. 3.13 Speed of the deep meridional circulation (*solid line*) and amplitude of the sunspot maximum (*broken line*) during the Maunder minimum. From [26]

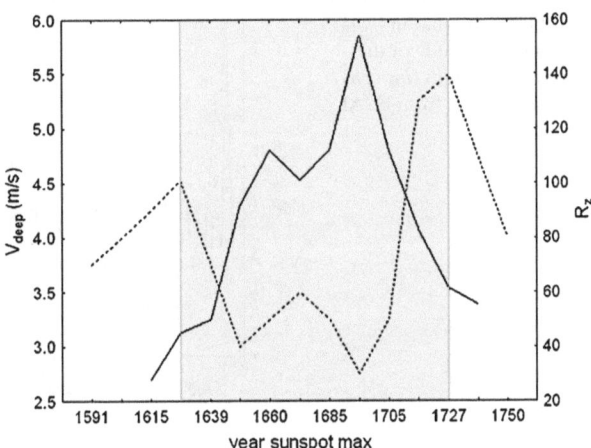

such as ^{10}Be in ice cores and ^{14}C in tree rings are produced in the Earth's atmosphere under the action of galactic cosmic rays which are modulated by the solar open flux related to the Sun's large-scale poloidal field and have proven to be a valuable tool in reconstructing past solar activity and changes in the geomagnetic field intensity over several millennia. At present, this is the only method to extend back the record of solar activity beyond the instrumental period [1]. Taking the years of minima in ^{10}Be [8] as a proxy for the changes in the open magnetic field of the Sun, the group sunspot number Rz [53], and the reconstructions of Schove [98, 99] for sunspot minima and maxima as a measure of sunspot activity, we can estimate the solar meridional circulation during the Maunder minimum in the same way as described in Sect. 3.4.2. Figure 3.13 presents the dependence of the sunspot maximum on the speed of the preceding deep meridional circulation during the Maunder minimum. In contrast to Fig. 3.5b, the correlation between V_{deep}

Fig. 3.14 Speed of the deep meridional circulation (*solid line*) and amplitude of the sunspot maximum (*dashed line*) since 1619 from ESAI database

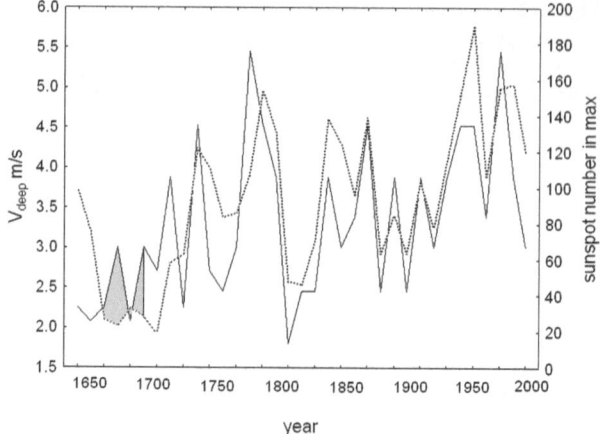

Fig. 3.15 Speed of the deep meridional circulation (*solid line*) and amplitude of the sunspot maximum (*broken line*) since 1099 from ESAI database

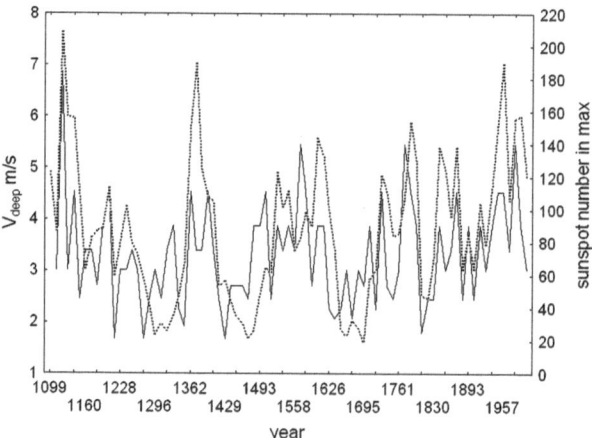

and R_z is negative: the faster the speed of the deep circulation, the lower the cycle amplitude.

The negative correlation between V_{deep} and the sunspot activity during the Maunder minimum is confirmed by reconstructed data from the ESAI, Extended time series of Solar Activity Indices, for geomagnetic activity aa-index and yearly sunspot areas since 1619 [75–77] (Fig. 3.14). An even longer ESAI reconstruction of the yearly Wolf numbers and the yearly aa-indices since 1099 [78] confirms that the correlation between V_{deep} and sunspot activity was negative during other grand minima of solar activity: Wolf minimum from about 1280 to 1350 and Spörer minimum from about 1450 to 1550 (Fig. 3.15).

Figures 3.14 and 3.15 demonstrate that during all grand minima since 1099 a faster deep equatorward meridional circulation was followed by a lower sunspot maximum. According to the model of Yeates et al. [115], this means that in these periods the time which the differential rotation has to generate the toroidal field at the base of the convection zone is more important than diffusion which acts on the

Fig. 3.16 Dependence of the maximum sunspot number in a cycle on the rise time from sunspot minimum to sunspot maximum during the Maunder minimum

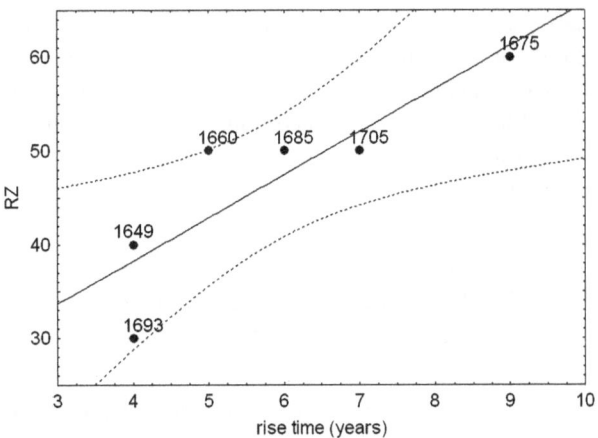

field during this time, or, in other words, this means during the Maunder minimum and other grand minima the solar dynamo operated in advection dominated regime at the base of the solar convection zone.

The Maunder minimum period can provide a test for the explanation of the Waldmeier rule, needed to prove the validity of the Babcock–Leighton-type dynamo model [101]. If the Waldmeier rule relating the sunspot cycle amplitude to the cycle rise time (the shorter cycle rise time, the higher sunspot maximum) is due to the solar dynamo operating in diffusion-dominated regime at the base of the convection zone, it should not hold during the Maunder minimum when the dynamo operated in advection-dominated regime. In advection-dominated regime, when diffusion is less important than advection, the shorter rise time means the shorter time for generation of toroidal field at the base of the convection zone and therefore lower sunspot maximum. Figure 3.16 demonstrates that this is exactly the case during the Maunder minimum: the slower the rise to sunspot maximum, the higher the maximum. The correlation between the rise time and the amplitude of the cycle is positive with $r = 0.84$ and $p = 0.037$.

Therefore, the Waldmeier rule about the relation between the cycle rise time and amplitude is a consequence of the dependence of the cycle amplitude on the speed of the deep meridional circulation, and its validity depends on the mode of operation of the solar dynamo in the lower part of the solar convection zone. The Waldmeier rule (faster rise = stronger cycle) is a manifestation of this relation in the case of diffusion-dominated regime, and it does not hold in the advection-dominated regime.

3.5 Factors Determining the Period and Amplitude of the Sunspot Cycle

In Sect. 3.4 it was demonstrated that, because of the relative importance of diffusion and advection, in the last 13 sunspot cycles the speed of the surface poleward meridional circulation has been anticorrelated with the polar field and with the amplitude of the following sunspot maximum. Further, it was shown that the speed of the deep equatorward circulation has been positively correlated with the amplitude of the following sunspot maximum. Now we are summarizing the sequence of relations in order to identify the factors responsible for the fluctuations in the sunspot cycles parameters.

3.5.1 Surface Meridional Circulation—Polar Field and Sunspot Maximum: Strong Correlation

The speed of the surface meridional circulation is important for the generation of the poloidal field from the toroidal field. As seen above, in moderately advection dominated regime in the upper part of the solar convection zone, a part of this newly generated poloidal field reaches the poles to cancel the polar field of the old cycle, to accumulate as the polar field of the new cycle, and to be carried down to the tachocline and equatorward, being transformed into toroidal field. Another part diffuses directly toward the tachocline at mid-latitudes where it joins the conveyor belt and is again carried equatorward and transformed into toroidal field. Therefore, the speed of the surface poleward circulation should correlate with both the polar field and with the amplitude of the sunspot maximum following it. Figure 3.4 demonstrates that a stronger polar field and a higher sunspot maximum follow after a slower surface poleward circulation V_{surf}, as it should be in the case where the advection time-scale is shorter than the diffusion time-scale: with slower surface circulation, more leading polarity flux diffuses across the equator to cancel with the leading polarity flux of the opposite hemisphere, and more uncanceled trailing polarity flux is left from which more poloidal field is generated, so more poloidal field is carried to the poles, and more poloidal field diffuses to the tachocline.

3.5.2 Polar Field—Deep Meridional Circulation: NO Correlation

There is controversy in the literature regarding the possible correlation between the maximum poloidal field around sunspot minimum and the speed of the deep equatorward circulation carrying this poloidal field equatorward to produce the following sunspot maximum. The basis of the expected correlation [15] is the fact that a large-scale magnetic field produces a Lorentz force which quenches the large-scale plasma motions. This back reaction of the magnetic field on the flow is known as the Malkus–Proctor mechanism [68].

Model studies have mainly dealt with the effects of the magnetic field on the solar differential rotation, and only few have focused on the effects on the meridional circulation. Rempel et al. [90] simulated the feedback of the Lorentz force on the meridional flow and found that the magnetic field will quench the transport by the

Fig. 3.17 Correlation between the magnitude of the solar polar field and the speed of the following deep meridional circulation. From [26]

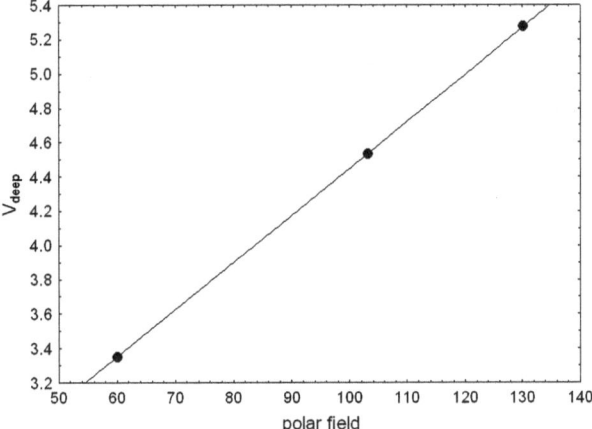

meridional circulation with a speed of a few m/s of toroidal field of more than 30 kG, while the transport of the much weaker poloidal field ($\sim$1–2 G) from which the toroidal field is produced will be unaffected.

Figure 3.17 demonstrates the correlation between the magnitude of the solar polar field and the speed of the following deep meridional circulation. Not only the magnetic field does not quench the large-scale motions as should be expected if the Malkus–Proctor mechanism were acting upon the poloidal field leading to anticorrelation between the field strength and the speed of the deep meridional circulation, but the correlation is opposite. We have only three cycles with measured polar fields and the speed of the deep equatorward meridional circulation following it, but the strong positive correlation is obvious. This means that the Malkus–Proctor effect is much weaker than the effect of some other factor, and the apparent correlation between the polar field and the deep circulation is a result of the influences of this factor on both of them.

3.5.3 Surface Poleward Circulation—Deep Equatorward Circulation: Strong Correlation

As it can be seen from Figs. 3.4 and 3.5, the higher the speed of the surface circulation, the lower the sunspot maximum following it; the higher the speed of the deep circulation, the higher the sunspot maximum following it. In other words, the speeds of the surface and deep circulation are anticorrelated: the higher the speed of the surface circulation, the lower the speed of the deep circulation following it: $r = 0.7$ with $p = 0.005$, see Fig. 3.18.

What is the physical meaning of the ratio of the speeds of the surface and deep meridional circulation? In periods of "normal" solar activity the speed of the surface circulation is much higher than the speed of the deep return flow, both as calculated from geomagnetic data and as assumed in models. This is because the density in

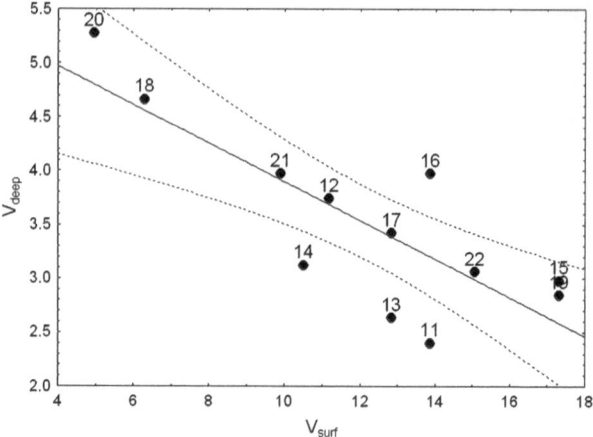

Fig. 3.18 Speed of the surface meridional circulation after a sunspot maximum and the following deep meridional circulation

the solar convection zone quickly increases with depth: 0.5 % of the solar mass is contained in the upper half of the convection zone versus 2.5 % in the lower half, so a slower deep flow is sufficient to balance the mass carried by the faster surface flow. The depth where the direction of the flow reverses from poleward to equatorward has not been measured directly, and indirect estimations are controversial. Giles et al. [32] from time-distance helioseismology evaluated a reversal depth of $0.94R_S$, but Braun and Fan [10] pointed out that the above result is not consistent with the observations of SOHO Solar Investigations Observation–Michelson Doppler Imager (SOI–MDI). Using the Fourier–Hankel method of spectral decomposition of the solar oscillation signal, Braun and Fan [10] found no reversal in the whole upper half of the convection zone. Later Giles [31], again from time-distance helioseismology, estimated a possible reversal depth of $0.8R_S$, while Duvall and Kosovichev [21], using local area helioseismology, found no reversal at all down to $0.725R_S$. Krieger et al. [63], applying the Fourier–Hankel spectral decomposition method of Braun and Fan [10] on data from SOHO SOI–MDI for 1999, also found a reversal depth of $0.8R_S$. On the other hand, applying a method using global seismology, Mitra-Kraev and Thompson [73] found a meridional flow that decreased with depth and became equatorward at a depth of only 40 Mm, but with big uncertainty, with a possible deeper counter-cell (however, Jouve and Brun [59] demonstrated that it is unlikely that multicellular meridional flows persist for a long period of time inside the Sun). Recently Hathaway [46], by tracking the motions of supergranules, found a return flow below 35 Mm, which is $0.95R_S$.

The ratio of the speeds of the surface and deep circulations provides another estimate of the reversal depth of the large-scale meridional circulation because this ratio depends on what part of the convection zone is involved in poleward and what in equatorward flow. Therefore, if we know the mass distribution in the Sun and the radius to which the whole circulation system penetrates, from the ratio of the speeds we can calculate the depth where the direction of the flow reverses. Different models suggest values of the penetration depth of the meridional circulation between $0.7R_S$

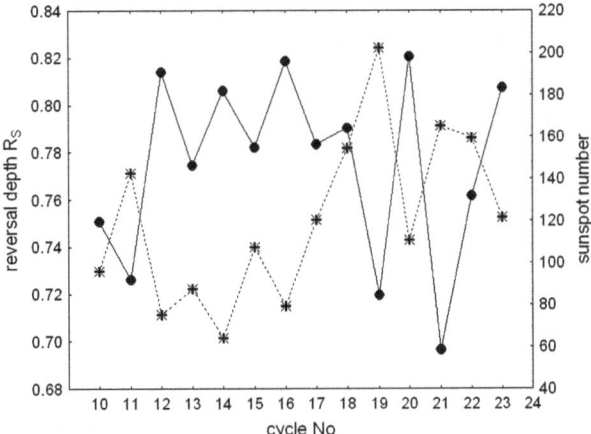

Fig. 3.19 *Solid line*: depth in the solar convection zone where the flow direction changes from poleward at the surface to equatorward below the reversal depth, in solar radii R_S; *dashed line*: amplitude of the following sunspot maximum

and $0.65R_S$, and even deeper [25, 33]. The difference between penetration depth of $0.7R_S$ and $0.65R_S$ leads to a difference in the reversal depth of up to $0.03R_S$.

Figure 3.19 presents the long-term variations in the estimated in this way reversal depth of the solar meridional circulation for penetration depth of $0.7R_S$, with the mass distribution in the Sun taken from the Standard Solar Model BS2005-AGS,OP [6] available online at http://www.sns.ias.edu/~jnb/SNdata/solarmodels.html. Also plotted for comparison is the amplitude of the following sunspot cycle. The calculated reversal depth varies between about $0.7R_S$ and $0.8R_S$, in clear anticorrelation with the amplitude of the following sunspot cycle: a higher sunspot maximum follows after a deeper reversal.

It should be noted that in solar dynamo simulations it is generally assumed that the speeds of the surface and the deep circulation increase and decrease simultaneously. This would mean a constant ratio of the speeds and a constant reversal depth. Figure 3.19 demonstrates that probably this is not the case: with increasing speed of the surface circulation, a smaller upper part of the convection zone is involved in poleward motion, the flow reverses direction closer to the surface, and the speed of the deep part of the convection zone decreases to compensate for the bigger mass transported equatorward.

3.5.4 Deep Meridional Circulation—Next Sunspot Maximum: Strong Correlation

As demonstrated in Figs. 3.5b and 3.13, the amplitude of the sunspot cycle is highly correlated with the speed of the deep equatorward circulation preceding it, positively during periods of "normal" activity and negatively during grand minima. The explanation, given by Yeates et al. [115] is that if diffusion dominates in the lower part of the solar convection zone, a faster circulation means less time for diffusive decay of the poloidal magnetic field during its equatorward transport along the tachocline

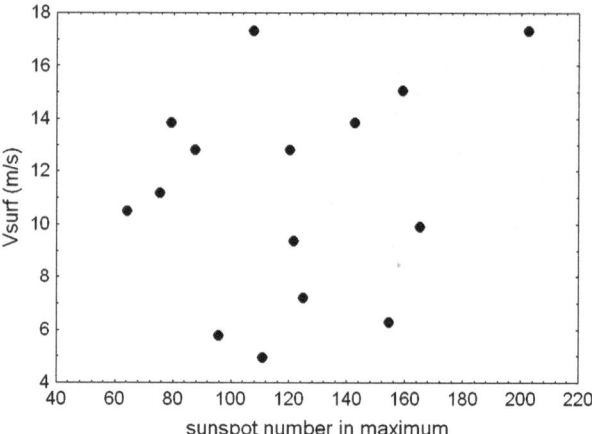

Fig. 3.20 Dependence of the surface meridional circulation V_{surf} on the maximum sunspot number preceding it. From [26]

where it is transformed by the differential rotation into toroidal field, while if the dominant transport mechanism is advection, the diffusion is not so important, and a faster circulation means less time for generation of toroidal field during the equatorward transport of the poloidal field. This is also the probable reason for the correlation between the depth where the flow reverses its direction from poleward to equatorward and the amplitude of the following sunspot cycle (Fig. 3.19): because, for mass conservation, with deeper reversal the speed of the equatorward circulation increases to balance the bigger mass transported poleward in the upper layers.

3.5.5 Sunspot Maximum—Surface Meridional Circulation: NO Correlation

There is no correlation at all between the amplitude of the sunspot maximum and the speed of the surface meridional circulation after it (Fig. 3.20). This finding is in agreement with the lack of correlation found by Jiang et al. [58] between the maximum sunspot number of a cycle and the polar field at the end of this cycle whose strength is determined by the surface circulation after the sunspot maximum.

The chain of influences is the following:

$$V_{surf} \Rightarrow B_{pol}, \qquad V_{deep} \Rightarrow B_{tor} \times V_{surf} \Rightarrow B_{pol},$$
$$V_{deep} \Rightarrow B_{tor} \times V_{surf} \Rightarrow B_{pol}, \qquad V_{deep} \Rightarrow B_{tor}, \ldots.$$

The speed of the surface circulation determines both the polar field and the speed of the deep circulation; the two of them determine the toroidal field and the amplitude of the sunspot cycle as its manifestation, but the toroidal field has no connection whatsoever with the speed of the following surface circulation. The chain breaks there. This finding is in agreement with the suggestion of Choudhuri [17] that while the production of toroidal field by the stretching of the poloidal field by differential rotation in the tachocline and the advection of poloidal field by the meridional

circulation first to high latitudes and then down to the tachocline are reasonably or-
dered and deterministic processes, the production of poloidal field by the Babcock–
Leighton mechanism of decay of tilted bipolar sunspots involves some randomness,
a "random kick" responsible for the fluctuations in the amplitudes and periods of
the solar cycles. According to Choudhuri [17], this randomness is due to the scatter
in the tilt angles of bipolar sunspot groups. As seen in Fig. 3.20, another possible
reason for the randomness is the lack of correlation between the toroidal field and
the following surface circulation responsible for the production of the poloidal field.

3.5.6 Sunspot Cycle "Memory"

The lack of correlation between the sunspot maximum and the speed of the following surface meridional circulation after it limits the sunspot cycle "memory"
assumed in some models, the correlations between the parameters of cycle n and
the amplitude of cycles $n + 1, n + 2$, etc.

Charbonneau and Dikpati [16] suggested that the duration of the Sun's memory of its own magnetic field in flux-transport dynamos is governed primarily by the
meridional flow speed and is no less than two solar cycles. This long memory in their
advection dominated model is achieved through very slow deep equatorward circulation—0.15 m/s versus 15 m/s for the surface poleward circulation—combined
with survival of multiple old-cycle polar fields thanks to the low diffusivity during the prolonged time for advective transport in the deep convection zone from
the poles to sunspot latitudes. However, as demonstrated by the negative correlation
between the time it takes the deep circulation to carry the flux to sunspot latitudes
and the amplitude of the following sunspot maximum, the diffusion is more important than the advective transport in the lower part of the solar convection zone,
so multiple old-cycle polar fields cannot survive a very slow transport. Indeed, as
demonstrated by Fig. 3.21, the sunspot cycle maximum is correlated to the speed of
the deep circulation preceding it, with a smaller contribution from the previous cycle and no contribution from earlier cycles. Therefore, the sunspot cycle has a short
memory.

According to the model of Yeates et al. [115], in the diffusion-dominated regime
the toroidal field for cycle $n + 1$ is produced primarily from the poloidal field of cycle n, with a small contribution from cycle $n - 1$, while in the advection-dominated
regime poloidal fields from cycles n, $n - 1$, and $n - 2$ combine to produce the
toroidal field for cycle $n + 1$. The reason for the correlations to persist from cycle
to cycle (and therefore to provide longer-than-one cycle memory) in the advection-
dominated regime and not in the diffusion-dominated regime according to Yeates
et al. [115] is that in the advection-dominated regime fluctuations are passed on in
both the poloidal-to-toroidal and in the toroidal-to-poloidal phases of the cycle by
means of the meridional circulation, while in the diffusion-dominated regime the
poloidal field is carried to the tachocline by means of downward diffusion "short-
circuiting" the meridional circulation, so the correlations are passed on only in the

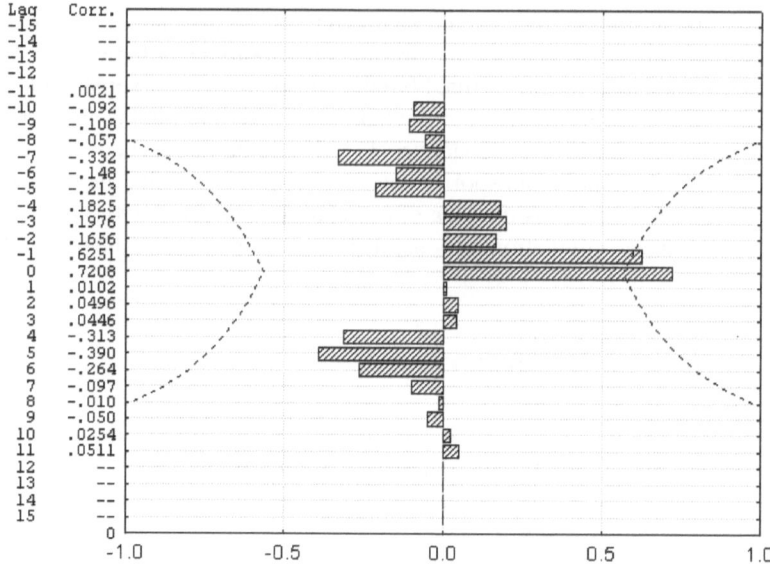

Fig. 3.21 Dependence of the maximum sunspot number in a cycle on the speed of the deep meridional circulation preceding it in the period 1868–2005

poloidal-to-toroidal phase. However, the cross-correlation between the speed of the deep meridional circulation and the maximum sunspot number for the period of the Maunder minimum (Fig. 3.22) demonstrates that the memory is equally short in both diffusion-dominated and advection-dominated regimes. The reason is the lack of connection of the toroidal field with the following surface circulation.

We can therefore conclude that a major factor important for the amplitude of the sunspot cycle is the speed of the surface poleward meridional circulation. In other words, the "random kick" suggested by Choudhuri [17], which solar dynamo receives at the time of transformation of the toroidal field into poloidal field, responsible for the cycle-to-cycle fluctuations of solar activity, acts on V_{surf}. On the other hand, the systematic variations in the sunspot cycles characteristics, the secular or Gleissberg cycle [34], imply that the "kick" may not be entirely random. The question is what process, if any, modulates V_{surf}.

3.6 Planetary Influences on Solar Activity

The dynamo theory outlined above explaining solar and stellar magnetic activity works for any Sun-like star with a convective envelope. If the star has planets, tidal effects exerted by the planets on the surface of the star can be described by the classical tidal theory. The tidal driving force is the gradient of the gravity field of

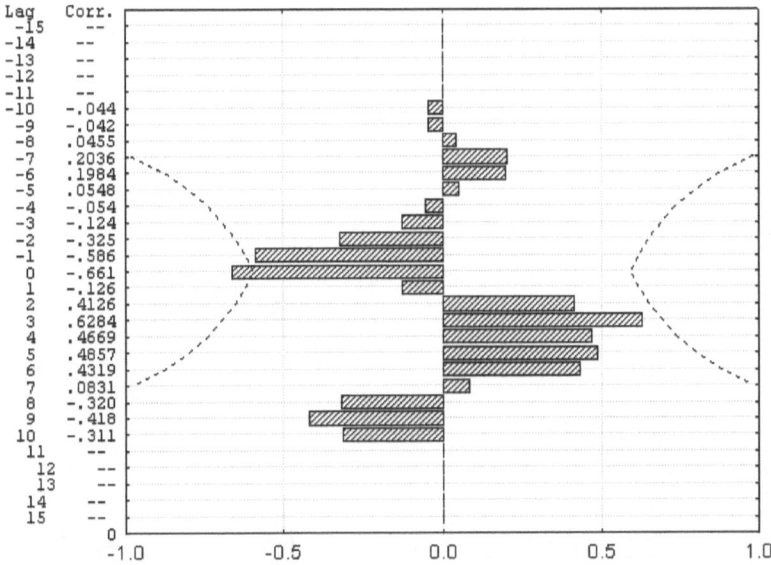

Lag	Corr.
-15	--
-14	--
-13	--
-12	--
-11	--
-10	-.044
-9	-.042
-8	.0455
-7	.2036
-6	.1984
-5	.0548
-4	-.054
-3	-.124
-2	-.325
-1	-.586
0	-.661
1	-.126
2	.4126
3	.6284
4	.4669
5	.4857
6	.4319
7	.0831
8	-.320
9	-.418
10	-.311
11	--
12	--
13	--
14	--
15	--

Fig. 3.22 Dependence of the maximum sunspot number in a cycle on the speed of the deep meridional circulation preceding it during the Maunder minimum

the planets. In the simplest case of only one planet orbiting in the stellar equatorial plane, the tide generating potential V is

$$V = -\frac{\gamma M r^2}{2R^3}\left(3\cos^2\varphi - 1\right),$$

where γ is the gravitational constant, M is the planet's mass, R is the distance between the centers of the star and the planet, r is the distance from the star's center, and φ is the latitude on the star's surface. The tide generating force has components perpendicular and parallel to the stellar surface. The horizontal component is

$$H = -\frac{1}{r}\frac{\partial V}{\partial \varphi} = \frac{2G}{r}\sin 2\varphi,$$

where $G = \frac{3}{4}\gamma M(\frac{r^2}{R^3})$ [14]. Figure 3.23 adapted from [19] illustrates the distribution of the horizontal component of the tidal force on the surface of the star when the tide-generating planet is above the subplanetary point SP at the equator. All vectors are directed to SP and its projection, with amplitudes depending on the planet's mass and distance from the star.

In the case of the Sun with a number of planets, the tidal forces depend on the distance and relative positions of the planets which change with time.

The tidal forces of the planets onto the solar surface were calculated by P.A. Semi [30], using JPL planetary and lunar ephemerides, version DE406 available at ftp://ssd.jpl.nasa.gov/pub/eph/export, which specify planets' positions at any time covered by their time-span, in ICRS Earth-centered reference frame. The reference

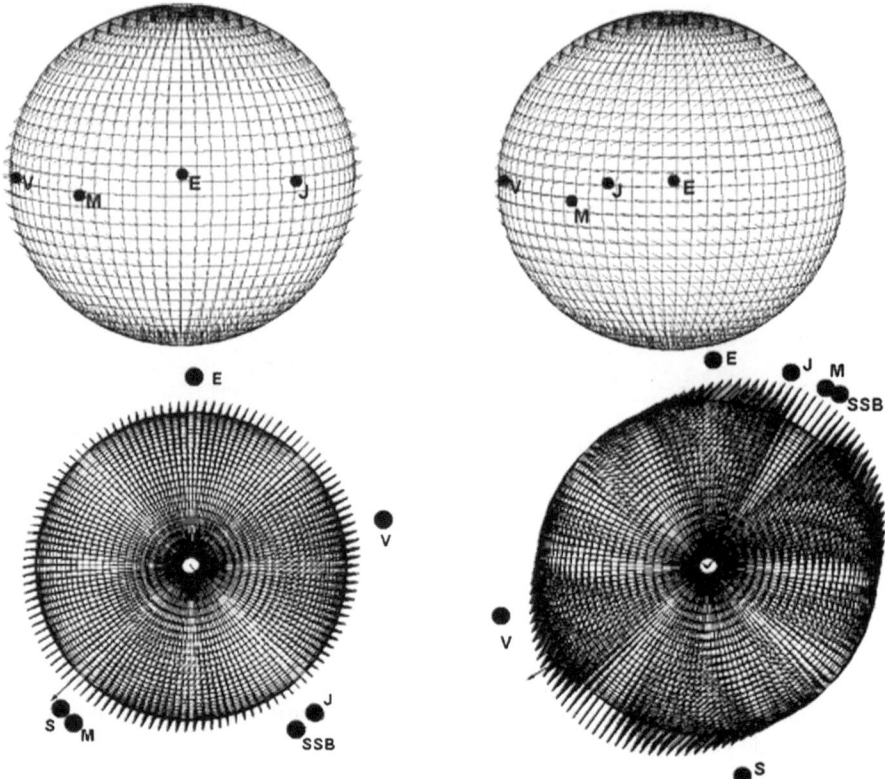

Fig. 3.23 Horizontal (*upper panels*) and vertical (*lower panels*) tidal forces due to the combined action of all planets, together with the positions of Mercury (M), Venus (V), the Earth–Moon system (E), Jupiter (J), Saturn (S), and the Solar system barycenter (SSB) in two periods, September 2005 (*left*) and September 2009 (*right*)

frame is first rotated to transfer it into Sun-centered. All nine planets are used including Pluto (as specified in the Ephemerides, excluding only the asteroids). The Solar system barycenter is also taken into account. Instead of the Earth alone, the Earth–Moon barycenter is used, with the combined mass of Earth and Moon. On the solar surface, the mesh of points is set up with 5° spacing, and the tidal force of each planet is evaluated at each mesh point, which are then summed over the latitude to get latitudinal averages of the tidal force. The daily values are calculated at midnight UTC in 1 Earth day steps, which are then averaged either to monthly or yearly averages.

Figure 3.23 shows the horizontal and the vertical tidal forces in two periods: September 2005 and September 2009, with the main tidal-forming planets.

Figure 3.23 shows that the tidal forces, both horizontal and vertical, vary strongly depending on the positions of the planets. Of course, the magnitudes of the vectors are strongly exaggerated, for the sake of illustration. In reality, the vertical tidal forces due to the combined action of all planets can cause elevation of the solar

Fig. 3.24 Distribution of the horizontal component of the tidal force on the surface of the star when the tide-generating planet is above the equator. SP is the subplanetary point

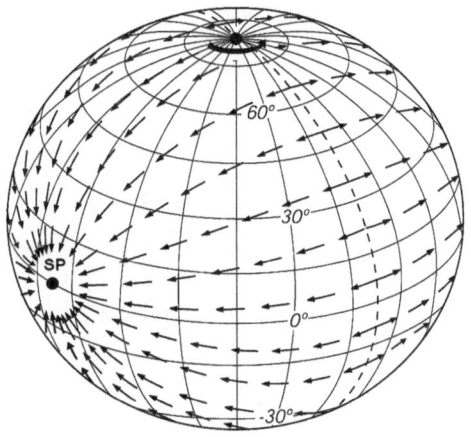

surface $\sim$1 mm [64]. Such an elevation is negligible, compared to the size of the Sun, and so this fact is often cited as an argument against the reality of the planetary influences on solar activity, in spite of the statistical correlations found between solar activity and planetary configurations.

As for the horizontal tidal force, it can be decomposed into zonal (east–west) and meridional (north–south) components. The zonal component could lead to modulation of the solar rotation rate, which is important for the generation of the toroidal field from the poloidal field at the base of the solar convection zone. The zonal velocity field which the planetary tidal forces can induce is up to $\pm$20 m/s [61] on the background of $\sim$2000 m/s average rotation velocity. However, as the force is directed toward the subplanetary point and its projection, its latitudinal average is zero (Fig. 3.24).

If the planet is above the star's equator, the meridional tidal force is always directed equatorward, and its effect is to slow down the surface poleward meridional circulation. In the case of the Sun, the tidal-forming planets are always close to the equatorial plane. The meridional tidal force varies periodically, and its average value does not change much (Fig. 3.25). However, what is important for the modulation of the solar cycle is not the meridional tidal force's overall average but rather its average magnitude during the period when the surface meridional circulation carries the flux from sunspot latitudes to the poles, that is, from sunspot maximum to the geomagnetic activity maximum on the sunspot declining phase. These periods, marked with thicker lines in Fig. 3.25, which covers the period 1750–2005, have different duration and come on different parts of the tidal force sinusoid.

In Fig. 3.26 the average meridional tidal force acting on V_{surf} during the poleward transport of the flux is compared to the maximum number of sunspots in the following solar cycle. Figure 3.26 demonstrates a very good correspondence between the planetary tidal force (solid line) and the amplitude of the sunspot cycle (dotted line), with the Dalton minimum (the beginning of 19th century) and Gleissberg minimum (end of 19th and beginning of 20th century) coinciding with low tidal forces during

Fig. 3.25 Meridional tidal
force in units 10^{-10} N/kg
due to the combined action of
all planets in the period
1750–2005

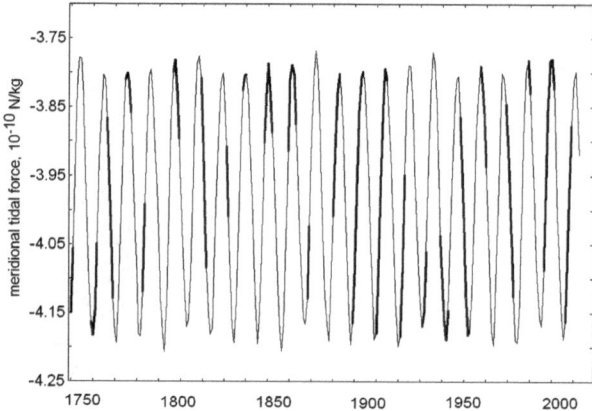

Fig. 3.26 *Solid line*: the
average meridional tidal force
acting on V_{surf} during the
poleward transport of the flux;
dotted line: the maximum
number of sunspots in the
following solar cycle

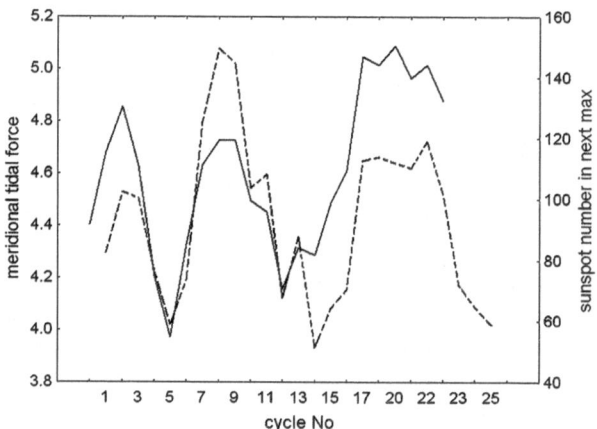

the surface flux transport, and the secular solar maxima in the 18th, 19th, and 20th
centuries, with maxima in the tidal forces during these periods.

We can make a rough estimation of the magnitude of the effect of the planetary
induced meridional tidal forces. The calculated magnitude of the tidal force is of
order $F \sim 10^{-10}$ N/kg. The acceleration caused by this force is $a = F/\rho$, where
the density ρ in the surface layer of the Sun is $\sim 10^{-5}$ gr/cm^3 = 10^{-2} kg/m^3. This
acceleration, acting on the flux during the time while it is carried from sunspot
latitudes to the poles (of order 10^8 s), can change the speed of the surface meridional
circulation by a few m/s, which corresponds to the observed variations in V_{surf}.

As seen in Fig. 3.26, the next sunspot cycle 24 is expected to be lower than cycle 23. Longer forecasts are difficult because we need to calculate the tidal force in
the period between the next sunspot maximum and the geomagnetic activity maximum following it, and these times are not known. If the next sunspot maximum is
in 2013, and the following geomagnetic activity maximum is in 2015, cycle 25 will

be even lower than cycle 24. The result is not much different if the periods of the maxima are shifted by ±1 year.

3.7 Long-Term Variations of Toroidal Field-Related and Poloidal Field-Related Solar Agents

According to the solar dynamo theory outlined above, the solar activity cycle is a result of the oscillation between the toroidal and poloidal components of the solar magnetic field. These two components are the two faces of the solar magnetism, so they are not independent. However, they are produced in two separate domains: at the base of the solar convection zone and in the surface layers, respectively, and their amplitudes strongly depend on the speed of the meridional circulation in the respective domains. As demonstrated above, the ratio between the speeds of the surface and deep meridional circulations is not constant; therefore, the ratio between the magnitudes of the solar poloidal and toroidal field may also be nonconstant.

The most obvious terrestrial manifestation of the solar activity is the geomagnetic activity, so it can be used to evaluate the evolution of the various manifestations of the solar magnetic field. As commented in Sect. 3.4.1, the disturbances in the geomagnetic field can be due to solar activity agents related to both the toroidal and poloidal solar magnetic field, coronal mass ejections (CME's) and high-speed solar wind streams from solar coronal holes, respectively; therefore the geomagnetic activity contains the contributions of solar agents which are manifestations of both types of solar magnetic activity. How can we evaluate their relative magnitude?

3.7.1 Two Components of the Geomagnetic Activity

Feynman [23] noted that if we plot a geomagnetic activity index (e.g., aa-index) as a function of the number of sunspots R, practically all points lie above a line (Fig. 3.27).

The equation of this line is

$$aa_R = a + b^* R,$$

where aa_R is the minimum geomagnetic activity for a given number of sunspots and, following Feynman [23], represents the contribution of sunspot related, or toroidal field related solar agents to geomagnetic activity. What is left is due to nonsunspot-related or poloidal field-related solar agents:

$$aa_P = aa - aa_R.$$

The source of the R component are shown to be CMEs and that of the P component high-speed solar wind from coronal holes. According to Feynman [23], the periodic variations of these two components are equally strong, each having the period of

Fig. 3.27 Dependence of the geomagnetic activity as measured by *aa*-index (since 1868) and *ak*-index (1844–1867) on the sunspot number *R*

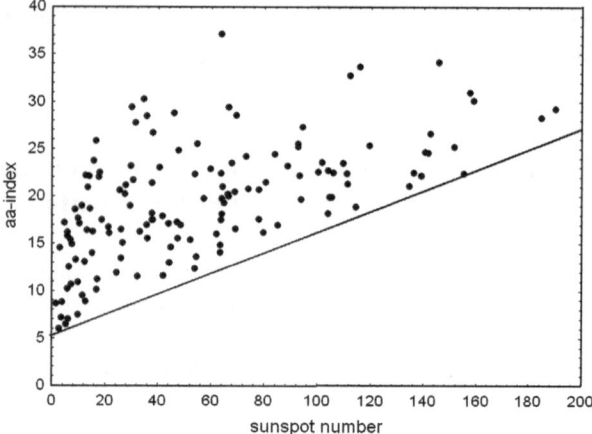

Fig. 3.28 Long-term variations of the aa_R-sunspot-related (*solid line*) and aa_P-nonsunspot-related (*dashed line*) geomagnetic activity, moving averages over 30 years with a step of 10 years (climatic normals)

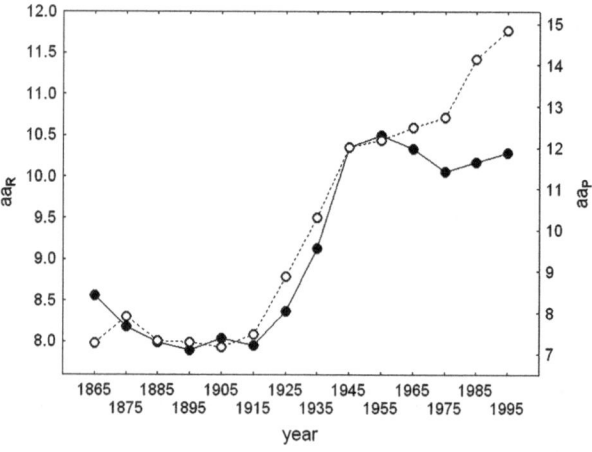

the sunspot cycle, but differing in phase. Ruzmaikin and Feynman [96] studied the long-term variations in aa_R and aa_P and found a secular trend in the strength and the relative phases of the toroidal and poloidal components of the solar dynamo. Moreover, their Fig. 5 hints at a change in the relative strength of the two components after the 1960s; however, the relative strength of the two components was outside the scope of the study, and the authors did not elaborate on it.

To evaluate the long-term variations in aa_R and aa_P, we calculate the so-called "climatic normals," first recommended by the World Meteorological Organization for the evaluation of climate changes [40] and also suitable for quantifying space climate changes: averages over 30 yearly means with a step of 10 years, e.g., 1901–1930, 1911–1940, etc. (Fig. 3.28). It is clearly seen that while both aa_R and aa_P have increased by about 30 % between the middle of the 19th century and the middle of the 20th century, beginning with the period 1951–1980, aa_R has been decreasing, while aa_P has continued increasing, in agreement with Ruzmaikun and Feynman's [96] Fig. 5.

3.7.2 Three Components of the Geomagnetic Activity

A closer examination of Fig. 3.27 reveals that aa_R contains not only the contribution of sunspot-related or toroidal field-related solar agents to geomagnetic activity as suggested by Feynman [23]. In fact, aa_R, which is the minimum geomagnetic activity for a given number of sunspots, consists of two parts: for any number of sunspots, even zero, there is some nonzero level of geomagnetic activity aa_{min} equal to the coefficient a in Feynman's [23] equation which is obviously not due to sunspot-related or toroidal field-related solar agents, plus there is a second part proportional to the number of sunspots equal to $aa_T = b^* R$, which actually reflects the contribution of the toroidal field-related solar agents to geomagnetic activity; therefore, $aa_R = aa_{min} + aa_T$. Further, there is the contribution to the overall level of geomagnetic activity aa of aa_P, so that Feynman's equation can be rewritten as

$$aa = aa_{min} + aa_T + aa_P.$$

The coefficient aa_{min} is a measure of the "floor" in geomagnetic activity for zero level of toroidal field-related solar activity, while the coefficient b reflects the sensitivity of the geomagnetic activity to the variations in the toroidal field-related solar activity (CMEs), and aa_P is the variable additional geomagnetic activity due to poloidal field-related solar activity (high-speed solar wind streams).

Figure 3.28 demonstrated that aa_P and $aa_R = aa_{min} + aa_T$ both vary in time, and the two have different long-term evolution with aa_R decreasing relative to aa_P after sunspot cycle 19, the secular sunspot maximum. As aa_R consists of two parts, we will now try to evaluate separately the long-term evolution of these two parts.

In Fig. 3.27, the relation between the geomagnetic activity and the sunspot number is presented for the whole period 1844–2009. If shorter periods are examined, it is found that in different periods, the coefficients aa_{min} and b are different, and the two of them also have different long-term evolution [29]. Figure 3.29(a)–(m) presents the scatterplots showing the dependence of the geomagnetic activity on the number of sunspots in consecutive periods each covering three full sunspot cycles from minimum to minimum, e.g., cycles 9–11, 10–12, 11–13, etc., up to cycles 21–23. It is clearly seen that the "floor" in geomagnetic activity and the rate of increase of the geomagnetic activity with increasing sunspot-related activity vary in time. In the beginning of the period, the floor in geomagnetic activity aa_{min} is low (the geomagnetic activity is low for zero or a low number of sunspots), and the minimum geomagnetic activity quickly increases with increasing sunspot-related (toroidal-field related) solar activity (the coefficient b is big, i.e., the minimum line is steep). In the later periods, the "floor" in geomagnetic activity begins increasing (much higher part of the geomagnetic activity is caused by nonsunspot-related, or nontoroidal field-related solar activity), while the impact of the sunspot–related,

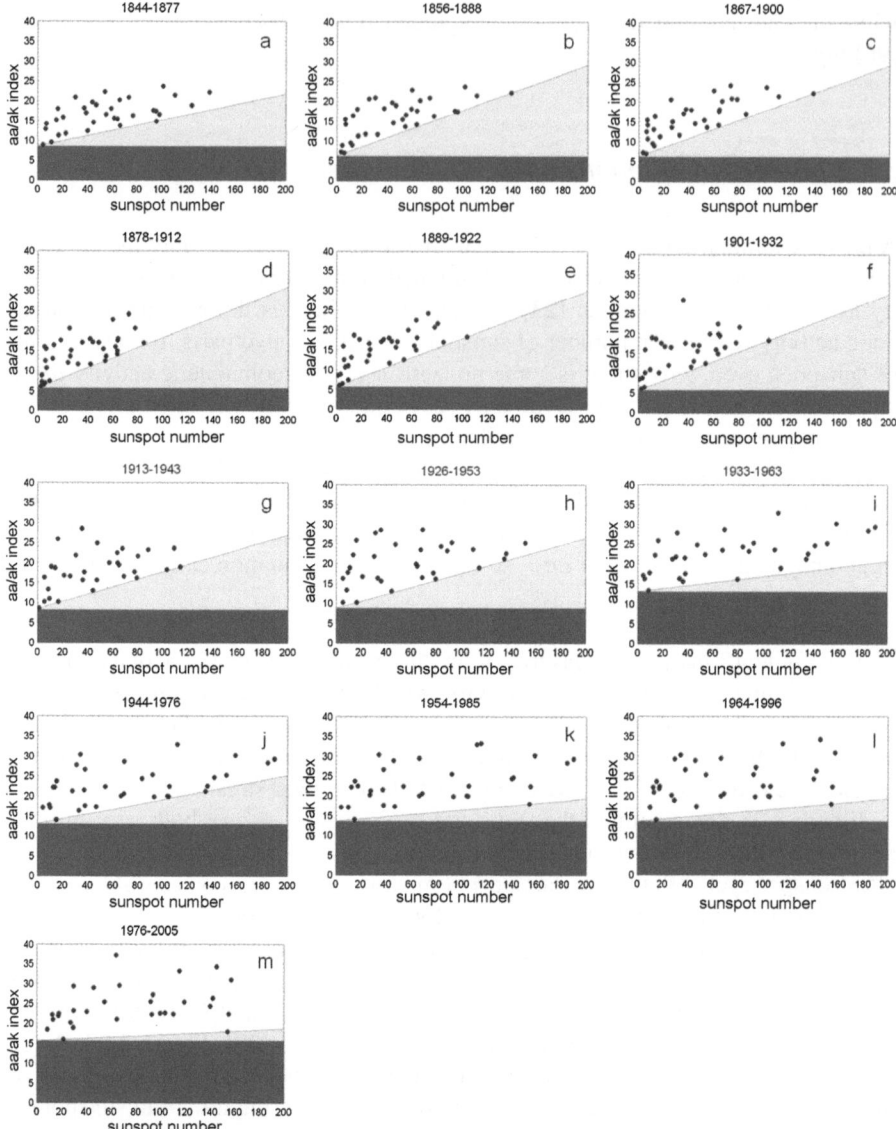

Fig. 3.29 a–m Scatterplots of the dependence of the geomagnetic activity on the sunspot number in consecutive ~30-year periods, each covering three sunspot cycles: (**a**) cycles 9–11; (**b**) cycles 10–12; (**c**) cycles 11–13, ... (**m**) cycles 21–23. From [29]

or toroidal field-related activity on geomagnetic activity is becoming progressively smaller to vanishing (geomagnetic activity is becoming less sensitive to the number of sunspots), especially in the last ~30-year period (see Fig. 3.29(m)).

Fig. 3.30 (**a**) Long-term variations in the coefficients a_{min} (*solid line*) and b (*dashed line*) in the period 1884–2005 based on aa and ak-index: values in consecutive periods each covering 3 full sunspot cycles (see text); from [29] (**b**) the same for the period 1712–1995 based on the data from ESAI

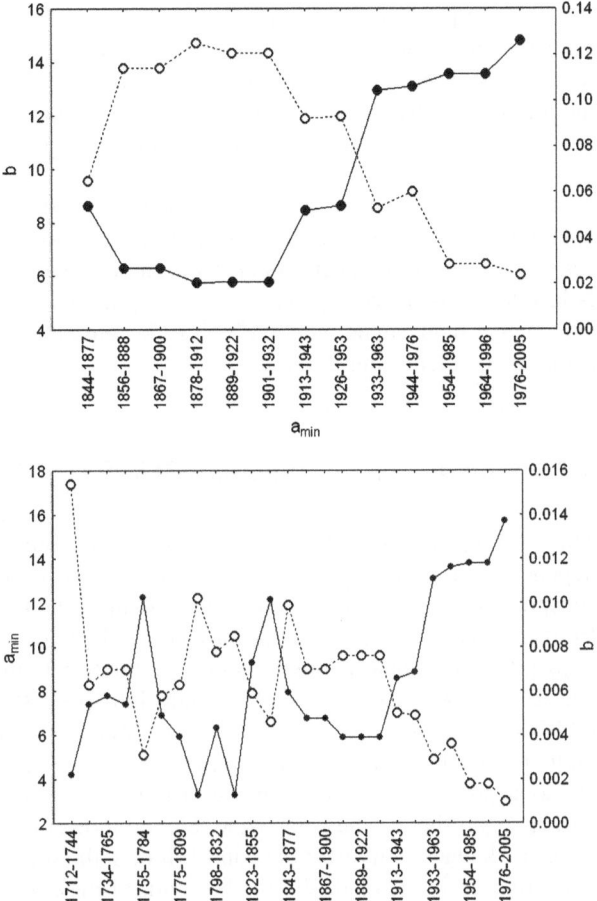

Figure 3.30(a) presents the long-term variations in the coefficients a_{min} and b calculated for these consecutive 3-cycle periods. The two coefficients are in antiphase and vary cyclically with a period of ∼100 years (the Gleissberg cycle). This cyclicity is confirmed by calculations based on the long-term reconstruction of solar and geophysical parameters ESAI (Fig. 3.30(b)). This is a new type of cyclicity in solar activity, not known before, which may shed some light on the theory of the solar dynamo and its long-term variations.

Another feature which can be noted in Fig. 3.29 is the behavior of aa_P, the additional geomagnetic activity above the minimum line due to high-speed solar wind from poloidal field-related long-lived coronal holes. During the 19th and early 20th century (up to the period 1901–1932), the impact of poloidal field-related geomagnetic activity to the overall activity is relatively small, and the maximum additional geomagnetic activity from poloidal field-related solar agents is almost constant for any level of sunspots and in all periods presented in Figs. 3.29(a)–(e): the maximum overall geomagnetic activity aa can be presented by a line parallel to the minimum line, and the level of aa_P in Fig. 3.28 is almost constant in this period. Later on, the impact of poloidal field-related geomagnetic activity to the overall activity becomes increasingly bigger in consecutive 3-cycle periods (see also Fig. 3.28), and within the individual periods, the maximum additional geomagnetic activity from poloidal field-related solar agents is not constant and does not seem to show any dependence on the sunspot number (Fig. 3.29(h)–(m)).

Therefore, not only aa_R has decreased relative to the poloidal field-related geomagnetic activity aa_P, but the contribution of the toroidal field-related geomagnetic activity aa_T to aa_R has also decreased. In other words, the decrease in the toroidal field-related solar activity relative to the poloidal field-related sunspot activity is even bigger than can be seen in Fig. 3.28 because a progressively bigger part of aa_R is due to the increasing floor.

3.7.3 Components of the Solar Wind

This decomposition of the geomagnetic activity into three parts (a floor $a_{\min}$ not dependent on the number of sunspots, a component proportional to the number of sunspots aa_T, and a component proportional to the high-speed solar wind streams from coronal holes aa_P) is in accordance with the decomposition of the solar wind transmitting the solar magnetic field from the Sun to the Earth into three parts [91–93, 103]: slow solar wind, CMEs, and high-speed streams. As mentioned above, CMEs originate from solar active regions, manifestation of the solar toroidal field; high-speed solar wind streams originate from solar coronal holes, manifestation of the solar poloidal field. What is the origin of the slow solar wind? In the classical Parker's [83] theory, the solar wind, a continuous outflow of plasma from the Sun, is due to the lack of hydrostatic equilibrium in the solar corona, leading to its expansion. To describe realistically the behavior of the coronal plasma as it expands into the interplanetary medium, it is necessary to take into account the interaction of this plasma with the magnetic field. The first to study the gas–magnetic field interaction were Pneuman and Kopp [89], who showed that slow solar wind can emanate not from the whole corona but from "helmet" streamers: regions of open magnetic loops with open field lines adjacent to and above the loops (Fig. 3.31(a)).

The source of the slow solar wind is has been identified as a belt of streamers encircling the Sun. It contains the magnetic neutral line that separates the two polarities of the global magnetic dipole and can therefore be regarded as the heliomagnetic equator [103]. The streamer belt's projection in the interplanetary medium is the heliospheric current sheet which rotates with the Sun and extends throughout the whole heliosphere like a giant ballerina skirt [3]. Around solar activity minimum, the coronal streamer belt is confined to low heliolatitudes and is almost flat, while at sunspot maximum it is strongly undulating and extends to high latitudes [9, 11, 22], among others). Figure 3.31(b) (from [89]) presents the geometry of the solar dipolar magnetic field with the equatorial streamer belt, and Figs. 3.32(a) and (b) are two examples of the solar coronal magnetic field for a period of minimum and maximum solar activity, respectively, from Wilcox Solar Observatory,

Fig. 3.31 (a) A helmet
streamer, a region of closed
magnetic loops with open
field lines adjacent and above
the loops (from [89]);
(b) a scheme of the solar
dipolar magnetic field with
the solar equatorial streamer
belt shaded

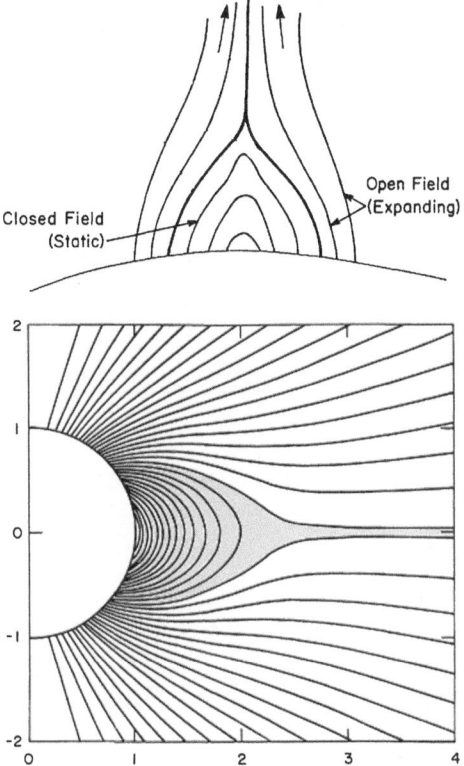

data obtained via the web site http://wso.stanford.edu, courtesy of J.T. Hoeksema.
The Wilcox Solar Observatory is currently supported by NASA. The association of
the slow solar wind with the equatorial streamer belt was unambiguously proved by
the data from the Ulysses high-latitude orbits which showed that near solar activity minimum low-speed flows are confined to a narrow band 40–45° wide centered
roughly on the solar equator, with remarkably uniform high-speed flow from the
superradially expanding polar coronal holes at higher heliolatitudes [39, 71], while
near solar maximum low-speed flows may dominate at all heliographic latitudes
consistent with the big tilt angle of the streamer belt at high solar activity as demonstrated in Fig. 3.32(b).

3.7.4 Long-Term Variations of the Components of the Solar Wind and Their Source Regions

In Sect. 3.7.2 it was shown that the geomagnetic activity consists of three parts:
a floor representing the geomagnetic disturbances at zero sunspot number; a contribution due to the toroidal field-related solar activity, increasing linearly with

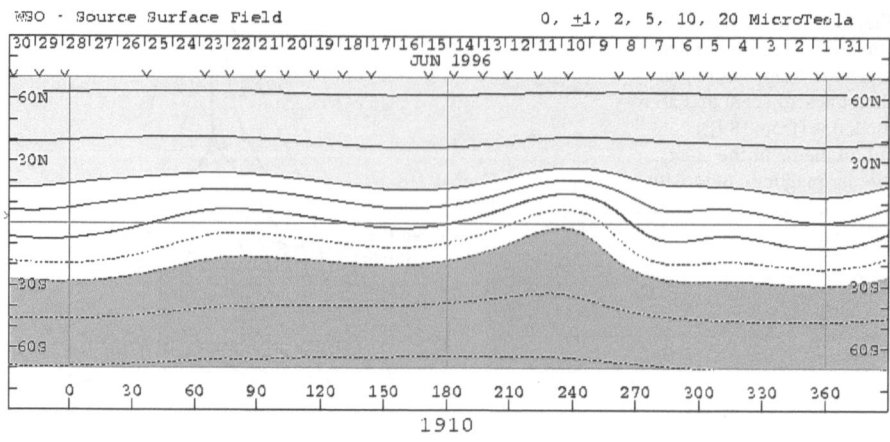

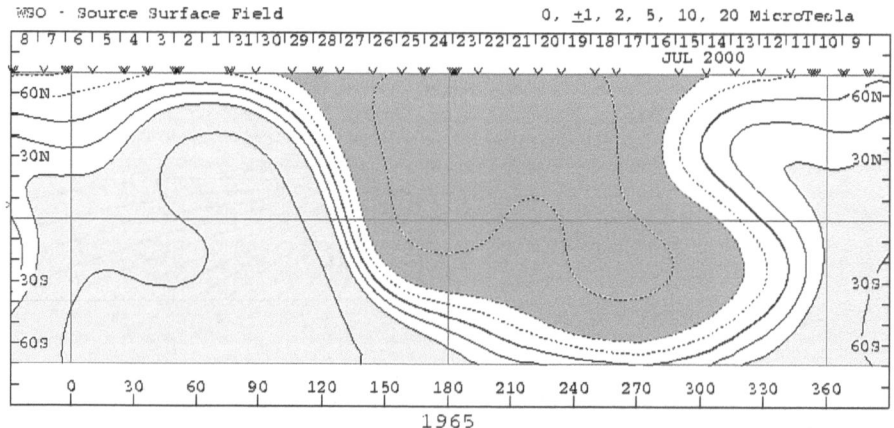

Fig. 3.32 Solar coronal magnetic field (**a**) at sunspot minimum: Carrington rotation 1910 (1916); (**b**) at sunspot maximum, Carrington rotation 1965 (2000). The negative and positive polarity magnetic field regions are represented as *dark grey* and *light grey shaded area*, respectively, while streamer belt is the band marked in white which separates them. Data from Wilcox Solar Observatory, data obtained via the web site http://wso.stanford.edu, courtesy of J.T. Hoeksema. The Wilcox Solar Observatory is currently supported by NASA

the number of sunspots; and a component due to the poloidal field-related solar activity. These three parts of the geomagnetic activity correspond to the three components of the solar wind: slow solar wind, CMEs, and high-speed solar wind from long-lived coronal holes. The long-term variations of the relative magnitudes of the three components of the geomagnetic activity reflect the long-term variations of the three components of the solar wind and can give information about the long-term variations in the operation of the solar dynamo.

3.7.4.1 The Floor in the Solar Wind and Geomagnetic Activity

The idea of a "floor" in the solar wind magnetic field and the association of this floor with the slow solar wind was first suggested by Svalgaard and Cliver [106], who, based on geomagnetic activity data, suggested that the solar wind magnetic field has a baseline value at the Earth's orbit (1 a.u.) of $B \sim 4.6$ nT, which is "approached but not broached" every sunspot minimum. They attributed the floor to a constant baseline solar open flux due to the slow solar wind, with the solar cycle variations of the interplanetary magnetic field strength, associated with CMEs and high-speed streams, riding on top of the floor.

However, in 2008 and 2009 the magnitude of the solar wind magnetic field at 1 a.u. dropped to ~ 4 nT, that is, below the alleged minimum level, which prompted a major revision in the level of the floor, as well as the view of its origin.

Cliver and Ling [18] defined the floor of the solar wind magnetic field in sunspot minima as consisting of two parts: a constant baseline of ≈ 2.8 nT due to the slow solar wind, plus a variable component that varies from cycle to cycle from ≈ 0 to ≈ 3 nT in concert with the solar polar fields which have maximum in sunspot minimum, and which is due to high-speed solar wind streams from the superradially expanding polar coronal holes. In other words, the solar wind at sunspot minimum consists not only of slow solar wind, but is actually a mixture of slow solar wind from the solar streamer belt and a higher speed supplement from outside the streamer belt which varies from minimum to minimum. The value 2.8 nT was obtained from two independent correlations: (i) between the solar polar field strength and the yearly averages of B for the last four 11-year minima for which measurements of the solar polar field and of B are available and (ii) between the peak sunspot number for cycles 14–23 and the minimum B at their preceding minima, calculated from IDV, the interdiurnal variability index of geomagnetic activity [105, 107].

The variable portion of higher-speed wind from outside the streamer belt at sunspot minimum suggested as an explanation of the variable floor of the solar wind magnetic field [18] is in agreement with the possible explanation of the overall increase of geomagnetic activity since the beginning of the 20th century suggested by Richardson et al. [94]. The authors studied the possible sources of the long-term increase in geomagnetic activity and came to the conclusion that the reason for the increase cannot be accounted for by the increase in the number of CMEs and high-speed solar wind streams, but rather the CMEs and recurrent high-speed streams at the beginning of the twentieth century must have been embedded in a background of slow solar wind that was less geoeffective (having, for example, lower interplanetary magnetic field strength and/or flow speed) than its modern counterpart. The increasing portion of high-speed component of the solar wind at consecutive sunspot minima since the beginning of the 20th century could have increased the geoeffectiveness of the slow solar wind.

What can be the explanation of the variations in the portion of the high-speed wind at sunspot minimum? A possible answer is suggested by Ouattara et al. [81], who used the number of geomagnetically "quiet" and "very quiet" days in the period 1900–2000 to quantify the long-term variations in geomagnetic activity and the related long-term evolution of the solar coronal field topology. Geomagnetic quiet and very quiet days are defined as days for which the daily average values of the geomagnetic aa-index are under certain thresholds (20 nT and 13 nT, respectively) and correspond to periods during which the Earth encounters slow solar wind streams flowing in the heliosheet from the solar coronal streamer belt when the solar magnetic field has a dipolar geometry (e.g., in solar minimum). The authors speculated that the smaller number of the geomagnetically quiet and very quiet days in the first half of the 20th century is a result of the shorter time spent by the Earth in the heliosheet (the interplanetary projection of the solar coronal streamer belt). Similarly, Cliver and Ling [18] showed that the decrease in geomagnetic activity in the last three sunspot minima corresponds to the increase in the percentage of time per year which the Earth spends in slow solar wind: 43.7 % in 1976, 45.1 in 1986, 51.8 % in 1995, and 71.3 % for 2009.

The percentage of time per year during sunspot minimum which the Earth spends in slow solar wind is indicative of the heliosheet thickness: the thicker the heliosheet, the longer time the Earth is immersed in it [104]. A proof of the long-term variations in the heliosheet thickness can be found in a study by Tlatov [109], who investigated the long-term change in the coronal large-scale structure at periods of minimum solar activity from 1878 to 2008. As a quantitative measure of the large-scale coronal structure at minimum solar activity, a parameter γ was introduced that characterizes the latitudinal extend of the equatorial streamer belt during solar activity minima and is therefore a measure of the thickness of the equatorial streamer belt. The variations of γ in cycles from 11 to 24 show that the thickness of the coronal streamer belt had a maximum in cycle 13, smoothly decreased to a minimum during cycle 19, which was the period when the global magnetic field of the Sun was closest to a dipole configuration, and has been increasing since then.

The relation between the thickness of the heliospheric current sheet, respectively the time the Earth spends in its slow solar wind at sunspot minimum periods, and the level of the floor in geomagnetic activity is supported by the studies cited above, and so the variations in $aa_{\min}$ are indicative of the coronal configuration. We should note, however, that this is not the only possible reason for the variations in the geomagnetic activity floor. Another factor can be the variations in the solar magnetic field, as will be demonstrated in the next section.

3.7.4.2 Relative Variations of the Solar Poloidal and Toroidal Magnetic Fields

The thickness of the equatorial streamer belt and its interplanetary projection, the heliosheet, is related to the solar poloidal magnetic field component, and therefore a change in the heliosheet thickness indicates a change in the solar poloidal field component [104]. The solar coronal holes, from which the high-speed solar wind streams originate, are another manifestation of the solar poloidal field. A comparison of Fig. 3.28 and Fig. 3.30(a) confirms that the long-term variations of aa_P (the geomagnetic activity due to high speed solar wind) and aa_{min} (the floor in geomagnetic activity due to the slow solar wind from the coronal streamer belt) are very similar, so we can assume that they both reflect the secular evolution of the poloidal component of the solar magnetic field. Important for the magnitude of the poloidal component are the speed of the surface poleward meridional circulation and the diffusivity in the upper part of the solar convection zone.

The toroidal component of the magnetic field depends on the poloidal component from which it is generated: partly from the polar field formed as a result of the transport by the meridional circulation of the poloidal field generated at midlatitudes and transported to the poles, down to the tachocline and equatorward, and partly from the poloidal field generated at midlatitudes and diffused down to the tachocline before reaching the poles. Also important for the magnitude of the toroidal component are the speed of the deep equatorward circulation, the differential rotation, and the diffusivity in the lower part of the solar convection zone.

As demonstrated by Figs. 3.28 and 3.29, there is no one-to-one correspondence between the poloidal and toroidal components of the solar magnetic field, which points at the importance of local factors in the regions of the generation of the two components for their magnitude.

Figures 3.29 and 3.30 demonstrate that the rate of increase of the geomagnetic activity with increasing number of sunspots (the coefficient b in Feynman's [23] equation) is different in different periods and has long-term variations in antiphase to the variations in the level of the geomagnetic activity floor. What can be the reason for the different sensitivity of the geomagnetic activity to the number of sunspots in different solar cycles? Sunspots are themselves not geoeffective, but they are a manifestation of the solar toroidal field, and geoeffective CMEs whose number and intensity are proportional to the number and area of sunspots are another manifestation of this toroidal field. Therefore the variation in sunspots' characteristic parameters is a measure of the variation in the toroidal magnetic field strength [100] determining the geoeffectiveness of CMEs.

Albregtsen and Maltby [2] were the first to find solar cycle dependence of the sunspot brightness. Norton and Gilman [80] using the Advanced Stokes Polarimeter and the Michelson Doppler Imager (MDI) data during 1998–2003, Penn and Livingston [84] using the National Solar Observatory Kitt Peak Vacuum Telescope spectromagnetograph data from 1998 to 2006, and Penn and MacDonald [87] using the using the NSO Kitt Peak McMath-Pierce telescope data from 1992 to 2003, all found that sunspot umbrae appear brighter during the minimum of the sunspot cycle and darker during the maximum.

The magnetic field is the central quantity determining the properties of sunspots. It greatly reduces the convective transport of heat from below and is finally responsible for sunspot darkness. Martínez Pillet and Vázquez [69] demonstrated that a strong correlation exists between the sunspot's maximum brightness and its magnetic field: the darker the spot, the more intense its magnetic field.

Therefore, solar cycle variations in the maximum umbral intensity (darker umbra during sunspot maximum and brighter umbra during sunspot minimum) means sunspot cycle variation in the sunspot magnetic field strength (stronger fields during sunspot maximum and weaker fields during sunspot minimum).

Pevtsov et al. [88] employed historic synoptic data sets from seven observatories in the former USSR covering the period from 1957 to 2011. The main results of this study can be summarized as follows:

(1) The sunspot field strengths vary cyclically reaching maxima around sunspot maxima and minima around sunspot minima, although the synchronicity is not absolute; therefore, the solar dynamo produces sunspots with noticeably larger field strengths near the maximum of the sunspot cycle.

(2) In the last five sunspot maxima (cycles 19–23), the maxima in the sunspot field strength in the different cycles show no indication of a secular trend, either decreasing or increasing with time;

(3) The sunspot field strength in sunspot minimum has variations from cycle to cycle.

(4) The sunspot field strength in sunspot minimum B_{min} tends to be correlated with the number of the sunspots in the following cycle maxima: the stronger the magnetic field in sunspot minimum, the bigger the number of sunspots in the following sunspot maximum.

Figure 3.33(a) demonstrates the correlation between B_{min} and the following sunspot maximum, and Fig. 3.33(b) is a plot of the values of the sunspot number in consecutive sunspot maxima (dashed line) together with the sunspot field strength in the preceding sunspot minima. Although the statistical sample is very small, the correlation is very high and highly statistically significant: $R = 0.9988$ with $p = 0.01$. If this correlation is confirmed, it can serve as a prognostic tool for the amplitude of the sunspot cycle which can be estimated as early as the minimum in the magnetic field strength in sunspots is observed, that is, around sunspot minimum. Applying a linear regression to these numbers yields a prediction for the maximum amplitude of Cycle 24 of 67 ± 35.

(5) The decrease in the sunspot field strength during the declining phase of each cycle is about 500–600 G, with gradients between -47.1 ± 8.9 (cycle 20) and -118.7 ± 7.9 (cycle 23). The amplitudes of gradients of sunspot magnetic field strengths do not appear to correlate with the amplitude of a solar cycle or the steepness of the declining phase of the cycle.

(6) Not commented by the authors, but also noticeable in their Fig. 4 is that the rate of increase of the magnetic field in sunspots from sunspot minimum to sunspot maximum also varies from cycle to cycle and depends on the strength of the sunspot magnetic field in cycle minimum (negative correlation, $R = -0.9$ with $p = 0.037$): the weaker the magnetic field in sunspots in cycle maximum, the faster the magnetic field strength grows toward cycle maximum (Fig. 3.34).

It should be noted that the solar cycle variation in the sunspot magnetic field strength and the constant field strength in the sunspot maxima contradict the results of Penn and Livingston [85, 86], who found that the magnetic field strength in sunspots has been decreasing in time since the 1990s, independently of the solar cycle. The reason for these diverging results may be in the different methodology the two teams use. The most obvious difference is that while Pevtsov et al. [88]

Fig. 3.33 (**a**) Dependence of the number of sunspots in the cycle maximum on the average magnetic field in sunspots in the preceding cycle minimum. The forecasted amplitude of cycle 24 is derived by extrapolation. (**b**) Long-term variations in the sunspot cycle amplitude (*dashed line, asterisk*) and the average magnetic field in sunspots in the preceding cycle minimum (*solid line, points*)

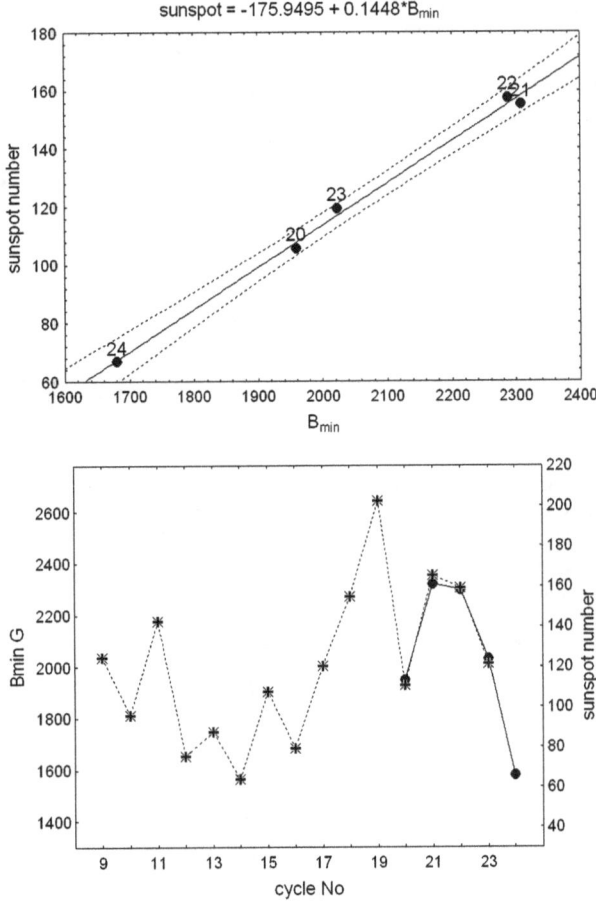

used only the biggest sunspots for the analysis, Penn and Livingston [85, 86] used all visible sunspots. Nevertheless, a solar cycle variation does show in the latter results, though less expressed, even though the authors are mainly concerned with the long-term trend and do not comment on it.

Pevtsov et al. [88] suggest that the sunspot cycle dependence of the magnetic field strength of the sunspots may be due to the different depths at which the sunspots originate. Javaraiah and Gokhale [56] compared the "initial" rotation velocity (the rotation velocity when they are first seen) of sunspot groups of different age, size and magnetic field strength, and the radial rotation profile in the solar convection zone derived by helioseismology, and found that sunspots with stronger magnetic fields originate deeper in the convection zone. Therefore, the solar cycle variation of the sunspot magnetic field strength may indicate a solar cycle variation of the depth at which sunspots originate.

A comparison of B_{min} and the reversal depth of the meridional circulation calculated as described in Sect. 3.5.3 shows that the sunspot magnetic field in cycle

Fig. 3.34 Dependence of the rate of increase of the magnetic field in sunspots from sunspot minimum to sunspot maximum on the strength of the sunspot magnetic field in cycle minimum

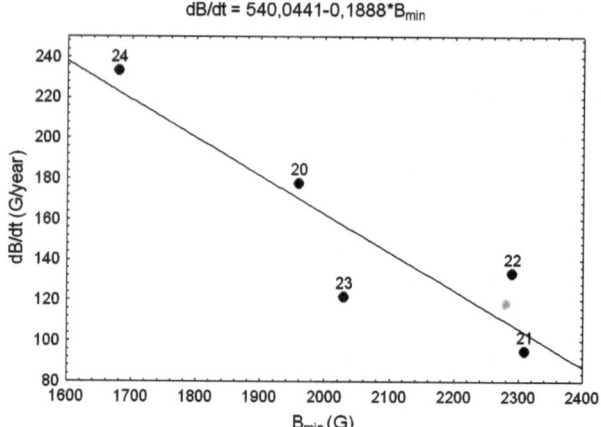

Fig. 3.35 Dependence of the sunspot magnetic field in cycle minimum B_{min} on the reversal depth of the meridional circulation

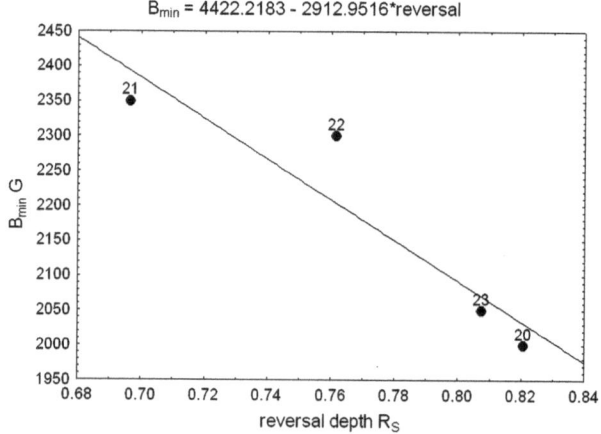

minimum is stronger when the meridional circulation reverses direction deeper in the convection zone (Fig. 3.35). This relation, together with the relation between B_{min} and the amplitude of the following sunspot maximum (Fig. 3.33), is in agreement with Fig. 3.19, which demonstrates that a higher sunspot maximum follows after a deeper reversal.

Another interesting result is the strong correlation between the sunspot field strength in sunspot minimum B_{min} and aa_{min}, the floor in geomagnetic activity (Fig. 3.36). This result is somewhat unexpected: the floor in geomagnetic activity is supposed to be related to the solar dipolar field and the thickness of the coronal streamer belt and its interplanetary projection the heliospheric current sheet, while the sunspots are manifestation of the solar toroidal field and the magnetic field in the sunspots is indicative of the magnitude of the toroidal component of the solar magnetic field. It can be speculated that the geomagnetic activity even in sunspot minimum is due not only to the poloidal component of the solar magnetic field but also to its toroidal component. Hassler et al. [44] showed a relationship between

Fig. 3.36 Relation between the floor in geomagnetic activity a_{min} and the magnetic field in sunspots in periods of cycle minima B_{min}

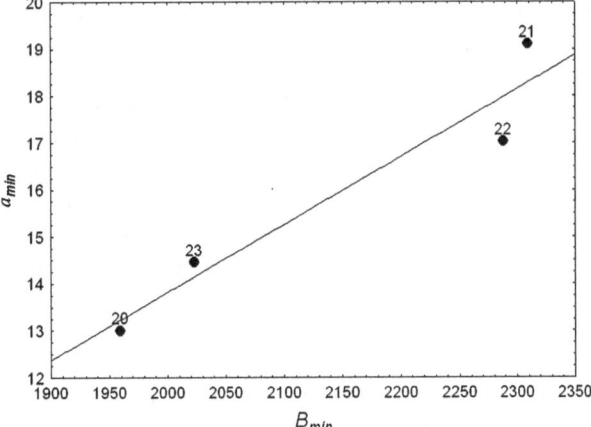

Fig. 3.37 Relation between the rate of increase of the magnetic field in sunspots from sunspot minimum to sunspot maximum grad B and the rate of increase of the geomagnetic activity with increasing number of sunspots b

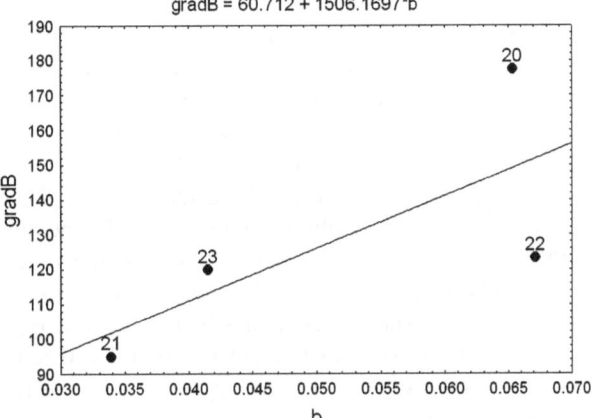

outflow velocity and chromospheric magnetic network structure, suggesting that the solar wind is rooted at its base to this structure, emanating from localized regions along boundaries and boundary intersections of magnetic network cells.

Figure 3.37 demonstrated the relation between the rate of increase of the magnetic field in sunspots from sunspot minimum to sunspot maximum grad B and the coefficient b in Feynman's [23] equation (Sect. 3.7.1). This provides an explanation of the different sensitivity of the geomagnetic activity to increasing number of sunspots in different cycles: with increasing number of sunspots, the toroidal magnetic field has different rates of increase in different solar cycles, and the magnetic field strength in CMEs is a main factor for their geoeffectiveness.

$aa_R = aa_{min} + aa_T$ in Feynman's equation is by definition proportional to the number of sunspots. As seen in Fig. 3.30, its two components, the floor in geomagnetic activity aa_{min} and the rate of increase of the geomagnetic activity with increasing number of sunspots $aa_T = b^*R$, are in antiphase. This negative correlation is the reflection of the dependence between strength of the sunspot magnetic field in cycle minimum and the rate of increase of the magnetic field in sunspots from cycle minimum to cycle maximum (Fig. 3.34).

The third component of geomagnetic activity, aa_P, which is due to high-speed solar wind streams from coronal holes, manifestation of the solar poloidal field, had been varying in accord with aa_R up to the 19th solar cycle, after which aa_R ceased increasing and began declining, while aa_P continued rising for three more cycles.

If we compare Fig. 3.28 and Fig. 3.11, it can be noted that the change in the relative evolution of the toroidal field-related and poloidal field-related geomagnetic activity coincides with a sharp drop in the diffusivity in the upper part of the solar convection zone. A lower diffusivity in the upper part of the solar convection zone means longer life-time of the low-latitude solar coronal holes, more recurrent high-speed solar wind streams.

The diffusivity in the upper part of the solar convection zone is proportional to the squired diameter of the supergranulation cells. McIntosh et al. [72] analyzed several thousand SOHO/EIT 304 A synoptic images taken over the last fourteen years including the "unusual" solar activity minimum of 2009. The observations of the STEREO spacecraft (in the same He II 304A channel) and ground-based observations from PSPT instrument were used to validate the observed variance, particularly through the 2009 solar minimum. The authors found a reduction of the mean supergranular cell radius, by about 0.5 Mm on average, between the cycle 22/23 and cycle 23/24 minima.

Tlatov [110] analyzed the characteristic size of the cells of the chromospheric network over the daily observations in K CaII line observatories Kodaiknal (1907–1999) and Meudon (1983–2011). It was found that the characteristic size of the chromospheric cells is ~36 Mm, and the variation associated with a solar activity is ~2 Mm. The largest cell size is observed typically ~1.5 years after solar maximum. There is a positive correlation ($R = 0.83$) between the cell size and the maximum amplitude of the next cycle of activity, in agreement with the positive correlation found between the coefficient of diffusivity and the next cycle amplitude (Sect. 3.4.2.6, Fig. 3.11). The biggest supergranular size was observed in the period 1950–1953, in agreement with Fig. 3.11. A sharp decrease in the cell size was observed in the next sunspot cycle, coinciding with the increased magnitude of aa_P compared to aa_T.

Acknowledgements I would like to thank the editors of this volume, Coralie Neiner and Jean-Pierre Rozelot, for their persistence and patience without which this chapter would have never been started and finished.

References

1. Abreau, H.A., Beer, J., Ferriz-Mas, A.: Past and future solar activity from cosmogenic radionuclides. In: Cranmer, S.R., Hoeksema, J.T., Kohl, J.L. (eds.) SOHO-23: Understanding a Peculiar Solar Minimum. ASP Conference Series, vol. 428, pp. 287–297 (2010)
2. Albregtsen, F., Maltby, P.: New light on sunspot darkness and the solar cycle. Nature **274**, 41–42 (1978)

3. Alfvén, H.: Electric currents in cosmic plasmas. Rev. Geophys. Space Phys. **15**, 271–284 (1977)
4. Antalova, A., Gnevyshev, M.N.: Principal characteristics of the 11-year solar activity cycle. Astron. Zh. **42**, 253–258 (1965)
5. Babcock, H.W.: The topology of the sun's magnetic field and the 22-year cycle. Astrophys. J. **133**, 572–587 (1961)
6. Bahcall, J.N., Serenelli, A.M., Basu, S.: New solar opacities, abundances, helioseismology, and neutrino fluxes. Astrophys. J. **621**, L85–L88 (2005)
7. Basu, S., Antia, H.M.: Changes in solar dynamics from 1995 to 2002. Astrophys. J. **585**, 553–565 (2003)
8. Beer, J., Tobias, S., Weiss, N.: An active sun throughout the Maunder minimum. Sol. Phys. **181**, 237–249 (1998)
9. Borrini, G.T., Gosling, J.T., Bame, S.J., Feldman, W.C., Wilcox, J.M.: Solar wind helium and hydrogen structure near the heliospheric current sheet: a signal of coronal streams at 1 AU. J. Geophys. Res. **86**, 4565–4573 (1981)
10. Braun, D.C., Fan, Y.: Helioseismic measurements of the subsurface meridional flow. Astrophys. J. Lett. **508**, L105–L108 (1998)
11. Bruno, R., Villante, U., Bassano, B., Schwenn, R., Mariani, F.: In-situ observations of the latitudinal gradients of the solar wind parameters during 1976 and 1977. Sol. Phys. **104**, 432–445 (1986)
12. Cameron, R., Schüssler, M.: Solar cycle prediction using precursors and flux transport models. Astrophys. J. **659**, 801–811 (2007)
13. Carrington, R.C.: On the distribution of the solar spots in latitudes since the beginning of the year 1854, with a map. Mon. Not. R. Astron. Soc. **19**, 1–3 (1985)
14. Cartwright, D.E.: Tides: A Scientific History. Cambridge University Press, Cambridge (1999)
15. Charbonneau, P.: Dynamo models of the solar cycle. Available via living reviews in solar physics. http://www.livingreviews.org/lrsp-2005. Cited 26 October 2011
16. Charbonneau, P., Dikpati, M.: Stochastic fluctuations in a Babcock–Leighton model of the solar cycle. Astrophys. J. **543**, 1027–1043 (2000)
17. Choudhuri, A.R.: The origin of the solar magnetic cycle. Pramana **77**, 77–96 (2011)
18. Cliver, E.W., Ling, A.G.: The floor in the solar wind magnetic field revisited. Sol. Phys. (2010). doi:10.1007/s11207-010-9657-6
19. Dietrich, G., Kalle, K., Krauss, W., Siedler, G.: General Oceanography, 2nd edn. Wiley-Interscience, New York (1980)
20. Dikpati, M., Gilman, P.A., de Toma, G., Ulrich, R.K.: Impact of changes in the Sun's conveyor-belt on recent solar cycles. Geophys. Res. Lett. **37**, L14107 (2010)
21. Duvall, T.L. Jr., Kosovichev, A.G.: New developments in local area helioseismology. In: Brekke P, P., Fleck, B., Gurman, J.B. (eds.) Recent Insights into the Physics of the Sun and Heliosphere: Highlights from SOHO and Other Space Missions. Proceedings of IAU Symposium, vol. 203, pp. 159–166. Astronomical Society of the Pacific, New York (2001)
22. Feldman, W.C., Asbridge, J.R., Bame, S.J., Fenimore, E.E., Gosling, J.T.: The solar origin of solar wind interstream flows: near equatorial coronal streamers. J. Geophys. Res. **86**, 5408–5416 (1981)
23. Feynman, J.: Geomagnetic and solar wind cycles, 1900–1975. J. Geophys. Res. **87**, 6153–6162 (1982)
24. Fisher, G.H., Fan, Y., Longcope, D.W., Linton, M.G., Abbett, W.P.: Magnetic flux tubes inside the sun. Phys. Plasmas **7**, 2173–2179 (2000)
25. Garaud, P., Brummell, N.H.: On the penetration of meridional circulation below the solar convection zone. Astrophys. J. **674**, 498–510 (2008)
26. Georgieva, K.: Why the sunspot cycle is double peaked. ISRN Astron. Astrophys. (2011). doi:10.5402/2011/437838
27. Georgieva, K., Kirov, B.: Long-term variations in solar meridional circulation from geomagnetic data: implications for solar dynamo theory (2007). arXiv:physics/0703187v2 [physics.space-ph]

28. Georgieva, K., Kirov, B.: Solar dynamo and geomagnetic activity. J. Atmos. Sol.-Terr. Phys. **73**, 207–222 (2011)
29. Georgieva, K., Kirov, B., Gavruseva, E.: Geoeffectiveness of different solar drivers, and long-term variations of the correlation between sunspot and geomagnetic activity. Phys. Chem. Earth **31**, 81–87 (2006)
30. Georgieva, K., Semi, P.A., Kirov, B., Obridko, V.N., Shelting, B.D.: Planetary tidal effects on solar activity. In: Proceedings of the XIII Pulkovo International Conference on Solar and Solar-Terrestrial Physics, St. Peterburg, Russia, 5–11 July 2009, pp. 117–120 (2009). ISSN 0552-5829. Available at http://195.201.30.3/russian/publ-s/conf_2009/conf_2009.pdf
31. Giles, P.M.: Time-distance measurements of large-scale flows in the solar convection zone. PhD Thesis, Stanford University (2000)
32. Giles, P.M., Duvall, T.L. Jr., Scherrer, P.H.: Time-distance measurements of subsurface rotation and meridional flow. In: Proceedings of the SOHO 6/GONG 98 Workshop Structure and Dynamics of the Interior of the Sun and Sun-like Stars, Boston, MA, 1–4 June 1998, pp. 775–780 (1998)
33. Gilman, P.A., Miesch, M.S.: Limits to penetration of meridional circulation below the solar convection zone. Astrophys. J. **611**, 568–574 (2004)
34. Gleissberg, W.: Evidence for a long solar cycle. Observatory **65**(282), 123–125 (1945)
35. Gnevyshev, M.N.: The corona and the 11-year cycle of solar activity. Sov. Astron. **AJ 7**, 311–318 (1963)
36. González Hernández, I., Komm, R., Hill, F., Howe, R., Corbard, T., Haber, D.A.: Meridional circulation variability from large-aperture ring-diagram analysis of global oscillation network group and Michelson Doppler imager data. Astrophys. J. **638**, 576–583 (2006)
37. Gonzalez, W.D., Tsurutani, B.T., Clúa de Gonzalez, A.L.: Interplanetary origin of geomagnetic storms. Space Sci. Rev. **88**, 529–562 (1999)
38. Gonzalez, W.D., Tsurutani, B.T., Clúa de Gonzalez, A.L.: Geomagnetic storms contrasted during solar maximum and near solar minimum. Adv. Space Res. **30**, 2301–2304 (2002)
39. Gosling, J.T.: Physical nature of the low-speed solar wind. Scientific basis for robotic exploration close to the sun. AIP Conf. Proc. **385**, 17–24 (1997)
40. Guttman, N.B.: Statistical descriptors of climate. Bull. Am. Meteorol. Soc. **70**, 602–607 (1989)
41. Hale, G.E.: On the probable existence of a magnetic field in sun-spots. Astrophys. J. **28**, 315–348 (1908)
42. Hale, G.E., Nicholson, S.B.: The law of sun-spot polarity. Mon. Not. R. Astron. Soc. **85**, 270–300 (1925)
43. Hale, G.E., Ellerman, F., Nicholson, S.B., Joy, A.H.: The magnetic polarity of sun-spots. Astrophys. J. **49**, 153–186 (1919)
44. Hassler, D.M., Dammasch, I.E., Lemaire, P., Brekke, P., Curdt, W., Mason, H.E., Vial, J.-C., Wilhelm, Kl.: Solar wind outflow and the chromospheric magnetic network. Science **283**, 810–813 (1999)
45. Hathaway, D.: Doppler measurements of the Sun's meridional flow. Astrophys. J. **460**, 1027–1033 (1996)
46. Hathaway, D.: The Sun's shallow meridional circulation (2011). arXiv:1103.1561v2 [astro-ph.SR]
47. Hathaway, D.H., Rightmire, L.: Variations in the Sun's meridional flow over a solar cycle. Science **237**, 1350–1352 (2010)
48. Hathaway, D.H., Wilson, R.M.: What the sunspot record tells us about space climate. Sol. Phys. **224**, 5–19 (2004)
49. Hathaway, D., Wilson, R.M., Reichmann, E.J.: The shape of the sunspot cycle. Sol. Phys. **151**, 177–190 (1994)
50. Hathaway, D., Nandy, D., Wilson, R., Reichmann, E.: Evidence that a deep meridional flow sets the sunspot cycle period. Astrophys. J. **589**, 665–670 (2003)

51. Hathaway, D., Nandy, D., Wilson, R., Reichmann, E.: Erratum: "Evidence that a deep meridional flow sets the sunspot cycle period". Astrophys. J. **602**, 543 (2004)
52. Hotta, H., Yokoyama, T.: Importance of surface turbulent diffusivity in the solar flux-transport dynamo. Astrophys. J. **709**, 1009–1017 (2010)
53. Hoyt, D.V., Schatten, K.H.: Group sunspot numbers: a new solar activity reconstruction. Sol. Phys. **181**, 491–512 (1998)
54. Ivanov, E.V., Obridko, V.N., Shelting, B.D.: Meridional drifts of large-scale solar magnetic fields and meridional circulation. In: Proc. 10th European Solar Physics Meeting "Solar Variability: from Core to Outer Frontiers", Prague, Czech Republic, 9–14 September 2002, pp. 851–854 (2002). ESA SP-506
55. Javaraiah, J.: Long-term variations in the mean meridional motion of the sunspot groups. Astron. Astrophys. **509**, A30 (2010)
56. Javaraiah, J., Gokhale, M.H.: Estimation of the depths of initial anchoring and the rising-rates of sunspot magnetic structures from rotation frequencies of sunspot groups. Astron. Astrophys. **327**, 795–799 (1997)
57. Javaraiah, J., Ulrich, R.K.: Solar-cycle-related variations in the solar differential rotation and meridional flow: a comparison. Sol. Phys. **237**, 245–265 (2006)
58. Jiang, J., Chatterjee, P., Choudhuri, A.R.: Solar activity forecast with a dynamo model. Mon. Not. R. Astron. Soc. **381**, 1527–1542 (2008)
59. Jouve, L., Brun, A.S.: On the role of meridional flows in flux transport dynamo models. Astron. Astrophys. **474**, 239–250 (2007)
60. Kitchatinov, L.L.: Solar differential rotation: origin, models, and implications for dynamo. In: Choudhuri, A.R., Rajaguru, S.P., Banerjee, D. (eds.) Proceedings of 1st Asia-Pacific Solar Physics Meeting. ASI Conference Series, pp. 1–10 (2011). arXiv:1108.1604v2 [astro-ph.SR]
61. Klvaňa, M., Švanda, M., Bumba, V.: Temporal changes of the photospheric velocity fields. Hvar Obs. Bull. **29**, 89–98 (2005)
62. Kopecky, M.: When did the latest minimum of the 80-year sunspot period occur? Bull. Astron. Inst. Czechoslov. **42**, 158–160 (1991)
63. Krieger, L., Roth, M., von der Lühe, O.: Estimating the solar meridional circulation by normal mode decomposition. Astron. Nachr. **328**, 252–256 (2007)
64. Krivtsov, A.M., Klvana, M., Bumba, V.: Photosphere velocity field generated by tidal forces. In: Wilson, A. (ed.) Solar Variability as an Input to the Earth'S Environment. International Solar Cycle Studies (ISCS) Symposium, Tatranská Lomnica, Slovak Republic, 23–28 June 2003, pp. 121–123. ESA Publications Division, Noordwijk (2003). ISBN 92-9092-845-X, ESA SP-535
65. Leighton, R.A.: Transport of magnetic fields on the Sun. Astrophys. J. **140**, 1547–1562 (1964)
66. Leighton, R.A.: Magneto-kinematic model of the solar cycle. Astrophys. J. **156**, 1–26 (1969)
67. Makarov, V.I., Tlatov, A.G., Sivaraman, K.R.: Does the poleward migration rate of the magnetic fields depend on the strength of the solar cycle? Sol. Phys. **202**, 11–26 (2001)
68. Malkus, W.V.R., Proctor, M.R.E.: The macrodynamics of α-effect dynamos in rotating fluids. J. Fluid Mech. **67**, 417–443 (1975)
69. Martínez, P.V., Vázquez, M.: The continuum intensity-magnetic field relation in sunspot umbrae. Astron. Astrophys. **270**, 494–508 (1993)
70. Mayaud, P.-N.: The aa indices: a 100-year series characterizing the magnetic activity. J. Geophys. Res. **77**, 6870–6874 (1972)
71. McComas, D.J., Barraclough, B.L., Funsten, H.O., Gosling, J.T., Santiago-Munoz, E., Skoug, R.M., Goldstein, B.E., Neugebauer, M., Riley, P., Balogh, A.: Solar wind observations over Ulysses' first full polar orbit. J. Geophys. Res. **105**, 10419–10433 (2000)
72. McIntosh, S.W., Leamon, R.J., Hock, R.A., Rast, M.P., Ulrich, R.K.: Observing evolution in the supergranular network length scale during periods of low solar activity. Astrophys. J. Lett. **730**, L3 (2011)
73. Mitra-Kraev, U., Thompson, M.J.: Meridional flow profile measurements with SOHO/MDI. Astron. Nachr. **328**, 1009–1012 (2007)

74. Mursula, K., Usoskin, I.G., Maris, G.: Introduction to space climate. Adv. Space Res. **40**, 885–887 (2007)
75. Nagovitsyn, Yu.A.: Solar and geomagnetic activity on a long time scale: reconstructions and possibilities for predictions. Astron. Lett. **32**, 344–352 (2006)
76. Nagovitsyn, Yu.A., Ivanov, V.G., Miletsky, E.V., Volobuev, D.M.: Extended time series of Solar Activity Indices (ESAI): new possibilities for complex description of magnetic cycle. In: Stepanov, A.V., Benevolenskaya, E.E., Kosovichev, A.G. (eds.) IAU Symposium, vol. 223, pp. 555–556. Cambridge University Press, Cambridge (2004)
77. Nagovitsyn, Yu.A., Ivanov, V.G., Miletsky, E.V., Volobuev, D.M.: ESAI database and some properties of solar activity in the past. Sol. Phys. **224**, 103–112 (2004)
78. Nagovitsyn, Yu.A., Miletsky, E.V., Ivanov, V.G., Guseva, S.A.: Reconstruction of space weather physical parameters on 400-year scale. Cosm. Res. **46**, 283–293 (2008)
79. Nevanlinna, H., Kataja, E.: An extension of the geomagnetic activity index series aa for two solar cycles (1844–1868). Geophys. Res. Lett. **20**, 2703–2706 (1993)
80. Norton, A.A., Gilman, P.A.: Magnetic Field-minimum intensity correlation in sunspots: a tool for solar dynamo diagnostics. Astrophys. J. **603**, 348–354 (2004)
81. Ouattara, F., Amory-Mazaudier, C., Menvielle, M., Simon, P., Legrand, J.-P.: On the long term change in the geomagnetic activity during the 20th century. Ann. Geophys. **27**(5), 2045–2051 (2009)
82. Parker, E.: Hydromagnetic dynamo models. Astrophys. J. **122**, 293–314 (1955)
83. Parker, E.N.: Dynamics of the interplanetary gas and magnetic fields. Astrophys. J. **128**, 664–676 (1958)
84. Penn, M.J., Livingston, W.: Temporal changes in sunspot umbral magnetic fields and temperatures. Astrophys. J. **649**, L45–L48 (2006)
85. Penn, M.J., Livingston, W.: Are sunspots different during this solar minimum? Eos **90**, 257–258 (2009)
86. Penn, M.J., Livingston, W.: Long-term evolution of sunspot magnetic tields (2010). arXiv:1009.0784v1 [astro-ph.SR]
87. Penn, M.J., MacDonald, R.K.D.: Solar cycle changes in sunspot umbral intensity. Astrophys. J. **662**, L123–L126 (2007)
88. Pevtsov, A.A., Nagovitsyn, Yu.A., Tlatov, A.G., Rybak, A.L.: Long-term trends in sunspot magnetic fields. Astrophys. J. Lett. **742**, L36–L39 (2011)
89. Pneuman, G.W., Kopp, R.A.: Interaction of coronal material with magnetic fields. Sol. Phys. **18**, 258–270 (1971)
90. Rempel, M., Dikpati, M., MacGregor, K.: Dynamos with feedback of $j \times B$ force on meridional flow and differential rotation. In: Favata, F., Hussain, G., Battrick, B. (eds.) Proc. 13th Cool Stars Workshop, Hamburg, 5–9 July 2004, pp. 913–916 (2005). ESA SP-560
91. Richardson, I.G., Cliver, E.W., Cane, H.V.: Sources of geomagnetic activity over the solar cycle: relative importance of coronal mass ejections, high-speed streams, and slow solar wind. J. Geophys. Res. **105**, 18203–18214 (2000)
92. Richardson, I.G., Cliver, E.W., Cane, H.V.: Sources of geomagnetic storms for solar minimum and maximum conditions during 1972–2000. Geophys. Res. Lett. **28**, 2569–2572 (2001)
93. Richardson, I.G., Cane, H.V., Cliver, E.W.: Sources of geomagnetic activity during nearly three solar cycles (1972–2000). J. Geophys. Res. **107**, SSH12-1 (2002). doi:10.1029/2001JA000507
94. Richardson, I.G., Cliver, E.W., Cane, H.V.: Long-term trends in interplanetary magnetic field strength and solar wind structure during the twentieth century. J. Geophys. Res. **107**, SSH8-1 (2002). doi:10.1029/2001JA000504
95. Rozelot, J.P., Lefebvre, S.: Advances in understanding elements of the Sun–Earth links. In: Solar and Heliospheric Origins of Space Weather Phenomena. LNP, vol. 699, Springer, Berlin, pp. 1–24 (2006). doi:10.1007/3-540-33759-8_2
96. Ruzmaikin, A., Feynman, J.: Strength and phase of the solar dynamo during the last 12 cycles. J. Geophys. Res. **106**, 15783–15790 (2001)

97. Ruzmaikin, A., Molchanov, S.A.: A model of diffusion produced by a cellular surface flow. Sol. Phys. **173**, 223–231 (1997)
98. Schove, D.J.: Sunspot turning-points and aurorae since A.D. 1510. Sol. Phys. **63**, 423–432 (1979)
99. Schove, D.J.: Sunspot Cycles. Benchmark Papers in Geology, vol. 68. Hutchinson Ross, Stroudsburg (1983)
100. Schüssler, M.: Flux tube dynamo approach to the solar cycle. Nature **288**, 150–152 (1980)
101. Schüssler, M.: Are solar cycles predictable? Astron. Nachr. **328**, 1087–1091 (2007)
102. Schwabe, H.: Sonnenbeobachtungen im Jahre 1843. Astron. Nachr. **21**, 233–236 (1844)
103. Schwenn, R.: Solar wind sources and their variations over the solar cycle. Space Sci. Rev. **124**, 51–76 (2006)
104. Simon, P.A., Legrand, J.P.: Some solar cycle phenomena related to the geomagnetic activity from 1868 to 1980, III: quiet-days, fluctuating activity or the solar equatorial belt as the main origin of the solar wind flowing in the ecliptic plane. Astron. Astrophys. **182**, 329–336 (1987)
105. Svalgaard, L., Cliver, E.W.: The IDV index: its derivation and use in inferring long-term variations of the interplanetary magnetic field strength. J. Geophys. Res. **110**, A12103 (2005). doi:10.1029/2005JA011203
106. Svalgaard, L., Cliver, E.W.: A floor in the solar wind magnetic field. Astrophys. J. **661**, L203–L206 (2007)
107. Svalgaard, L., Cliver, E.W.: Heliospheric magnetic field 1835–2009. J. Geophys. Res. **115**, A09111 (2010). doi:10.1029/2009JA015069
108. Švanda, M., Kosovichev, A.G., Zhao, J.: Speed of meridional flows and magnetic flux transport on the Sun. Astrophys. J. **670**, L69–L72 (2007)
109. Tlatov, A.G.: The centenary variations in the solar corona shape in accordance with the observations during the minimal activity epoch. Astron. Astrophys. **522**, A27 (2010)
110. Tlatov, A.G.: Long-period variations in the solar supergranulation size from observations in K CaII line. In: Proceedings of the Yearly Conference on Solar and Solar-Terrestrial Physics, St. Petersburg, 3–7 October 2011, pp. 93–98 (2011) (in Russian). Available online at http://www.gao.spb.ru/russian/publ-s/conf_2011/conf_2011.pdf
111. Ulrich, R.K.: Solar meridional circulation from Doppler shifts of the FeI line at λ5250 Å as measured by the 150-foot solar tower telescope at the Mt. Wilson observatory. Astrophys. J. **725**, 658–669 (2010)
112. Wang, Y.-M., Sheeley, N.R. Jr.: Magnetic flux transport and the sunspot-cycle evolution of coronal holes and their wind streams. Astrophys. J. **365**, 372–386 (1990)
113. Wang, Y.-M., Sheeley, N.R. Jr., Nash, A.G.: A new solar cycle model including meridional circulation. Astrophys. J. **383**, 431–442 (1991)
114. Wang, Y.-M., Sheeley, N.R. Jr., Lean, J.: Meridional flow and the solar cycle variation of the Sun's open magnetic flux. Astrophys. J. **580**, 1188–1196 (2002)
115. Yeates, A.R., Nandy, D., Mackay, D.H.: Exploring the physical basis of solar cycle predictions: flux transport dynamics and persistence of memory in advection versus diffusion dominated solar convection zones. Astrophys. J. **673**, 544–556 (2008)
116. Yoshimura, H.: The solar-cycle period-amplitude relation as evidence of hysteresis of the solar-cycle nonlinear magnetic oscillation and the long-term /55 year/ cyclic modulation. Astrophys. J. **227**, 1047–1058 (1979)
117. Zhao, J., Kosovichev, A.: Torsional oscillation, meridional flows, and vorticity inferred in the upper convection zone of the Sun by time-distance helioseismology. Astrophys. J. **603**, 776–784 (2004)

Part II
Tides in Planetary Systems
and Massive Stars

Chapter 4
Tides in Planetary Systems and in Multiple Stars: a Physical Picture

Stéphane Mathis and Françoise Remus

Abstract Many stars belong to close binary or multiple stellar systems. Moreover, since 1995, a large number of extrasolar planetary systems have been discovered where planets can orbit very close to their host star. Finally, our own Solar system is the seat of many interactions between the Sun, the planets, and their natural satellites. Therefore, in such astrophysical systems, tidal interactions are one of the key mechanisms that must be studied to understand the celestial bodies' dynamics and evolution. Indeed, tides generate displacements and flows in stellar and planetary interiors. The associated kinetic energy is then dissipated into heat because of internal friction processes. This leads to secular evolution of orbits and of spins with characteristic time-scales that are intrinsically related to the properties of dissipative mechanisms, the latter depending both on the internal structure of the studied bodies and on the tidal frequency. This lecture is thus aimed to recall the basics of the tidal dynamics and to describe the different tidal flows or displacements that can be excited by a perturber, the conversion of their kinetic energy into heat, the related exchanges of angular momentum, and the consequences for astrophysical systems evolution.

S. Mathis (✉) · F. Remus
Laboratoire AIM Paris-Scalay, CEA/DSM-CNRS-Université Paris Diderot, IRFU/SAp, CEA, 91191 Gif-sur-Yvette, France
e-mail: stephane.mathis@cea.fr

F. Remus
e-mail: francoise.remus@obspm.fr

S. Mathis
LESIA, Observatoire de Paris, CNRS, Université Paris Diderot, Université Pierre et Marie Curie, Observatoire de Paris, 5 place Jules Janssen, 92195 Meudon, France

F. Remus
LUTH, Observatoire de Paris, CNRS, Université Paris Diderot, Observatoire de Paris, 5 place Jules Janssen, 92195 Meudon Cedex, France

F. Remus
IMCCE, Observatoire de Paris, UMR 8028 du CNRS, Université Pierre et Marie Curie, Observatoire de Paris, 77 avenue Denfert-Rochereau, 75014 Paris, France

J.-P. Rozelot, C. Neiner (eds.), *The Environments of the Sun and the Stars*,
Lecture Notes in Physics 857,
DOI 10.1007/978-3-642-30648-8_4, © Springer-Verlag Berlin Heidelberg 2013

4.1 Introduction

Many stars belong to binary or multiple stellar systems (see, for example, [138]). Moreover, since 1995, a large number of extrasolar planetary systems have been discovered and characterised [104]. Finally, our own Solar system is the seat of many interactions between the Sun, the planets and their natural satellites. In such astrophysical systems, bodies can orbit close to the others, and thus tidal interactions are one of the key physical processes that must be studied to understand orbital and rotational evolution. In stellar systems, the main related key question concerns the difference of evolution between multiple stars and single ones. In planetary systems, such evolution is crucial for the hability question whether they could host the development of life; determining factors are the presence of liquid water and of a protective magnetosphere which are closely linked to the values of the planets' orbital elements and rotation rate.

To answer such important astrophysical problematics, we have thus to study in details the action of tides. Then, once a given binary or multiple system (star–star or star–planet) is formed, its fate is determined by the initial conditions and the mass ratio between its components. Through tidal interaction between each of them, the system evolves either to a stable state of minimum energy (where all spins are aligned, the orbits are circular and the rotation of each body is synchronised with the orbital motion), or the companion tends to spiral into the parent body. Indeed, by converting kinetic energy into heat through internal friction, tidal interactions modify the orbital and rotational properties of the components of the considered system and their internal structure through internal heating [1, 7, 23, 24, 33, 50, 65, 67, 68]. This mechanism depends sensitively on the internal structure (rocky, icy or fluid) and dynamics of the perturbed body (asynchronism, orbital eccentricity and inclination, obliquity).

In this context, the main goal of this review is to give a detailed state-of-the-art review on tidal interactions modelling in stellar and planetary interiors. First, in Sect. 4.2, we recall the basics on the tidal dynamics, i.e. the derivation of the tidal potential and the related evolution equations for orbital elements and spins where the dissipative processes are introduced. Then, in Sect. 4.3, we show how the properties of the tidal dissipation are strongly related to bodies' internal structure. We thus review such dissipative processes first in fluid regions, then in solid or icy ones, and finally at their interfaces. Next, in Sect. 4.4, we show the impact of tidal interactions on stars' internal angular momentum exchanges and magnetic activity, and we discuss their couplings with MHD interactions of stars with their environment.

4.2 Tidal Dynamics

4.2.1 The Tidal Potential

First, we consider the studied component A submitted to a tidal force exerted by the perturber B, which derives from a perturbing time dependent potential $U(\mathbf{r}, t)$

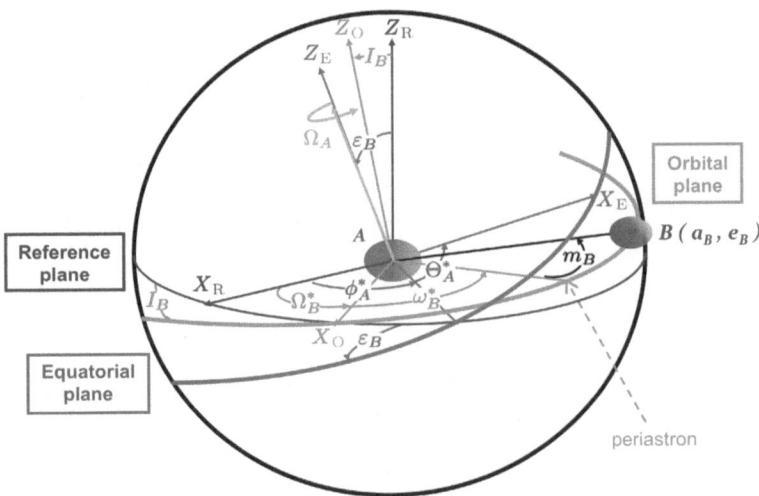

Fig. 4.1 Inertial reference ($\mathcal{R}_R$), orbital ($\mathcal{R}_O$), and equatorial ($\mathcal{R}_E$) rotating frames, and associated Euler angles of orientation (adapted from [73], courtesy Astronomy & Astrophysics)

where **r** and t are respectively the position vector and the time. Following [129] and generalising it by using [49, 58] and [73] (hereafter MLP09) in the general case of a close binary system where spins are not aligned, the components are not synchronised with the orbital motion and where the orbit is not circular, we expand the tidal potential U in spherical harmonics $Y_l^m(\theta, \varphi)$, where θ and φ are the angular spherical coordinates in the A equatorial reference frame ($\mathcal{R}_E$).

Before we proceed, we need to define the Euler angles that link the spin equatorial frame $\mathcal{R}_E : \{A, \mathbf{X}_E, \mathbf{Y}_E, \mathbf{Z}_E\}$ of the central body A, on one hand, and the orbital frame $\mathcal{R}_O : \{A, \mathbf{X}_O, \mathbf{Y}_O, \mathbf{Z}_O\}$, on the other hand, to the quasi-inertial frame $\mathcal{R}_R : \{A, \mathbf{X}_R, \mathbf{Y}_R, \mathbf{Z}_R\}$ whose axis $\mathbf{Z}_R$ has the direction of the total angular momentum of the whole system.

We need the three following Euler angles to locate the orbital reference frame $\mathcal{R}_O$ with respect to $\mathcal{R}_R$:

- I, the inclination of the orbital plane of B;
- ω^*, the argument of the orbit pericenter;
- Ω^*, the longitude of the orbit ascending node.

The equatorial reference frame $\mathcal{R}_E$ is defined by three other Euler angles with respect to $\mathcal{R}_R$:

- ε, the obliquity of the rotation axis of A;
- Θ^*, the mean sideral angle defined by $\Omega = d\Theta^*/dt$;
- ϕ^*, the general precession angle.

Refer to Fig. 4.1 for an illustration of the relative position of these three reference frames and the associated angles. For convenience, all the following developments will be done in the spin equatorial frame $\mathcal{R}_E$ of A (as it has been done in [73]).

All following results are derived from the Kaula's transform [49], used to explicitly express the whole generic multipole expansion in spherical harmonics of the perturbing potential U in terms of the Keplerian orbital elements of B in the equatorial A-frame:

$$\frac{1}{r_B^{l+1}} P_l^m(\cos\theta_B) e^{im\varphi_B} = \frac{1}{a^{l+1}} \sum_{j=-l}^{l} \sum_{p=0}^{l} \sum_{q\in\mathbb{Z}} \left\{ \sqrt{\frac{2l+1}{4\pi} \frac{(l-|j|)!}{(l+|j|)!}} \right.$$

$$\left. \times d_{j,m}^l(\varepsilon) F_{l,j,p}(I) G_{l,p,q}(e) e^{i\Psi_{l,m,j,p,q}} \right\}, \quad (4.1)$$

where θ_B and φ_B are respectively the colatitude and the longitude of the point mass perturber B, and where the phase argument is given by

$$\Psi_{l,m,j,p,q}(t) = \sigma_{l,m,p,q}(n,\Omega)t + \tau_{l,m,j,p,q}(\omega^*,\Omega^*,\phi^*). \quad (4.2)$$

We have defined here the tidal frequency

$$\sigma_{l,m,p,q}(n,\Omega) = (l-2p+q)n - m\Omega \quad (4.3)$$

and the phase $\tau_{2,m,j,p,q}$,

$$\tau_{l,m,j,p,q} = (l-2p)\omega^* + j(\Omega^* - \phi^*) + (l-m)\frac{\pi}{2}. \quad (4.4)$$

We study here binary systems close enough for the tidal interaction to play a role, but we also consider that the companion is far (or small) enough to be treated as a point mass (i.e. $a \geq 5\overline{R}_A$, where $\overline{R}_A$ is the mean A radius (see [73])). We then are allowed to assume the *quadrupolar approximation*, where we only keep the first mode of the potential, $l = 2$:

$$U(r,\theta,\varphi,t) = \Re\left[\sum_{m=-2}^{2} \sum_{j=-2}^{2} \sum_{p=0}^{2} \sum_{q\in\mathbb{Z}} U_{m,j,p,q}(r) P_2^m(\cos\theta) e^{i\Phi_{2,m,j,p,q}(\varphi,t)} \right], \quad (4.5)$$

where

$$\Phi_{2,m,j,p,q}(\varphi,t) = m\varphi + \Psi_{2,m,j,p,q}(t). \quad (4.6)$$

The functions $U_{m,j,p,q}(r,\theta)$ may be expressed in terms of the Keplerian elements (the semi-major axis a of the orbit, its eccentricity e and its inclination I) and the obliquity ε of the rotation axis of A as

$$U_{m,j,p,q}(r) = (-1)^m \sqrt{\frac{(2-m)!(2-|j|)!}{(2+m)!(2+|j|)!}}$$

$$\times \frac{\mathcal{G}M_B}{a^3} \left[d_{j,m}^2(\varepsilon) F_{2,j,p}(I) G_{2,p,q}(e) \right] r^2, \quad (4.7)$$

where $\mathcal{G}$ is the gravitational constant.

Table 4.1 Values of the obliquity function $d^2_{j,m}(\varepsilon)$ in the case where $j \geq m$ obtained from Eq. (4.8); (see [73])

j	m	$d^2_{j,m}(\varepsilon)$
2	2	$(\cos\frac{\varepsilon}{2})^4$
2	1	$-2(\cos\frac{\varepsilon}{2})^3(\sin\frac{\varepsilon}{2})$
2	0	$\sqrt{6}(\cos\frac{\varepsilon}{2})^2(\sin\frac{\varepsilon}{2})^2$
1	1	$(\cos\frac{\varepsilon}{2})^4 - 3(\cos\frac{\varepsilon}{2})^2(\sin\frac{\varepsilon}{2})^2$
1	0	$-\sqrt{6}\cos\varepsilon\,(\cos\frac{\varepsilon}{2})(\sin\frac{\varepsilon}{2})$
0	0	$1 - 6(\cos\frac{\varepsilon}{2})^2(\sin\frac{\varepsilon}{2})^2$

The obliquity function $d^2_{j,m}(\gamma)$ is defined, for $j \geq m$, by

$$d^2_{j,m}(\gamma) = (-1)^{j-m}\left[\frac{(2+j)!(2-j)!}{(2+m)!(2-m)!}\right]^{\frac{1}{2}}$$

$$\times \left[\cos\left(\frac{\gamma}{2}\right)\right]^{j+m}\left[\sin\left(\frac{\gamma}{2}\right)\right]^{j-m}P^{(j-m,j+m)}_{2-j}(\cos\gamma), \qquad (4.8)$$

where $P^{\alpha,\beta}_l(x)$ are the Jacobi polynomials. The values of these functions, for indices $j < m$, are deduced from $d^2_{j,m}(\pi + \gamma) = (-1)^{2-j}d^2_{-j,m}(\gamma)$ or from their symmetry properties: $d^2_{j,m}(\gamma) = (-1)^{j-m}d^2_{-j,-m}(\gamma) = d^2_{m,j}(-\gamma)$; moreover, we have $d^2_{j,m}(0) = \delta_{j,m}$. The values are given in Table 4.1.

We also define, the inclination function $F_{2,j,p}(I)$:

$$F_{2,j,p}(I) = (-1)^j\frac{(2+j)!}{4p!(2-p)!}$$

$$\times \left[\cos\left(\frac{I}{2}\right)\right]^{j-2p+2}\left[\sin\left(\frac{I}{2}\right)\right]^{j+2p-2}P^{(j+2p-2,j-2p+2)}_{2-j}(\cos I),$$

$$(4.9)$$

with the symmetry property

$$F_{2,-j,p}(I) = \left[(-1)^{2-j}\frac{(2-j)!}{(2+j)!}\right]F_{2,j,p}(I); \qquad (4.10)$$

the values are given in Table 4.2.

The eccentricity functions $G_{2,p,q}(e)$ are polynomial functions having e^q for argument (see [49]). Their values for the usual sets $\{2, p, q\}$ are given in Table 4.3, knowing that in the case of weakly eccentric orbits, the summation over a small number of values of q is sufficient ($q \in [-2, 2]$). In the following, let us denote by $\mathbb{I} \in [-2, 2] \times [-2, 2] \times [0, 2] \times \mathbb{Z}$ the set in which the quadruple $\{m, j, p, q\}$ takes its values.

If we simplify the expansion of the potential in the case where spins are aligned and perpendicular to the orbital plan, where obliquity ε and orbital inclination I are zero, Eq. (4.5) reduces to the expression of the potential given by [133].

Table 4.2 Values of the inclination function $F_{2,j,p}(I)$. Values for indices $j < 0$ can be deduced from Eq. (4.10) (see [73])

j	p	$F_{2,j,p}(I)$
0	0	$\frac{3}{8}\sin^2 I$
0	1	$-\frac{3}{4}\sin^2 I + \frac{1}{2}$
0	2	$\frac{3}{8}\sin^2 I$
1	0	$\frac{3}{4}\sin I (1 + \cos I)$
1	1	$-\frac{3}{2}\sin I \cos I$
1	2	$-\frac{3}{4}\sin I (1 - \cos I)$
2	0	$\frac{3}{4}(1 + \cos I)^2$
2	1	$\frac{3}{2}\sin^2 I$
2	2	$\frac{3}{4}(1 - \cos I)^2$

Table 4.3 Values of the eccentricity function $G_{2,p,q}(e)$; (see [73])

p	q	p	q	$G_{2,p,q}(e)$
0	−2	2	2	0
0	−1	2	1	$-\frac{1}{2}e + \cdots$
0	0	2	0	$1 - \frac{5}{2}e^2 + \cdots$
0	1	2	−1	$\frac{7}{2}e + \cdots$
0	2	2	−2	$\frac{17}{2}e^2 + \cdots$
1	−2	1	2	$\frac{9}{4}e^2 + \cdots$
1	−1	1	1	$\frac{3}{2}e + \cdots$
		1	0	$(1 - e^2)^{-3/2}$

4.2.2 Equations of the Dynamical Evolution

Mass redistribution due to the tide is at the origin of a tidal torque of non-zero average which induces an exchange of angular momentum between each component and the orbital motion. As shown for example in [73, 92], this tidal torque is proportional to the tidal dissipation k_2/Q (see [20, 21]), where k_2 is the Love number that describes the adiabatic response of A to the tidal excitation [66], and Q the tidal quality factor, which scales as the inverse of the ratio of the energy losses during an orbital period to the total energy of system [38]. For a perfectly adiabatic response to the tidal excitation, A will be elongated in the direction of the line of centres, inducing a torque with periodic variations of zero average, so that no secular exchanges of angular momentum will be possible (see [92, 129]). Next, if dissipative processes are taken into account, the deformation of A presents a time delay Δt with respect to the tidal forcing, which may be measured also by the tidal lag angle

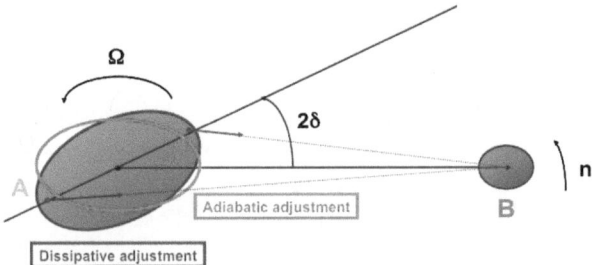

Fig. 4.2 Tidal interaction between two bodies A and B. In an ideal adiabatic view, the A response to the tidal potential exerted by B is directed along the line of centers. However, since there is internal friction, A responds with a time delay Δt with respect to the tidal forcing which may be also measured by the tidal angle 2δ which is related to the tidal quality factor Q (see Eq. (4.11))

2δ or equivalently by the quality factor Q (see [33]):

$$\tan[\Delta t \times \sigma_{2,m,p,q}] = \tan\left[2\delta(\sigma_{2,m,p,q})\right] = \frac{1}{Q(\sigma_{2,m,p,q})}. \qquad (4.11)$$

Thus, the tidal bulge is no more aligned with the line of centers, as shown in Fig. 4.2. The resulting tidal angle is at the origin of a torque of non-zero average which causes exchange of spin and orbital angular momentum between the components of the system.

Then, the evolution of the semi-major axis a, of the eccentricity e, of the inclination I, of the obliquity ε and of the angular velocity Ω ($\bar{I}_A$ denotes the mean moment of inertia of A) is governed by the following equations [30, 73, 92]:

$$\frac{d(\bar{I}_A \Omega)}{dt} = -\frac{8\pi}{5} \frac{\mathcal{G} M_B^2 \overline{R}_A^5}{a^6} \sum_{(m,j,p,q)\in\mathbb{I}} \left\{ \frac{k_2(\sigma_{2,m,p,q})}{Q(\sigma_{2,m,p,q})} \left[\mathcal{H}_{2,m,p,q}(e,I,\varepsilon)\right]^2 \right\}, \qquad (4.12)$$

$$\bar{I}_A \Omega \frac{d(\cos\varepsilon)}{dt} = \frac{4\pi}{5} \frac{\mathcal{G} M_B^2 \overline{R}_A^5}{a^6}$$
$$\times \sum_{(m,j,p,q)\in\mathbb{I}} \left\{ (j + 2\cos\varepsilon) \frac{k_2(\sigma_{2,m,p,q})}{Q(\sigma_{2,m,p,q})} \left[\mathcal{H}_{2,m,j,p,q}(e,I,\varepsilon)\right]^2 \right\}, \qquad (4.13)$$

$$\frac{1}{a}\frac{da}{dt} = -\frac{2}{n}\frac{4\pi}{5} \frac{\mathcal{G} M_B \overline{R}_A^5}{a^8}$$
$$\times \sum_{(m,j,p,q)\in\mathbb{I}} \left\{ (2 - 2p + q) \frac{k_2(\sigma_{2,m,p,q})}{Q(\sigma_{2,m,p,q})} \left[\mathcal{H}_{2,m,p,q}(e,I,\varepsilon)\right]^2 \right\}, \qquad (4.14)$$

$$\frac{1}{e}\frac{de}{dt} = -\frac{1}{n}\frac{1-e^2}{e^2}\frac{4\pi}{5}\frac{\mathcal{G}M_B\overline{R}_A^5}{a^8}$$

$$\times \sum_{(m,j,p,q)\in\mathbb{I}}\left\{\left[(2-2p)\left(1-\frac{1}{\sqrt{1-e^2}}\right)+q\right]\frac{k_2(\sigma_{2,m,p,q})}{Q(\sigma_{2,m,p,q})}\right.$$

$$\left.\times\left[\mathcal{H}_{2,m,p,q}(e,I,\varepsilon)\right]^2\right\},\tag{4.15}$$

$$\frac{d(\cos I)}{dt} = \frac{1}{n}\frac{1}{\sqrt{1-e^2}}\frac{4\pi}{5}\frac{\mathcal{G}M_B^2\overline{R}_A^5}{a^8}$$

$$\times \sum_{(m,j,p,q)\in\mathbb{I}}\left\{\left[j+(2q-2)\cos I\right]\frac{k_2(\sigma_{2,m,p,q})}{Q(\sigma_{2,m,p,q})}\right.$$

$$\left.\times\left[\mathcal{H}_{2,m,j,p,q}(e,I,\varepsilon)\right]^2\right\},\tag{4.16}$$

where the functions $\mathcal{H}_{m,j,p,q}(e,I,\varepsilon)$ are defined by

$$\mathcal{H}_{2,m,j,p,q}(e,I,\varepsilon) = \sqrt{\frac{5}{4\pi}\frac{(2-|j|)!}{(2+|j|)!}}d_{j,m}^2(\varepsilon)F_{2,j,p}(I)\,G_{2,p,q}(e),\tag{4.17}$$

and $\overline{R}_A$ accounts for the equatorial radius of the core of body A. Note that these expressions (derived from [73]) are valid for high eccentricities if enough terms are considered in the expansion [105].

From these equations one may derive the characteristic times of synchronization, circularization and spin alignment:

$$\frac{1}{t_{\text{sync}}} = -\frac{1}{\Omega-n}\frac{d\Omega}{dt} = -\frac{1}{\overline{I}_A(\Omega-n)}\frac{d(\overline{I}_A\Omega)}{dt},\tag{4.18}$$

$$\frac{1}{t_{\text{circ}}} = -\frac{1}{e}\frac{de}{dt},\tag{4.19}$$

$$\frac{1}{t_{\text{align}_A}} = -\frac{1}{\varepsilon}\frac{d\varepsilon}{dt} = \frac{1}{\varepsilon\sin\varepsilon}\frac{d(\cos\varepsilon)}{dt},\tag{4.20}$$

$$\frac{1}{t_{\text{align}_{\text{Orb}}}} = -\frac{1}{I}\frac{dI}{dt} = \frac{1}{I\sin I}\frac{d(\cos I)}{dt}.\tag{4.21}$$

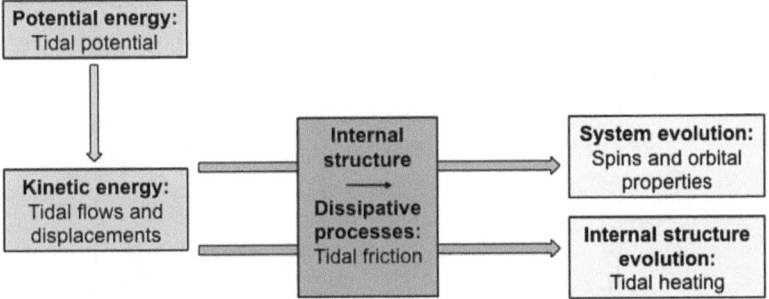

Fig. 4.3 The tidal energy dissipation: first, the gravific tidal potential energy generates tidal flows with a given kinetic energy. Next, because of internal friction related to the properties of the internal structure of the studied body, this kinetic energy is dissipated into heat that leads to the system evolution (spins and orbital properties) and to the modification of the internal structure because of the tidal heating [63]

4.3 Tidal Dissipation

4.3.1 The Tidal Kinetic Energy Dissipation: The Engine of Systems Evolution

As it has been emphasised in the previous section, the tidal interaction can be understood in two steps. First, if we adopt an ideal adiabatic view of the action of the tidal potential exerted by the secondary on the primary, the latter becomes elongated along the line of centers. However, dissipative processes such as viscous friction and heat diffusion that convert the tidal kinetic energy into heat have to be taken into account (see Fig. 4.3). Then, the response of the studied body to the tidal excitation presents a delay (often called the tidal lag), which translates into an angle, the tidal angle, between the tidal bulge and the line of centres. Those tidal delay and angle are thus directly related to the dissipative mechanisms and their dependence on the tidal frequency and as we will see in the next sections. Moreover, this shift between the tidal bulge and the line of centers is responsible of the net tidal torque of non-zero average over an orbital period that governs the secular evolution of the system (see for example [129]).

Then, if a theoretical isolated two-body system is considered, two evolutions are possible. In the most common case, provided that the system does not loose angular momentum, it tends to a state of minimum energy in which the orbits are circular, the rotation of the components is synchronised with the orbital motion, and the spins are aligned. However, in very close systems, such final state cannot be achieved: instead, the secondary spirals towards the primary and may be engulfed by it [46, 64]. To predict the fate of a binary system, one has thus to identify and to model the dissipative processes that achieve the conversion of kinetic energy into thermal energy in fluid and solid layers and at their interfaces, from which one may then draw the characteristic times of circularisation, synchronisation and spins' alignment.

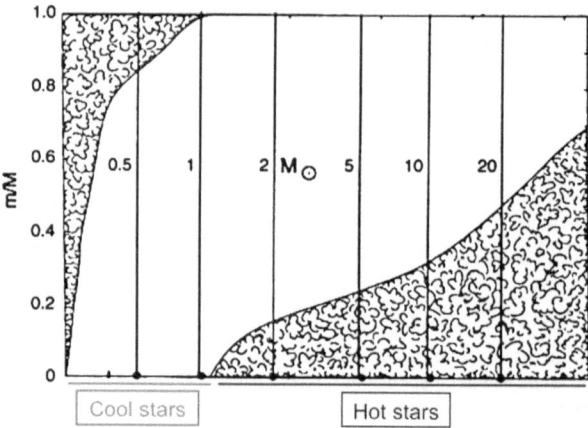

Fig. 4.4 Internal structure of main-sequence stars as a function of their mass represented using Kippenhahn diagram: stellar masses are given in the *horizontal axis*, while the fraction in mass and the position of convection zones (cloudy regions) and of radiation zones are given along the *vertical axis* (adapted from [52], courtesy Springer-Verlag). Very low-mass stars are entirely convective. Next, as the stellar mass grows, the radiative core becomes more and more important. Finally, a transition occurs because hydrogen is converted into helium through the CNO cycle and a convective core takes place with an external radiative envelope

4.3.2 Tidal Dissipation in Fluid Bodies

While stars are fluid bodies, planets host gaseous and liquid layers such as the deep envelopes of giant planets, and the internal core, the atmosphere and the potential ocean of telluric ones (see Figs. 4.4 and 4.5). Therefore, one has to obtain a complete understanding of flows that are tidally excited and dissipated in such regions in planetary systems.

4.3.2.1 Type of Fluid Tides

Two types of tides operate in stars and in fluid planetary layers: the equilibrium and dynamical tides. On one hand, the equilibrium tide designates the large-scale flow induced by the hydrostatic adjustment of studied fluid layers in response to the gravitational force exerted by the companion [129, 130]. On the other hand, the dynamical tide corresponds to the fluid eigenmodes that are excited by the tidal potential. Let us now detail the different types of eigenmodes that should be studied in stellar and fluid planetary regions. First, if Fig. 4.6 is considered, four characteristic frequencies are introduced: the Alfvén frequency ($\omega_A = \frac{B}{\sqrt{\mu \rho r \sin \theta}}$, where B is the field amplitude, and μ is the magnetic permeability), the inertial frequency (2Ω), the Brunt–Väisäla frequency (N), and the Lamb frequency (f_L). These delimit the frequency domain of corresponding Alfvén waves ($\omega < \omega_A$), inertial waves ($\omega < 2\Omega$), gravity waves (for which $\omega < N$ and that are also called internal waves),

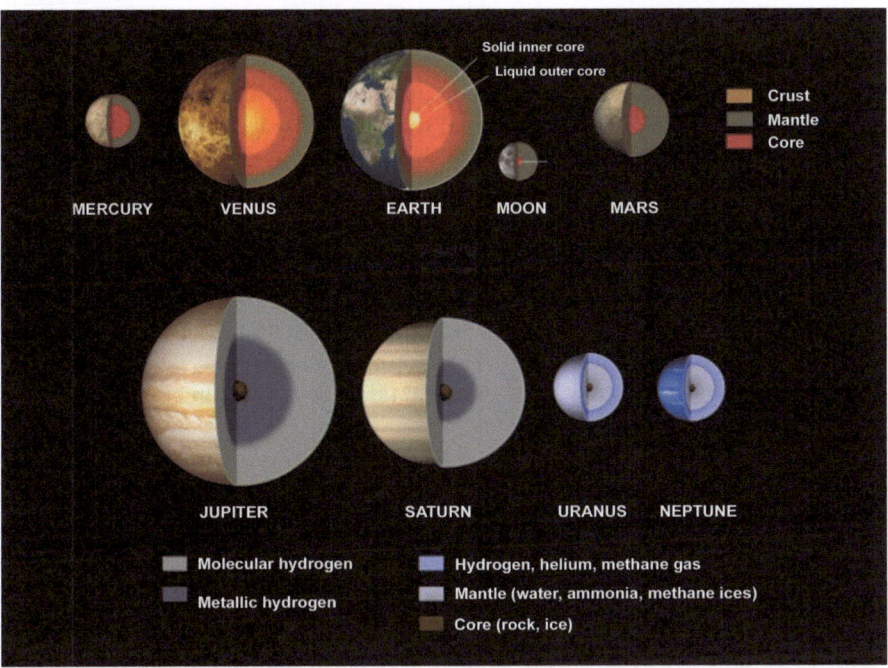

Fig. 4.5 *Top*: Internal structure of telluric planets in the Solar system and of the Moon. *Bottom*: Internal structure of gaseous giant planets (Jupiter and Saturn) and of icy giant planets (Uranus and Neptune) in the Solar system (see [2, 34, 41, 42]). (Courtesy NASA/JPL-Caltech)

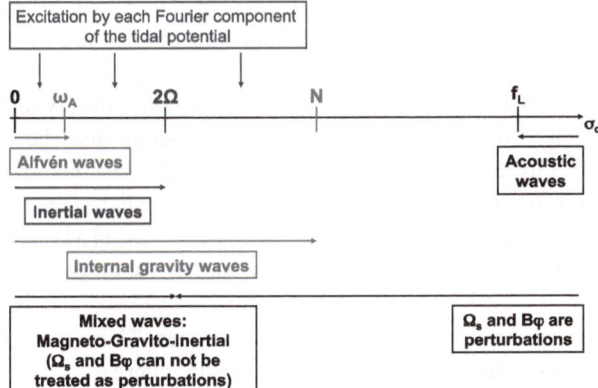

Fig. 4.6 Main wave types in stellar interiors and in fluid planetary layers. Tidal interactions excite low-frequency waves such as inertial waves in convective regions and gravito-inertial waves in stably stratified zones (and associated magneto-inertial and magneto-gravito-inertial waves if magnetic fields are taken into account); high-frequency acoustic waves are only perturbed by the tidal potential. (adapted from [71], courtesy Astronomy & Astrophysics)

and acoustic waves ($\omega > f_L$); these are respectively driven by the magnetic tension force, the Coriolis acceleration, the buoyancy force and the compressibility of the studied layers. If we now focus on low-frequency waves, inertial waves are propagating in convective regions while internal waves are propagating in stably stratified regions. For the latter, if considered frequencies are of the same order of magnitude as the Alvén and the inertial frequencies, these become gravito-inertial waves if we add the action of the Coriolis acceleration to the one of buoyancy and magneto-gravito-inertial waves if the magnetic field is taken into account. In this picture, tidal excitation is mostly efficient for low-frequency eigenmodes and thus for inertial and internal waves. Moreover, for acoustic waves which are high-frequency waves, the action of tides is only a perturbation. Therefore, one has to focus on inertial and on gravito-inertial waves to study the dynamical tide respectively in convective and in stably stratified regions.

Next, dissipative processes that convert kinetic energy of tidally excited fluid velocities into heat have to be identified.

First, stellar and planetary convective layers host strong turbulent flows because of the high value of the Reynolds number in such celestial bodies. In such regions, the action of turbulence on the tidal flows (the equilibrium tide and the dynamical tide, i.e. the inertial waves excited by the tidal potential) can be modelled as a viscous force with turbulent viscosity coefficient (see for example [35, 130]). This implicitly assumes that the respective length scales of tidal and convective flows allow one to distinguish one from the other. Such rough modelling has now been confirmed using direct numerical simulations of the interaction between a highly turbulent convection and a tidal velocity (see for example [89, 90]); note that in these works, the prescription given by [130] where the turbulent viscosity scales linearly with the tidal period (and thus the inverse of the tidal frequency) has been confirmed). So, in convective regions, the kinetic energy of tidal flows is dissipated into heat because of the turbulent viscous friction.

Next, in stably stratified stellar and planetary regions, the dynamical tide (i.e. the gravito-inertial waves) is dissipated through viscous and thermal diffusions (see for example [36, 37]). Then the ratio between the viscous and the radiative damping is govern by the Prandtl number ($Pr = \nu/K$, where ν and K are respectively the viscosity and the thermal diffusivity). Moreover, [131] has demonstrated that the dissipation of the equilibrium tide in stellar radiation zones can be neglected.

Finally, planetary interiors host solid/fluid interfaces. There, viscous friction occurs that contributes to the dissipation of tides kinetic energy.

All of this underlines the importance of the internal structure of the studied bodies, because each stellar or planetary layer dissipates the kinetic energy in function of its fluid or solid nature and of its stability with respect to the convective instability for fluid regions. A summary of the dissipation mechanisms that contribute to the tidal friction in liquid and gaseous regions is given in Fig. 4.7.

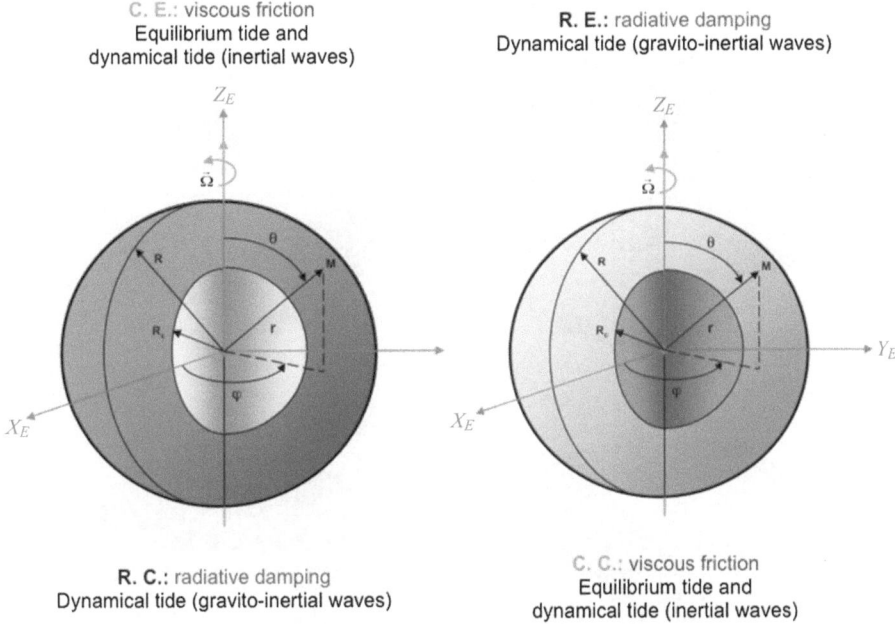

Fig. 4.7 Dissipative processes acting in solar-type stars (*left*) and in massive stars and in Jupiter- (and Saturn)-type giant planets (*right*). CE, RE, CC, RC are respectively convective and radiative envelopes and cores. For giant planets, a rocky/icy core could exist [2, 34, 41, 42]

4.3.2.2 The Fluid Equilibrium Tide

Let us first consider the equilibrium tide. As it has been explained in previous sections, the studied component adjusts in a quasi-hydrostatic way when this is submitted to the tidal potential exerted by the companion. Then, a large-scale flow in phase with the tidal potential is excited as a response to such structural adjustment with an amplitude scaling with the tidal frequency; this is the equilibrium tide (see for example [47, 92, 129, 133]). Since the tidal force is derived from a potential, the density is constant on an isobar, which is also an equipotential of the total potential (the sum of the self-gravitation potential and of the tidal one). Then, by the definition of the equilibrium tide, the tidal deformation and the related structural variables (the total gravific potential and the density) and the equilibrium tide velocity field are time-independent in a frame rotating with the studied Fourier component of the tidal potential; because of that property, the equilibrium tide velocity field is divergence-free (see [92]), contrary to the claim of [31, 109]. It is also important to point that since this velocity field represented in Fig. 4.8 verifies the momentum equation in this rotating frame, the Coriolis acceleration must be taken into account in its derivation (from a technical point of view, the equilibrium tide velocity field has both a poloidal component and a toroidal one). Next, the view of the equilibrium tide we give above is an adiabatic view, and we must now introduce the associated friction mechanism. For the equilibrium tide, this is the turbulent friction in con-

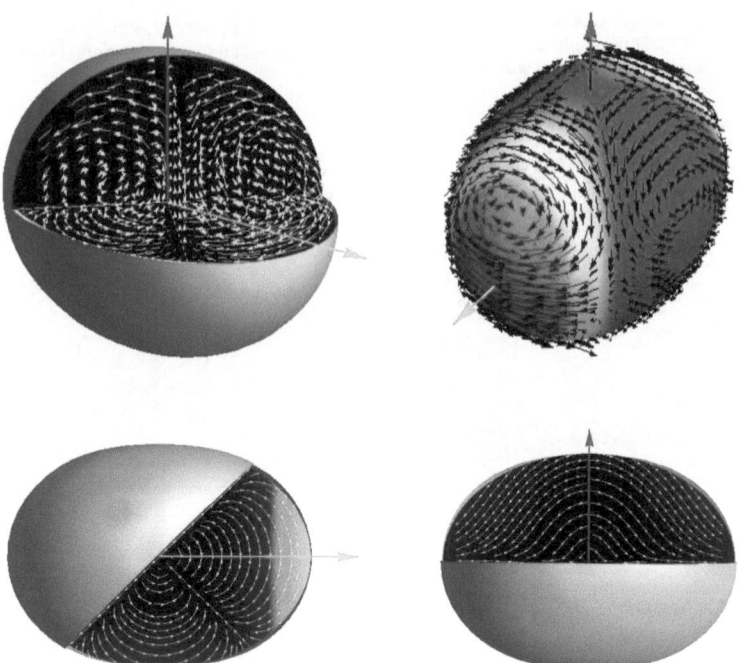

Fig. 4.8 *Top-left*: 3D view of the total (poloidal and toroidal) adiabatic equilibrium tide velocity field (*white arrows*). The *red* and *orange arrows* indicate the direction of the primary's rotation axis and the line of centers respectively. *Top-right*: Representation of this velocity field at the surface of the primary (*black arrows*); the *color-scaled background* represents the normalised tidal potential intensity (*blue* and *red* for the minimum and maximum values respectively). *Bottom-left*: View of the velocity field (*white arrows*) in its equatorial plane of symmetry; the *color-scaled background* represents the velocity value (*black* and *orange* for the minimum and maximum values respectively). *Bottom-right*: View of the velocity field (*white arrows*) in its meridional plane of symmetry; the *color-scaled background* represents the velocity value (*black* and *purple* for the minimum and maximum values respectively (taken from [92], courtesy Astronomy & Astrophysics)

vective regions that will act to convert its kinetic energy into heat (note that [131] has demonstrated that its dissipation in stably stratified regions is negligible). Then, the interaction between turbulent convective flows and the equilibrium tide velocity field has to be examined. From now on, two main assumptions are assumed: first, we consider that the respective scales of the turbulent convection and of the equilibrium tide are different enough to be separated; next, we assume that the action of the turbulent convection on the tidal flow can be modelled as a viscous force, where the used viscosity is a "turbulent viscosity" which is enhanced compared to the molecular viscosity of the plasma. Next, since the adiabatic equilibrium tide has both poloidal and toroidal components, the related viscous force has also the two corresponding components that sustain a secondary toroidal flow, called the "convective" or dissipative equilibrium tide in quadrature with the tidal potential that redistributes, because of the induced movement, the density. This gives a "dissipative" perturbation of the gravific potential that drives the secular evolution of the orbit and of the spins of the components of the system.

To describe the dissipation of the kinetic energy of the adiabatic equilibrium tide, its amplitude and its dependence on the tidal frequency, the key physical ingredient is the assumed prescription for the turbulent viscosity coefficient. Then, two different regimes can be drawn: the slow tide regime and the fast tide one. In the first one, the orbital period of the perturber is longer than the characteristic convective turn-over time. Then, the turbulent friction can be efficient to dissipate the kinetic energy into heat. In the opposite, in the fast tide regime, the orbital period is shorter than the characteristic convective turn-over time and the turbulent friction losses part of its efficiency to convert the kinetic energy into heat that lead to a saturation of the associated energy dissipation. The way in which the dissipation becomes less efficient when the tidal period becomes shorter remains one of the unsolved question in the treatment of the equilibrium tide. Two main prescriptions have been given today in the literature: those by [130, 134] and by [35]. In the first one the turbulent viscosity scales linearly with the tidal period, while in the second one this scales as the squared tidal period. The most efficient way to probe such prescriptions on the action of turbulence is then to use three-dimensional numerical simulations of highly turbulent convective flows submitted to a periodic forcing, which is often modelled as a shear that oscillates with time. The most recent numerical simulations have been achieved with such setup in Cartesian coordinates [88–90] that tends to confirm the prescription by [130]. In a near future, more simulations have to be computed in order to reach flows that are more turbulent and to take into account the spherical (ellipsoidal) geometry of the problem. The conclusion which can be then draw on the dependence on the tidal frequency of the viscous dissipation of the equilibrium tide is that is will vary as the tidal frequency (this is the dependence of the velocity field) multiplied by the frequency dependence of the turbulent viscosity. In the case of the linear prescription given by [130], we shall note that in the fast tide case the turbulent viscosity scales as the inverse of the tidal frequency that leads to a constant tidal dissipation while in the slow tide regime the turbulent viscosity is constant that gives a dissipation scaling with the tidal frequency (see Fig. 4.9).

To conclude this part on the equilibrium tide, we shall note that in a near future both the differential rotation and the magnetic field have to be taken into account since these modify both the equilibrium tide velocity field and the convective flows and the associated turbulence properties.

4.3.2.3 The Fluid Dynamical Tide: Inertial and Gravito-Inertial Waves

Inertial Waves   Once the equilibrium tide has been studied, it is then necessary to focus on the dynamical tide, i.e. the eigenmodes of the studied body, which are excited by the tidal potential. To achieve this aim, let us first consider convective zones in stellar and planetary interiors. In those regions, if we neglect magnetic field, two types of waves are propagating, the acoustic and the inertial waves. However, as it has been explained above, acoustic waves are high-frequency waves and are thus

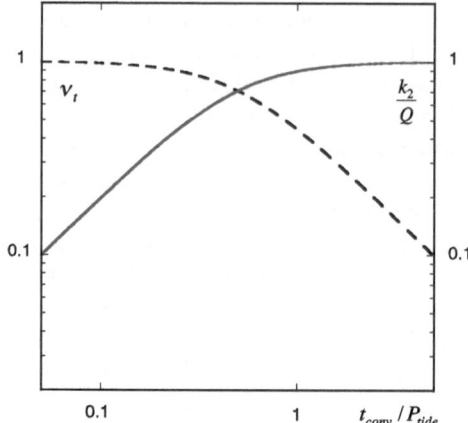

Fig. 4.9 The two regimes of the turbulent dissipation of the equilibrium tide according to the prescription by [130]. As long as the local convective turn-over time remains shorter than the tidal period ($t_{\mathrm{conv}} < P_{\mathrm{tide}}$), the turbulent viscosity ν_t (in *black dashed line*) is independent of the tidal frequency, and the inverse quality factor k_2/Q (in *red continuous line*) varies proportionally to the tidal frequency (σ_l) (so does also the tidal lag angle). When $t_{\mathrm{conv}} > P_{\mathrm{tide}}$, ν_t varies proportionally to the tidal period, whereas k_2/Q does no longer depend on the tidal frequency. ν_t and k_2/Q have been scaled by the value they take respectively for $t_{\mathrm{conv}}/P_{\mathrm{tide}} \to 0$ and $\to \infty$ (taken from [92], courtesy Astronomy & Astrophysics)

only weakly perturbed by the tidal potential. This is not the case of inertial waves for which the Coriolis acceleration is the restoring force. Then, inertial waves can be efficiently coupled with the tidal potential (see [48, 81–85, 97, 127, 128]). Such coupling can then lead to an important tidal dissipation in the convective envelopes of low-mass stars and giant planets and may be in the convective core of massive stars and telluric planets. Because of inertial waves properties, the necessary condition to get such dissipation is that the tidal frequency is such that $|\sigma_T| \in [0, 2\Omega]$, where from now on σ_T is the tidal frequency.

Let us now examine the properties of tidally excited inertial waves. First, two configurations can occur in stellar and planetary interiors; first, the tidal potential can be coupled with inertial waves that propagate in a full sphere (the case of an entirely convective star or fluid planet or of a stellar convective core); next, inertial waves can propagate between concentric spheres with stress-free or no-slip boundary conditions depending on if we are studying stellar or planetary interiors (the case of the convective envelopes of low-mass stars and of giant planets if those have an heavy element rocky or icy core). Because of the cylindrical geometry related to the Coriolis acceleration, boundary conditions then strongly influence the excited flows and the related dissipation.

The "Full sphere" Configuration   The first case of entirely convective sphere has been studied for example by [127, 128] (Fig. 4.10). In this work, the original solution for inertial waves in a full sphere with a constant density derived by [10] has been generalised to the stratified case. This leads to global modes which

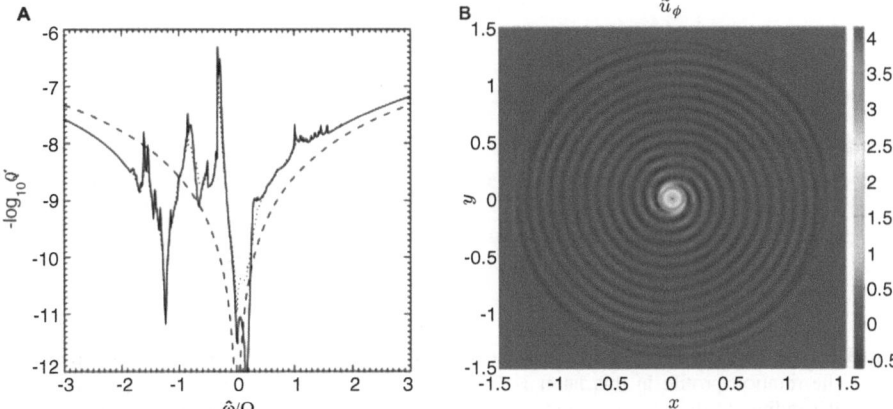

Fig. 4.13 **A**: Tidal dissipation by gravito-inertial waves in a solar-type star and the corresponding quality factor as a function of the tidal frequency computed by [83] (courtesy The Astrophysical Journal). **B**: Breaking of large-amplitude tidally excited internal waves at the centre of a solar-type star computed by [4]. There, waves deposit their angular momentum that can accelerate the centre of the star and make evolve the orbit of the planetary companion (courtesy Monthly Notices of the Royal Astronomical Society)

stars) and in stably stratified planetary layers (for example in non-convective layers just below the surface of giant planets in our solar system or of giant extra-solar planets, where those layers can be created because of the heating of the surface by the close star [40], and in stably stratified regions in telluric planets [16]). Then, the displacement as in the case of convective regions is the sum of the equilibrium tide and of the dynamical tide, which are here gravito-inertial waves. Let us now consider the properties of the tidal dissipation related to gravito-inertial waves. We must here recall that the main dissipative mechanism acting on such waves is the thermal diffusion. Then, as in the case of inertial waves, the tidal dissipation can be increased by several orders of magnitude compared to the one of the equilibrium tide, in particular in resonances that occurs for gravity waves it has been shown by [82, 83, 86, 101–103, 106, 108, 118, 132]. Moreover, because of such increase of the tidal dissipation during resonances, its behaviour is highly dependent on the tidal frequency (see Fig. 4.13).

Let us now discuss the modification of gravito-inertial waves propagation by rotation and magnetic field. First, for the rotation, gravito-inertial wave propagation strongly depends on the value of the tidal frequency compared to the inertial frequency (2Ω) where we recall that Ω is the rotation of the studied body. First, in the super-inertial regime ($\sigma_T > 2\Omega$), gravito-inertial waves are propagating in the whole sphere. However, in the sub-inertial one ($\sigma_T < 2\Omega$), waves become trapped in an equatorial region, propagating only above a given so-called critical latitude [69, 76] (see Fig. 4.14). Such kind modification of gravito-inertial propagation is very important for their coupling with tidal-induced displacements in adjacent convective regions. Moreover, it has been discovered by [107, 123–126] that the Coriolis acceleration can lead to the so-called "tapping in resonance" where

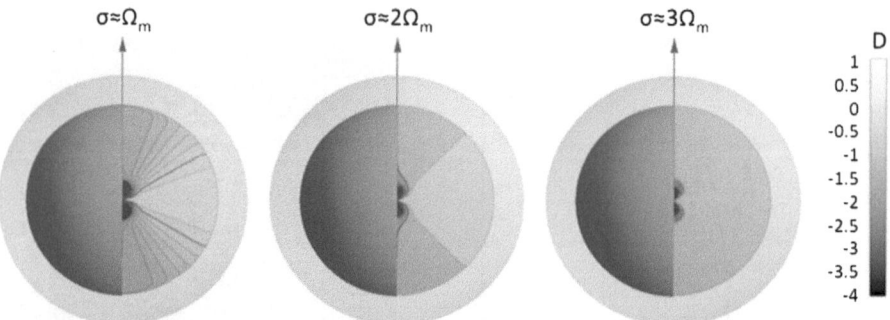

Fig. 4.14 We consider here a solar-twin star with an external convective envelope and a radiative core. The rotation profile in this latter is flat (i.e. $\Omega = \Omega_m$) for $r \in [0.2R, 0.7R]$, where R is the stellar radius. In the central region, Ω increases until $\Omega = 5\Omega_m$ in the center. Finally, at the radiation/convection border we choose the same differential rotation that in the Solar tachocline. *Regions with reds contours* correspond to those where waves are propagative, while polar and central regions with blue contours correspond to "dead" zones for waves propagation. We choose three frequencies ($\sigma = \{\Omega_m, 2\Omega_m, 3\Omega_m\}$), which shows that for frequency below $2\Omega_m$, equatorial trapping phenomena appear. The *black line* corresponds to the critical surface (the critical latitude in the case of uniform rotation) at the level of which the wave propagation regime changes. The central region is always a non-propagative region because of the central rapid rotation. D is a function of σ, Ω, and the gradients of Ω. Its definition is given in [69]

retrograde and prograde waves exert respectively a negative and a positive torque that act to block the studied system in a resonant state where the tidal dissipation is very efficient. Then, the tidal dissipation is also dependent on the rotation rate and the associated Coriolis acceleration. Next, for example in the case of solar-type stars, gravito-inertial wave propagation is modified by the presence of magnetic field, for example at the bottom of the convective envelope. Then, waves become magneto-gravito-inertial waves where the Lorenz force has to be taken into account. Such waves have been studied for example by [71, 72]. First, waves excited with frequencies close to the Alvén frequency will be vertically trapped. Then, as in the gravito-inertial case, an equatorial trapping can occur depending both on ω_A and 2Ω.

As in the case of inertial waves, we can conclude that the tidal dissipation strongly depends on the tidal frequency, on the rotation and on potential impact of magnetic fields.

Interactions with Shear Flows and Instabilities   Finally, let us discuss the interaction of the dynamical tide with differential rotation with the dynamical tide in stably stratified layers. First, as it has been explained by [36, 37], gravity waves can transfer angular momentum only if these are damped by a dissipative mechanism (which is mostly the thermal diffusion) or if these meet corotation resonances during their propagation (i.e. if we consider a "shellular" rotation that depends only on r, radius where $\Omega(r_c)$ is proportional to the tidal frequency). Let us first examine the thermal diffusion effect: an important point is that the thermal damping depends on the prograde or retrograde behaviour of the wave because of the Doppler shift. Then, in the case of a differentially rotating body, the synchronisation of each layer will progress from the surface to deeper regions [37]. We must point here that such mechanism will be coupled with other transport processes as it will be discussed in Sect. 4.3.1. Let us now focus on corotation regions, which are also called the critical layers. There, these are strong interactions between internal waves and the shear of the differential rotation. We can summarise such type of exchanges as follows: first, if the studied layer is stable with respect to the shear instabilities, waves deposit their angular momentum, the damping rate being dependent on the so-called Richardson number, which compares the strength of the stabilisation by the stratification and the destabilisation by the shear gradient; then, if the layer is already turbulent, internal waves can be reflected and transmitted by such layers with an amplitude greater than their initial one because waves take energy for the turbulent flows. In this context, it is important to study the possible instabilities that could affect internal waves dynamics. First, if waves are excited with a large amplitude, waves will break, and then, these could overturn the stable stratification (see for example [4, 5] and Fig. 4.13 for dynamical tide dynamics at the centre of solar-type star). Then, even for weak amplitude, internal waves can undergo parametric instabilities where a "parent" wave gives birth to "daughter" waves that could be then also dissipated [122]. Thus, as in the case of inertial waves, the interaction with the differential rotation as well as their own instabilities could strongly modify the value of the tidal dissipation.

4.3.3 Tidal Dissipation in Rocky or Icy Planetary Regions

As it has been shown in previous sections, the tidal potential is able to excite several types of velocity fields in fluid stellar and planetary layers leading to possible high values of the quality factor Q which is function of the tidal frequency both for the equilibrium and the dynamical tides. However, planets (and associated natural satellites) are composed both of fluid and solid layers, and tidal dissipation in the latter and at their interfaces should also be treated.

In this way, the treatment of what is often called "the bodily tide", in other word the solid tide, has been one of the first studies of tidal dissipation using continuous media mechanics (see for example [66]). These studies were of course motivated by the Earth case where both tidal interactions with the Moon and the Sun have to be taken into account [79]. In solid layers, tidal physical mechanisms are similar to those occurring in fluids. First, a solid equilibrium tide is generated that consists on a permanent large-scale displacement (with a zero velocity field that constitutes a difference with the fluid case). In an adiabatic modelling, this displacement is allowed by the elasticity of the material and directed along the line of centres. However, as in the fluid case, solid layers host dissipation because of their anelasticity, and the tidal energy is dissipated into heat that leads to internal heating, to a net applied torque, and to a small delay between the tidal bulge and the line of centers. This anelasticity,

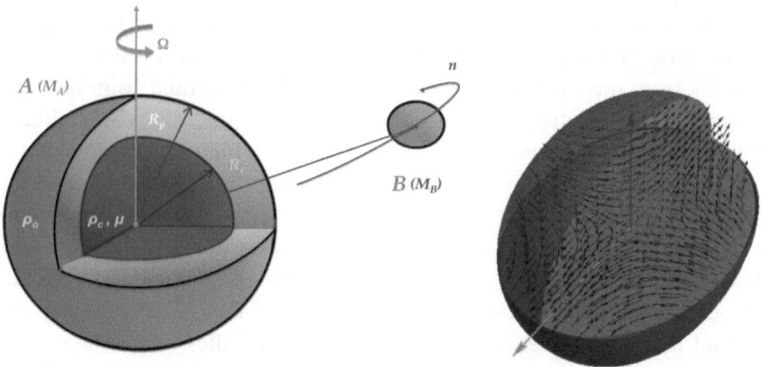

Fig. 4.15 *Left*: Two-layer model of planetary internal structure (body A with mass M_A) formed by an internal rocky core surrounded by a fluid envelope perturbed by a companion (B with mass M_B). R_c and R_p are respectively the core and the planetary radii, ρ_o and ρ_c the density of the fluid envelope and of the core, and μ is the shear modulus of the core. *Right*: tidal displacement in the solid core of the two-layer model in the adiabatic limit; the *red* and *orange arrows* indicate respectively the symmetry axis of the planetary core and the direction of the companion (taken from [93]; courtesy Astronomy & Astrophysics)

which is often modelled as a viscous behaviour that adds to the elasticity as in the Maxwell's body model, depends on the intrinsic properties of the considered material (for example silicates or ices), which are described by its rheology [78]. Then, the rheological relation gives the relation between the stress tensor ($\overline{\overline{\sigma}}$) and the strain one $\overline{\overline{\varepsilon}}$): $\overline{\overline{\sigma}} = \widetilde{\mu}(\sigma_T)\overline{\overline{\varepsilon}}$, where $\widetilde{\mu}$ is the complex rigidity, which depends on the tidal frequency (σ_T). Its real and imaginary parts represent respectively the energy storage and the energy losses of the system. Then, using the correspondence principle (see [6, 78]), one can calculate the tidal dissipation and the associated quality factor Q for any linear rheology.

Such type of computation has already been computed for the Moon [87], for rocky core of giant planets [26], for icy natural satellites (see for example [119, 120]), and for telluric extrasolar planets as Earth like planets or Super-Earth [44].

Such type of solid tide can be illustrated if we examine a two-layer planets with an internal rocky part with an external fluid envelope as studied by [26, 93] (see Fig. 4.15). This corresponds to the cases of a telluric planet with an external ocean or atmosphere or of a gaseous or icy giant planet with a potential rocky/icy core born during the planetary formation surrounding by the external fluid envelope [41, 42]. Then, if we consider the obtained tidal dissipation due to the solid layers for example for a Jupiter or a Saturn-like planet, this can reach values greater by several orders of magnitude than those due to tidal fluid velocities described in the previous sections for realistic values of the rigidity and viscosity (Figs. 4.16 and 4.17). This can be compared in our Solar system to values of the quality factor Q determined for example through astrometric measurements [56, 57] that are now confirmed by realistic scenarii for natural satellites formation, for example in the Saturnian system [17]. Moreover, if the dependence on the tidal frequency of such dissipation is studied (Fig. 4.17), a smooth behaviour is obtained compared to the case of inertial

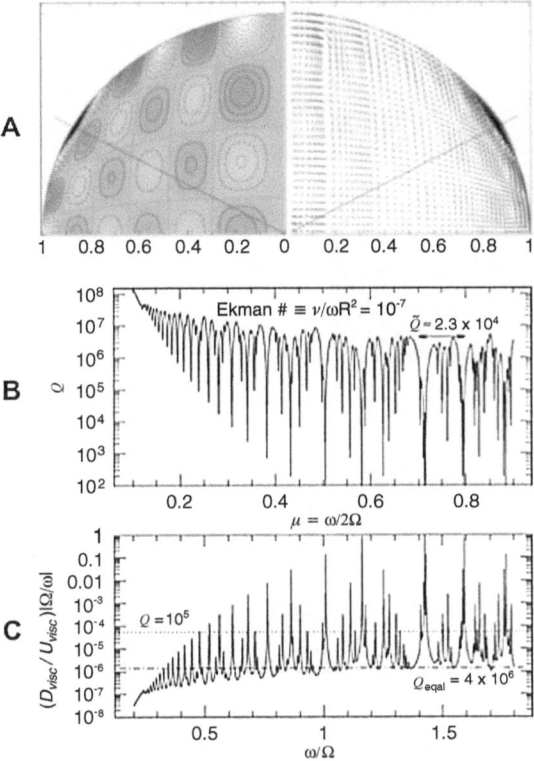

Fig. 4.10 **A**: Density perturbation (*left*) and radial and latitudinal components (*right*) of a tidally excited retrograde inertial mode with $|m| = 2$. The mode amplitude and the wavevector remain relatively uniform over much of the planet and rise sharply towards the surface; this rise is most striking near the critical latitude $\theta_c = \arccos(\sigma_T/2\Omega)$ marked by straight lines (Ω is the planet rotation) (taken from [127], courtesy The Astrophysical Journal). **B & C**: Viscous dissipation of such inertial waves excited by the tidal potential in a fully convective planet (**B**) and the associated tidal quality factor Q (**C**) as a function of the tidal frequency for the Eckman number $E = \nu/\Omega R^2 = 10^{-7}$, where ν is the viscosity, and R the radius of the studied planet (taken from [128], courtesy The Astrophysical Journal)

then couple to the tidal potential as in the academic case of a forced oscillator. To study the dissipation of their kinetic energy, the same assumptions that in the case of the equilibrium tide (i.e. the spatial scales separation between the convective flows and the tidal inertial waves and the modelling of the friction using a viscous force with a turbulent viscosity) are assumed. Numerous resonances between eigenmodes and the tidal potential are obtained and identified. Then, the main important properties to study is the related amplitude of the dissipation and its dependence on the tidal frequency. First, the tidal dissipation can be increased by several order of magnitudes when a resonance is encountered compared to the case of the equilibrium tide. Next, the main difference with the equilibrium tide is the strong dependence on the tidal frequency of the tidal dissipation because of the resonances.

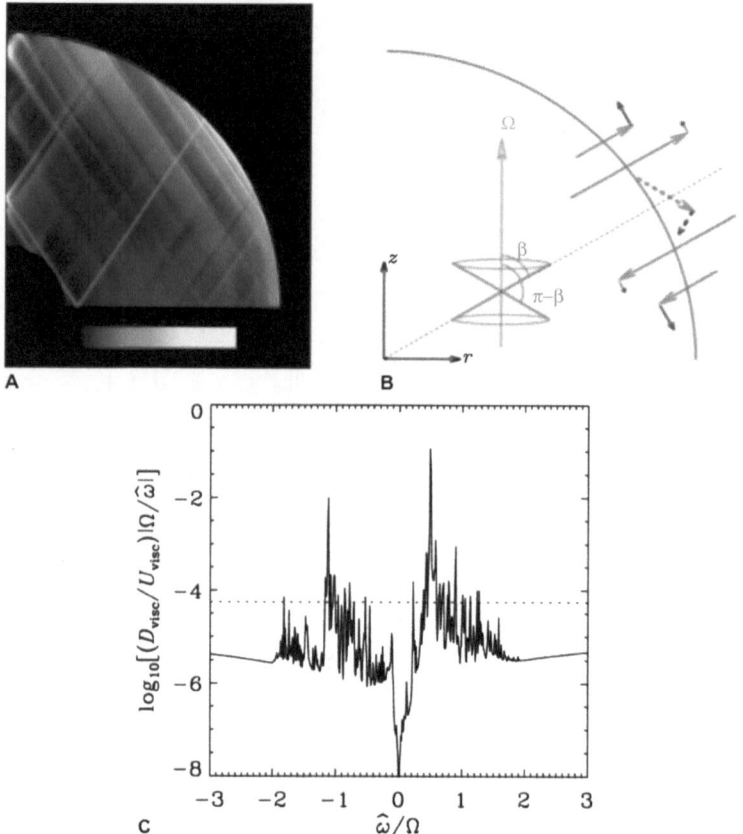

Fig. 4.11 **A**: High-resolution calculation ($E = 10^{-9}$) of the tidal response of a uniformly rotating planet with an internal core. The r.m.s. velocity of the total tide (equilibrium and dynamical) is plotted in a meridional slice through the convective region (the velocity scale is linear, black representing zero). Attractors with associated inertial wave beams can be identified. The forcing tidal frequency is chosen to be near the peak of an inertial mode resonance (taken from [82], courtesy The Astrophysical Journal). **B**: Production of short inertial waves by scattering of the equilibrium tide off the core at critical latitudes to explain results obtained by [82] (taken from [39], courtesy The Astrophysical Journal). **C**: Viscous dissipation of such inertial waves excited by the tidal potential as a function of the tidal frequency for the Eckman number $E = 10^{-7}$; the *dotted line* corresponds to $Q = 10^5$ (taken from [82], courtesy The Astrophysical Journal)

The Cored Configuration   The second case of inertial waves propagating in spherical shells has been studied in [82] for the planetary case and in [83] for the stellar case. Then, because of the conflict between the spherical geometry of the problem and the one related to the Coriolis acceleration, the propagation of inertial waves becomes more complex (Fig. 4.11). First, inertial waves propagate along characteristic rays that are inclined to the rotation axis at a certain angle, which depends on the wave frequency. This angle is necessarily preserved in reflections of the waves from boundaries so that a beam is focused of defocused in such a reflection. When propagating in such spherical annulus (in opposite to the first full sphere case), inertial waves are focused onto attractors where an intense dissipation occurs [81, 96, 99, 100]. Then, [80] demonstrated mathematically that the mean dissipation rate associated with waves attractors (in a simplified wave equation) becomes independent of the viscosity in the limit of a very small Ekman number (i.e. for very small viscosity). This constitutes a remarkable behaviour compared for example to the case of the equilibrium tide dissipation, which is directly proportional to the viscosity. In the complete global modelling by [82], a similar behaviour is observed in the limit of small viscosity, and obtained solutions indicate that the dissipation is typically concentrated along the rays that emanate from the critical latitude on the inner boundary (see also [97]). Reference [39] have proposed a physical interpretation of such phenomena: using WKB methods, they demonstrate the production of short inertial waves by scattering of the equilibrium tide off the core at critical latitudes. The tidal dissipation rate associated with these waves scales as the fifth power of the core radius. They also find that even if the core of rock or ice is unlikely to be rigid, Ogilvie and Lin's mechanism should still operate if the core is substantially denser than its immediate surroundings.

As a partial conclusion, we must point that the viscous dissipation of inertial waves are one of the most important processes to take into account in stellar and in planetary interiors. Let us now draw some perspectives on what should be done to improve the modelling of such processes. First, as in the case of the equilibrium tide, the differential rotation and the magnetic field have to be taken into account. First, convective flows are those that establish the differential rotation that depends both on radius and on latitude in convective regions [8]. Next, such regions host dynamo-generated magnetic fields [9] that are generated because of the simultaneous action of differential rotation and of convective turbulence that are themselves modified by the magnetic field because of the Lorentz force feed-back in the momentum equation. Then, inertial waves propagation and dissipation will be modified both by the differential rotation and the magnetic fields. Moreover, the dissipation may be modified both by corotation resonances between inertial waves and the sheared rotation (see also the case of gravito-inertial waves) and by the Ohmic heating that constitute supplementary dissipation sources.

Inertial Waves Instabilities   Finally, it is important to point the possibility of tidal inertial waves' related instabilities. In the case of inertial waves, main instabilities can come for the interaction with differential rotation (see also the case of gravito-inertial waves) and from the tidal elliptic instability. The latter corresponds to the astrophysical version of the generic elliptical instability, which affects all rotating fluids with elliptically deformed streamlines [51, 53, 54, 61]. In the astrophysical case, the origin of such elliptic instability is a resonance between inertial waves in rotating stars and planets and the tidal wave, i.e. the underlying strain field responsible for the elliptic deformation [121]. This instability is able to generate and sustain large-scale flows (for example the so-called spin-over mode) that superpose to basic flows such as differential rotation or convection in planetary and in stellar interiors (see Fig. 4.12). Then, as in the case of inertial waves, viscous forces can act to dissipate the generated kinetic energy that leads to potential important evolution of the

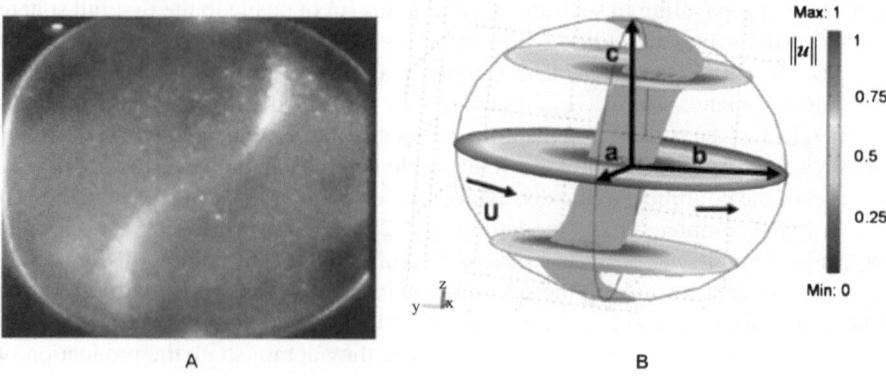

Fig. 4.12 Spin-over mode of the tidal elliptic instability observed in the laboratory (**A**) and computed in numerical simulations (**B**) (adapted from [12, 13], courtesy D. Cébron and Physics of the
Earth and Planetary Interiors)

considered system. Recent studies of the tidal elliptic instability have been recently
achieved in the context of binary stars [62], planetary cores [13, 14] and extra-solar
planetary systems [15]. Then, interesting behaviour of the elliptic instability has
been isolated. First, this can develop both in convective or stably stratified regions
(where inertial waves become gravito-inertial because of the supplementary stabil-
ising buoyancy force) [14, 43, 59, 60]. In the case of convective regions, the elliptic
instability can thus develop with a growth-rate that diminishes with the intensity of
the convection; thus, the flow generated with the tidal instability can superpose to
the convective one. Next, such tidal flow may play an important role in the induction
of a magnetic field leading to a "tidal dynamo" [45, 55] in planetary interiors and
may be in stellar ones. This last point constitutes one of the must important question
to examine in a near future to see a possible impact of tidal interactions on celestial
bodies magnetic activity (see Sect. 4.4.2). Finally, we must point that the dissipation
related to the elliptic instability depends on boundary conditions that are applied (see
Sect. 4.3.4), that can lead to important differences between no-slip boundary conditions (in planetary cores of telluric planets and at the interface between a central
core and a surrounding fluid envelope in giant planets) and stress-free conditions (as
at giant planets' and stars' surfaces.)

Internal Waves   Let us now consider the case of stably stratified zones in stellar
and planetary interiors. As it has been described in Sect. 2.2.1, gravity (and gravito-
inertial waves) are propagating in such regions. These are excited at the border with
adjacent convective regions both by turbulent movements and in the case where
there is a close companion, by pressure fluctuations induced by tidally excited inertial waves (for example those of inertial waves attractors in the case of external
convective envelopes). Then, in the case of binary or multiple systems, gravito-
inertial waves will be forced in stellar radiation zones (the core of low-mass main
sequence stars and the envelope of intermediate-mass or massive main sequence

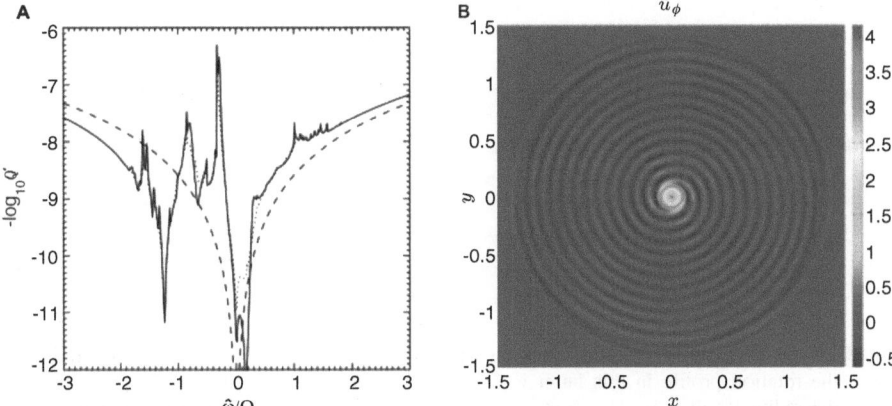

Fig. 4.13 **A:** Tidal dissipation by gravito-inertial waves in a solar-type star and the corresponding quality factor as a function of the tidal frequency computed by [83] (courtesy The Astrophysical Journal). **B:** Breaking of large-amplitude tidally excited internal waves at the centre of a solar-type star computed by [4]. There, waves deposit their angular momentum that can accelerate the centre of the star and make evolve the orbit of the planetary companion (courtesy Monthly Notices of the Royal Astronomical Society)

stars) and in stably stratified planetary layers (for example in non-convective layers just below the surface of giant planets in our solar system or of giant extra-solar planets, where those layers can be created because of the heating of the surface by the close star [40], and in stably stratified regions in telluric planets [16]). Then, the displacement as in the case of convective regions is the sum of the equilibrium tide and of the dynamical tide, which are here gravito-inertial waves. Let us now consider the properties of the tidal dissipation related to gravito-inertial waves. We must here recall that the main dissipative mechanism acting on such waves is the thermal diffusion. Then, as in the case of inertial waves, the tidal dissipation can be increased by several orders of magnitude compared to the one of the equilibrium tide, in particular in resonances that occurs for gravity waves it has been shown by [82, 83, 86, 101–103, 106, 108, 118, 132]. Moreover, because of such increase of the tidal dissipation during resonances, its behaviour is highly dependent on the tidal frequency (see Fig. 4.13).

Let us now discuss the modification of gravito-inertial waves propagation by rotation and magnetic field. First, for the rotation, gravito-inertial wave propagation strongly depends on the value of the tidal frequency compared to the inertial frequency (2Ω) where we recall that Ω is the rotation of the studied body. First, in the super-inertial regime ($\sigma_T > 2\Omega$), gravito-inertial waves are propagating in the whole sphere. However, in the sub-inertial one ($\sigma_T < 2\Omega$), waves become trapped in an equatorial region, propagating only above a given so-called critical latitude [69, 76] (see Fig. 4.14). Such kind modification of gravito-inertial propagation is very important for their coupling with tidal-induced displacements in adjacent convective regions. Moreover, it has been discovered by [107, 123–126] that the Coriolis acceleration can lead to the so-called "tapping in resonance" where

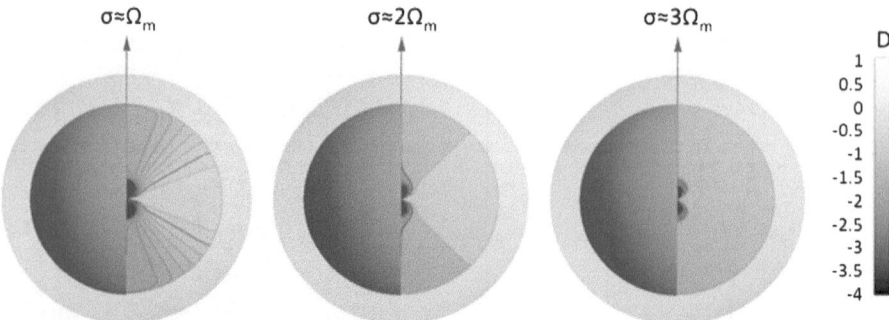

Fig. 4.14 We consider here a solar-twin star with an external convective envelope and a radiative core. The rotation profile in this latter is flat (i.e. $\Omega = \Omega_m$) for $r \in [0.2R, 0.7R]$, where R is the stellar radius. In the central region, Ω increases until $\Omega = 5\Omega_m$ in the center. Finally, at the radiation/convection border we choose the same differential rotation that in the Solar tachocline. *Regions with reds contours* correspond to those where waves are propagative, while polar and central regions with blue contours correspond to "dead" zones for waves propagation. We choose three frequencies ($\sigma = \{\Omega_m, 2\Omega_m, 3\Omega_m\}$), which shows that for frequency below $2\Omega_m$, equatorial trapping phenomena appear. The *black line* corresponds to the critical surface (the critical latitude in the case of uniform rotation) at the level of which the wave propagation regime changes. The central region is always a non-propagative region because of the central rapid rotation. D is a function of σ, Ω, and the gradients of Ω. Its definition is given in [69]

retrograde and prograde waves exert respectively a negative and a positive torque that act to block the studied system in a resonant state where the tidal dissipation is very efficient. Then, the tidal dissipation is also dependent on the rotation rate and the associated Coriolis acceleration. Next, for example in the case of solar-type stars, gravito-inertial wave propagation is modified by the presence of magnetic field, for example at the bottom of the convective envelope. Then, waves become magneto-gravito-inertial waves where the Lorenz force has to be taken into account. Such waves have been studied for example by [71, 72]. First, waves excited with frequencies close to the Alvén frequency will be vertically trapped. Then, as in the gravito-inertial case, an equatorial trapping can occur depending both on ω_A and 2Ω.

As in the case of inertial waves, we can conclude that the tidal dissipation strongly depends on the tidal frequency, on the rotation and on potential impact of magnetic fields.

Interactions with Shear Flows and Instabilities Finally, let us discuss the interaction of the dynamical tide with differential rotation with the dynamical tide in stably stratified layers. First, as it has been explained by [36, 37], gravity waves can transfer angular momentum only if these are damped by a dissipative mechanism (which is mostly the thermal diffusion) or if these meet corotation resonances during their propagation (i.e. if we consider a "shellular" rotation that depends only on r, radius where $\Omega(r_c)$ is proportional to the tidal frequency). Let us first examine the thermal diffusion effect: an important point is that the thermal damping depends on the prograde or retrograde behaviour of the wave because of the Doppler shift. Then, in the case of a differentially rotating body, the synchronisation of each layer will progress from the surface to deeper regions [37]. We must point here that such mechanism will be coupled with other transport processes as it will be discussed in Sect. 4.3.1. Let us now focus on corotation regions, which are also called the critical layers. There, these are strong interactions between internal waves and the shear of the differential rotation. We can summarise such type of exchanges as follows: first, if the studied layer is stable with respect to the shear instabilities, waves deposit their angular momentum, the damping rate being dependent on the so-called Richardson number, which compares the strength of the stabilisation by the stratification and the destabilisation by the shear gradient; then, if the layer is already turbulent, internal waves can be reflected and transmitted by such layers with an amplitude greater than their initial one because waves take energy for the turbulent flows. In this context, it is important to study the possible instabilities that could affect internal waves dynamics. First, if waves are excited with a large amplitude, waves will break, and then, these could overturn the stable stratification (see for example [4, 5] and Fig. 4.13 for dynamical tide dynamics at the centre of solar-type star). Then, even for weak amplitude, internal waves can undergo parametric instabilities where a "parent" wave gives birth to "daughter" waves that could be then also dissipated [122]. Thus, as in the case of inertial waves, the interaction with the differential rotation as well as their own instabilities could strongly modify the value of the tidal dissipation.

4.3.3 Tidal Dissipation in Rocky or Icy Planetary Regions

As it has been shown in previous sections, the tidal potential is able to excite several types of velocity fields in fluid stellar and planetary layers leading to possible high values of the quality factor Q which is function of the tidal frequency both for the equilibrium and the dynamical tides. However, planets (and associated natural satellites) are composed both of fluid and solid layers, and tidal dissipation in the latter and at their interfaces should also be treated.

In this way, the treatment of what is often called "the bodily tide", in other word the solid tide, has been one of the first studies of tidal dissipation using continuous media mechanics (see for example [66]). These studies were of course motivated by the Earth case where both tidal interactions with the Moon and the Sun have to be taken into account [79]. In solid layers, tidal physical mechanisms are similar to those occurring in fluids. First, a solid equilibrium tide is generated that consists on a permanent large-scale displacement (with a zero velocity field that constitutes a difference with the fluid case). In an adiabatic modelling, this displacement is allowed by the elasticity of the material and directed along the line of centres. However, as in the fluid case, solid layers host dissipation because of their anelasticity, and the tidal energy is dissipated into heat that leads to internal heating, to a net applied torque, and to a small delay between the tidal bulge and the line of centers. This anelasticity,

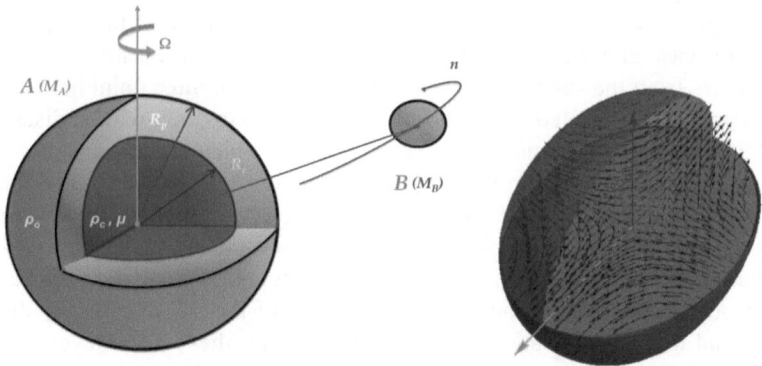

Fig. 4.15 *Left*: Two-layer model of planetary internal structure (body A with mass M_A) formed by an internal rocky core surrounded by a fluid envelope perturbed by a companion (B with mass M_B). R_c and R_p are respectively the core and the planetary radii, ρ_o and ρ_c the density of the fluid envelope and of the core, and μ is the shear modulus of the core. *Right*: tidal displacement in the solid core of the two-layer model in the adiabatic limit; the *red* and *orange arrows* indicate respectively the symmetry axis of the planetary core and the direction of the companion (taken from [93]; courtesy Astronomy & Astrophysics)

which is often modelled as a viscous behaviour that adds to the elasticity as in the Maxwell's body model, depends on the intrinsic properties of the considered material (for example silicates or ices), which are described by its rheology [78]. Then, the rheological relation gives the relation between the stress tensor ($\overline{\overline{\sigma}}$) and the strain one $\overline{\overline{\varepsilon}}$): $\overline{\overline{\sigma}} = \tilde{\mu}(\sigma_T)\overline{\overline{\varepsilon}}$, where $\tilde{\mu}$ is the complex rigidity, which depends on the tidal frequency (σ_T). Its real and imaginary parts represent respectively the energy storage and the energy losses of the system. Then, using the correspondence principle (see [6, 78]), one can calculate the tidal dissipation and the associated quality factor Q for any linear rheology.

Such type of computation has already been computed for the Moon [87], for rocky core of giant planets [26], for icy natural satellites (see for example [119, 120]), and for telluric extrasolar planets as Earth like planets or Super-Earth [44].

Such type of solid tide can be illustrated if we examine a two-layer planets with an internal rocky part with an external fluid envelope as studied by [26, 93] (see Fig. 4.15). This corresponds to the cases of a telluric planet with an external ocean or atmosphere or of a gaseous or icy giant planet with a potential rocky/icy core born during the planetary formation surrounding by the external fluid envelope [41, 42]. Then, if we consider the obtained tidal dissipation due to the solid layers for example for a Jupiter or a Saturn-like planet, this can reach values greater by several orders of magnitude than those due to tidal fluid velocities described in the previous sections for realistic values of the rigidity and viscosity (Figs. 4.16 and 4.17). This can be compared in our Solar system to values of the quality factor Q determined for example through astrometric measurements [56, 57] that are now confirmed by realistic scenarii for natural satellites formation, for example in the Saturnian system [17]. Moreover, if the dependence on the tidal frequency of such dissipation is studied (Fig. 4.17), a smooth behaviour is obtained compared to the case of inertial

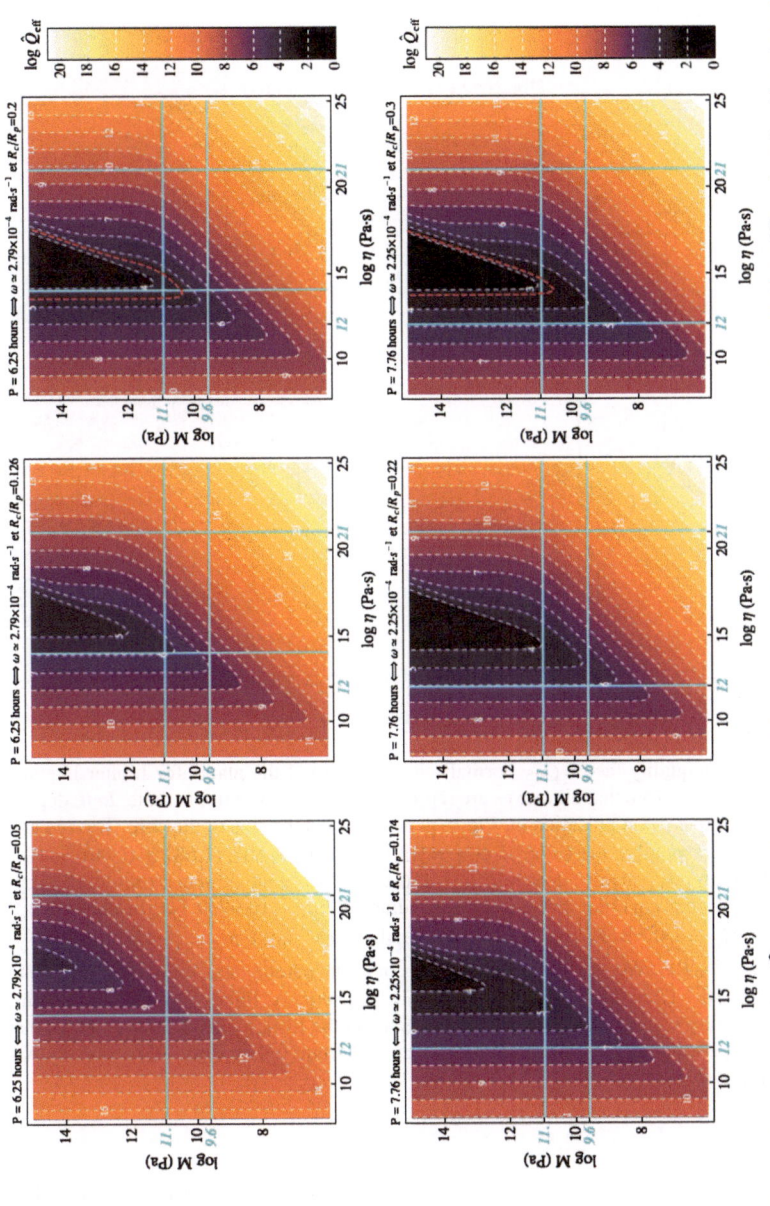

Fig. 4.16 Dissipation quality factor $\widehat{Q}_{\mathrm{eff}}$ of the Maxwell model in function of the viscoelastic parameters M (the stiffness) and η (the viscosity). *Top*: For a Jupiter-like two-layers planet tidally perturbed at the Io's frequency $\sigma_T = 2.79 \times 10^{-4}$ rad s^{-1}. *Bottom*: For a Saturn-like two-layers planet tidally perturbed at the Enceladus' frequency $\sigma_T = 2.25 \times 10^{-4}$ rad s^{-1}. The *red dashed line* corresponds to the value of $\widehat{Q}_{\mathrm{eff}}$ calculated by [56, 57] and needed by [17] to form the mid-sized satellites of Saturn. The *blue lines* corresponds to the lower and upper limits of the more realistic values taken by the viscoelastic parameters M and η at very high pressure for an unknown mix of ice and silicates (taken from [93]; courtesy Astronomy & Astrophysics)

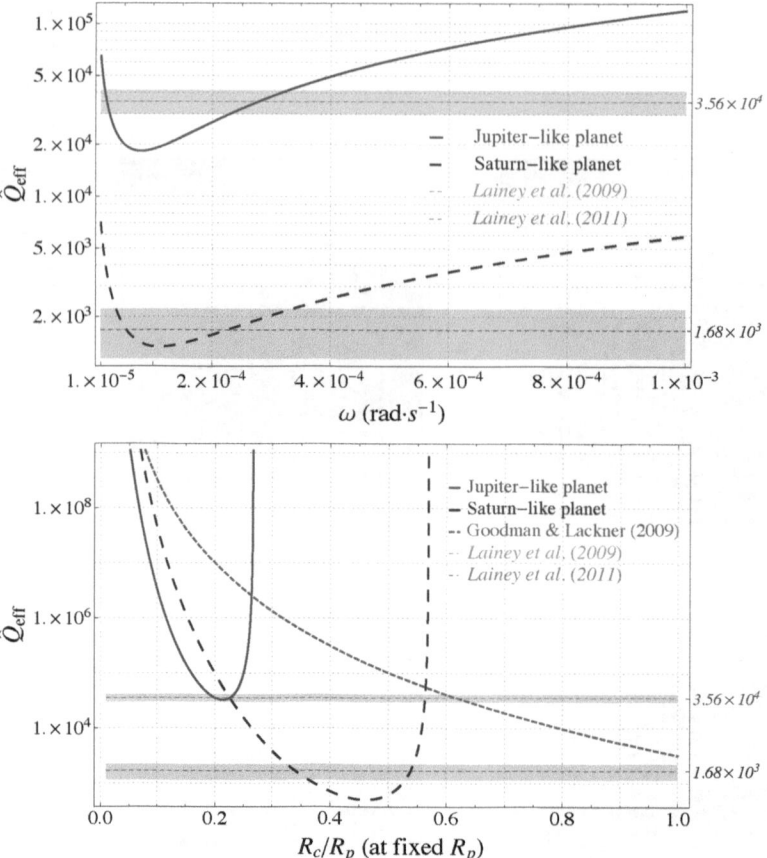

Fig. 4.17 Dissipation quality factor $\widehat{Q}_{\text{eff}}$ normalised to the size of the planet for Jupiter-like and Saturn-like giant planets. Note that all curves are represented with a logarithmic scale. *Left*: dependence to the perturbative strain pulsation ω. *Right*: dependence to the size of the core. The *red* and *blue dashed lines* correspond to the value of $\widehat{Q}_{\text{eff}} = \{(3.56 \pm 0.56) \times 10^4, (1.682 \pm 0.540) \times 10^3\}$ (for Jupiter and Saturn respectively) determined by [56, 57]. Their zone of uncertainty is also represented in the corresponding color. We assume the values of $R_p = \{10.97, 9.14\}$ (in units of $R_p^{\oplus}$, the Earth radius), $M_p = \{317.8, 95.16\}$ (in units of $M_p^{\oplus}$, the Earth mass), $M_c = \{6.42, 6.55\} \times M_p^{\oplus}$, and $R_c = \{0.20, 0.34\}R_p$. We also take for the viscoelastic parameters $M = 10^{11}$ (Pa), and $\eta = \{5.81, 6.02\} \times 10^{14}$ (Pa s^{-1}) for Jupiter and Saturn respectively. Finally, we achieve a comparison with the quality factor related to inertial waves using the [39] prescription that shows that anelastic tidal dissipation may be stronger and also very important (taken from [93]; courtesy Astronomy & Astrophysics)

and gravito-inertial waves for which variations of several orders of magnitude of the dissipation are obtained (c.f. Sect. 4.3.2). Furthermore, the dissipation is very sensitive to the size of the solid core as in the inertial waves case. This shows how this becomes crucial to get constraints on the size of the rocky core of giant planets from

Fig. 4.18 Schematic view of Eckman boundary layers that can be important at boundaries of planetary cores (taken from [98] in which details concerning boundary velocity fields $\mathbf{b}_1$ and $\mathbf{b}_2$ are given, courtesy The Astrophysical Journal)

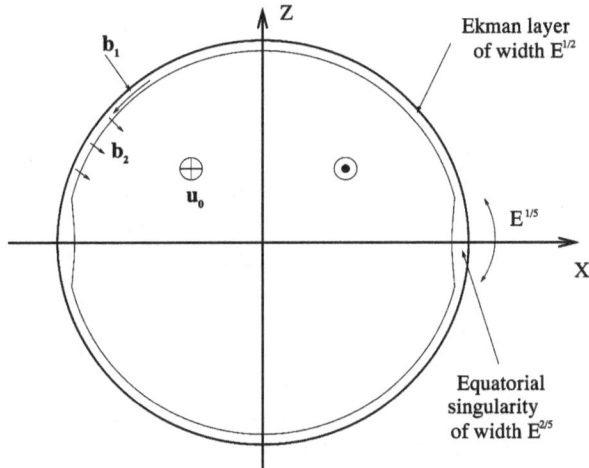

observations and theoretical predictions. Moreover, this large value, as in the fluid case, shows the strong need to go beyond phenomenological prescriptions [28, 29].

4.3.4 Boundary Conditions

As it has been shown previously, tidal interactions excite flows (i.e. the equilibrium tide, the dynamical tide, and fluid movements that result from their instabilities). Then, depending on the internal structure of the studied body, boundary conditions, and particularly at the surface and near solid/fluid interfaces (in planetary interiors) should be examined carefully. Indeed, boundary layer strongly sheared flows can develop there, which may lead to a strong dissipation. Let us first consider the case of the surface of stars or of planets. References [115–117] have proposed that Eckman boundary layers take place and lead to an important viscous dissipation. However, following [94, 98], we must point that such boundary constitutes a free surface with stress-free boundary conditions for the velocity field and thus that the dissipation related to associated boundary layers will be very weak. However, if we now consider the solid/fluid interfaces, we are in the case of no-slip boundary conditions that correspond to the classical Eckman boundary layers (see Fig. 4.18) where a strong viscous dissipation may occur, particularly if the studied region is turbulent. As a partial conclusion on boundary flows, one has thus to remember that solid/fluid regions (at the top of rocky/icy cores of giant planets and at boundaries of liquid cores in telluric planets) may host strong viscous dissipation while this will not be the case below the surface of stars or planets with a fluid envelope. At solid-fluid boundaries, we must also point that couplings between the fluid dynamical tide (inertial or gravito-inertial) and the (an)elastic tide in solid regions must be studied in a near future since these may modify the related tidal dissipation.

4.3.5 Hierarchy Between Dissipative Physical Processes and the Associated Obtained States

In this review, we have tried to give a complete review of dissipative processes acting on tidally excited velocity fields in stellar and in planetary interiors. It is thus interesting at that point to draw a hierarchy between the intensity of the related dissipations. First, for fluid regions, dynamical tides (inertial and gravito-inertial waves) and related instabilities (the elliptic instability for inertial waves and the convective and parametric instabilities for internal waves) lead to a stronger dissipation that can dominate by several orders of magnitude the one associated to the equilibrium tide. Next, if we study planetary interiors, we have isolated that tidal dissipation in rocky/icy regions can dominate the one in fluid regions depending on their respective size. Recall also that each type of tidal dissipation has a different dependence on the tidal frequency that can be constrained by observations to unravel the action of the different physical mechanisms. Finally, it is important to recall that once one has identified all the dissipation processes and derived the associated torques as shown in Sect. 2.2, equilibrium states for orbital and spin properties can be obtained (see for example [19–22] for telluric planets) and compared to observational constraints obtained on our Solar system and on exoplanetary systems.

Now, we will show how tidal interactions can modify star(s) in binary or multiple stellar systems and in planetary systems.

4.4 The Central Star

4.4.1 Tidal Action on the Internal Transport of Angular Momentum

First, the internal transport of angular momentum in multiple systems where stars host planetary, stellar, or compact objects companions is different from the case of single stars. In this way, stars in binary or multiple systems are submitted to tidal torques (and other interaction torques; see Sect. 4.4.3) that modify their internal redistribution of angular momentum and therefore their rotational history. Indeed, because of gravific (and magnetic) interactions with their environment, the evolution of such stars will be modified.

Therefore, if a star is submitted to efficient tidal interactions with its environment, both equilibrium and dynamical tide torques must be taken into account. First, the equilibrium tide generates a torque acting on stellar convective regions where the dissipation of its kinetic energy takes place because of the turbulent convection (see Sect. 4.3.2.1). References [130, 133] demonstrate that this dissipation mechanism is mainly efficient for stars having an external convective envelope where turbulent convective motions are able to lead to an effective friction acting on tidal flows like as for example in solar-type stars, while in the case of stars having a convective core as for example in massive stars, one must consider the action of the dynamical tide.

If we now consider the action of dynamical tides, the action of inertial waves and internal waves respectively excited in convective and stably stratified regions by the tidal potential must be described separately. Fist, the viscous dissipation of the inertial waves in convective regions leads, as in the case of the equilibrium tide, to net torques that are applied on the latter. As explained in Sect. 4.3.2.3, the efficiency of the torque then depends on the convective region geometry [39, 81–83, 85, 128]. Next, to understand the action of internal waves excited by the tidal potential in stably stratified regions [37, 83, 132], the importance of transport mechanisms which take place in such regions has to be recalled. In this manner, dynamical processes in stellar radiation zones are responsible both of the secular transport of angular momentum and of the associated chemicals mixing during stars' evolution. As described in [70, 111, 137], these impact the rotation history, the associated differential rotation profiles, and the stellar structure and evolution. In this framework, the dynamical tide, i.e. the internal waves excited by the tidal potential, redistribute the angular momentum in radiation zones in the same manner as those that are excited by the turbulent convective movement at the borders with convection regions [139]. Then, the tidal interaction is able to generate internal waves which propagate and are able to deposit/extract angular momentum as soon as they are dissipated through heat diffusion or when these reach critical layers [36]. Such action on the transport of angular momentum by tidal waves has been described for example by [113] in massive stars and by [4] at the centre of solar-type stars. This shows how tidal waves are able to modify the differential rotation profile in a star hosting a companion. Then, as observed in [112] and described in [72], the stresses induced by the simultaneous action of the dynamical tide and of internal waves excited by the convection combine with the torques applied by the equilibrium tide, the tidal inertial waves, and stellar winds to trigger large-scale meridional currents in stellar radiation zones [74, 135]. These movements then advect heat that leads to a baroclinic state responsible for the establishment of a new differential rotation profile as explained in [25]. This means that shear could be subject to instabilities that generate turbulence (see for example [75, 114]) leading to chemical mixing and viscous stresses that combine with waves stresses and applied torques to sustain the meridional circulation. This shows how tidal waves thus constitute one of the main important processes in the exchanges of angular momentum in stars hosting companions (see Fig. 4.19) and how this can modify the properties of such stars compared to those of single stars for example concerning their chemical properties (for example the metallicity of stars hosting planets; see Fig. 4.20 [136]).

4.4.2 Tidal Action on Magnetic Activity

The next main important topics that should be studied in a near future concerns the interaction between tidal interactions and the planetary and stellar magnetisms. The first question is about systems formation. Indeed, for intermediate or massive binaries (with components with an external radiative envelope and a potential associated

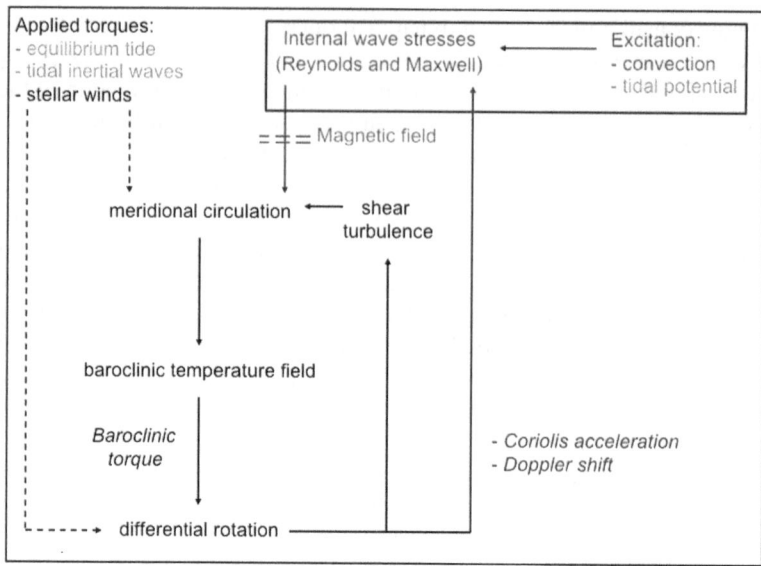

Fig. 4.19 Synopsis of the transport loop in stars hosting companions. Differences with single stars are emphasised in *purple*; these are the tidal torques applied on convection zones both by the equilibrium tide and the tidal inertial waves and the action of internal waves excited in radiation zones

with supposed fossil magnetic field), the first important question is to understand why when such a binary systems is formed, one of the components can be magnetic, while the second component can be non-magnetic. In such cases, is the absence (or the presence) of a magnetic field a consequence of an evolutive process? Is it a threshold for the intensity of the field? What could be the difference between the amplitude of the magnetic field of each component if magnetic fields are present? What is the geometry of the field of each component if magnetic fields are present (axisymmetric vs. non-axisymmetric, obliquity in the case of oblique rotators)? For low-mass stars, such type of questions may be relevant for fossil fields trapped in their radiative cores but not for dynamo fields that are generated by their external convective envelopes. Then, once close binary stellar or star-planet systems has been formed, one has to understand their magnetic properties. First, in the case of binary stars with almost one intermediate-mass or massive component (with an external radiative envelope and the associated potential magnetic field of fossil origin), is it some different magnetic signatures that appear in close binary systems compared to what is expected in the case of single stars, and if yes, what are their amplitude, their geometrical properties and their physical origin (coming from example from couplings of the internal dynamics of studied external radiative envelopes and tidal interactions). Next, in the case of binaries with almost one low-mass component (with an external convective envelope), the main questions are related to the modification of the magnetic activity and cyclicity because of binarity (see Fig. 4.21 and [11, 27, 32]). First, the binarity, because of tidal interactions, modifies the global

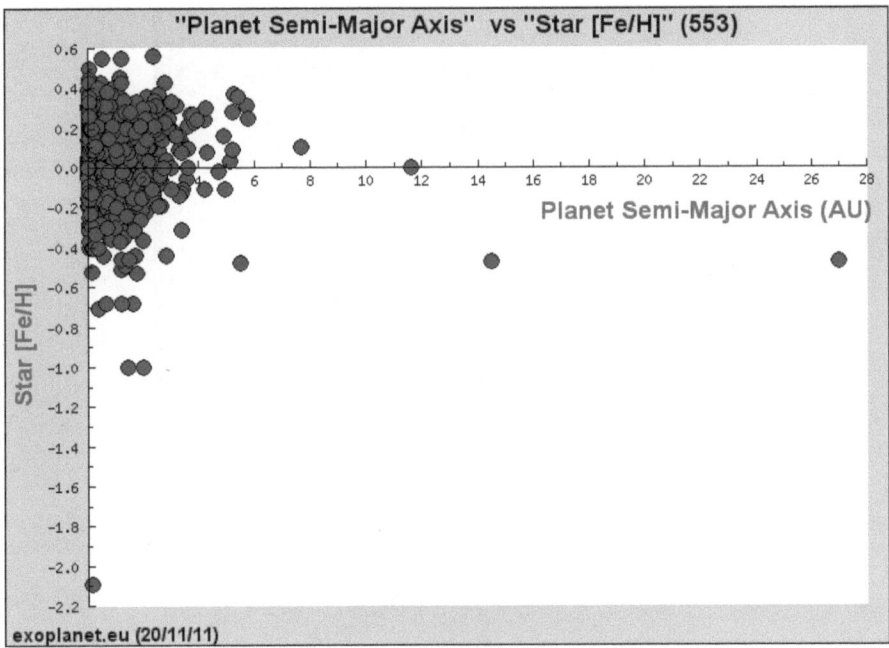

Fig. 4.20 Metalicity of stars hosting planets as a function of their semi-major axis (in A.U.) (courtesy the extrasolar planets encyclopedia by J. Schneider (http://exoplanet.eu/); see also [110]). The understanding of the coupling of tidal interactions with internal transport processes and the generated mixing constitutes one of the most important issue of our understanding of Star–Planets Interactions

angular velocity of the star; moreover, tidal interactions may modify the differential rotation in the star. This can modify the so-called Ω-effect and thus the activity of low-mass components in binary systems. Next, because tidal induced-flows are helical, these can combine with convection action to modify the so-called α-effect and, as a consequence, the dynamo behaviour. Finally, in the case of planetary interiors, in the same way as for low-mass stars, it will be important to study consequences of tidal interactions (on the spin and on excited internal flows) to see the modification of planets' magnetism and of related magnetospheres and the consequences for example on their habitability. All these exciting questions will stimulate theoretical MHD studies of tidal flows interaction with magnetic field and differential rotation for the equilibrium and the dynamical tide and their related instabilities.

In our purpose above, we have presented the possible impact of tidal interactions on the magnetic activity. However, as it has also been pointed at the end of the description of each fluid dissipative mechanism, the study of the interaction between magnetic fields and tidally excited flows is only in its infancy. Then, progresses have to be obtained in a near future on the description of tidal induced movements taking into account magnetic fields; as a consequence, this will leads to a complete MHD view of tidal interactions where we will be able to see if tides velocity fields are influenced by magnetic fields through the Lorentz force and if such MHD flows

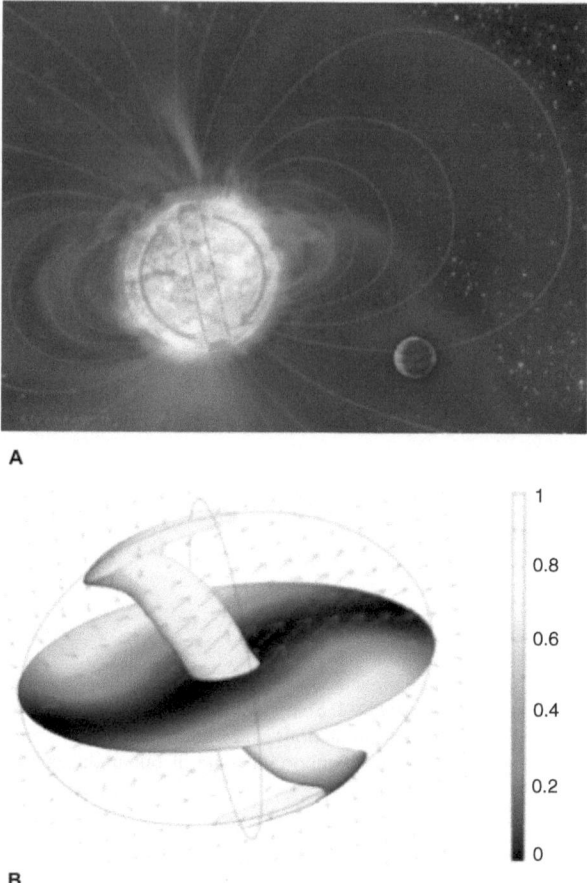

Fig. 4.21 **A**: the τ Bootis star-planet system [11, 27, 32]. This F star hosts a close giant planet ($\approx 6.5 M_J$, where M_J is the mass of Jupiter). The equatorial surface layers of the star have been synchronised by tidal forces; moreover, τ Bootis has very short magnetic cycles. Such type of systems constitutes an excellent laboratory to study magnetic Star–Planet Interactions and their couplings with tides (courtesy K. Teramura, University of Hawaii, Institute for Astronomy). **B**: MHD numerical simulation of the spin-over mode induced by the tidal elliptic instability with an imposed uniform magnetic field along the rotation axis. The induced magnetic field is represented by arrows which scale with its local value. Finally, the Ohmic dissipation is represented in the equatorial plane (a normalisation to its maximum value has been done) (taken from [12], courtesy D. Cébron)

are then able to sustain such magnetic fields through instabilities and dynamo processes. Moreover, tidal dissipation will be modified by the magnetic fields as well as related torques, and it will be interesting to study the modified evolution of the orbital and the spins' properties of the studied systems. Finally, if bodies' magnetic fields are modified by tidal interactions, their magnetic interactions (and the associated observational signatures) have to be studied (interactions between stellar winds

Fig. 4.22 3D numerical simulations of MHD Star–Planet Interactions between the stellar wind of a solar-type star and the magnetosphere of a close planet. Color contours represent plasma number density on the $y = 0$ plane, and yellow three-dimensional streamlines represent stellar magnetic field lines that originate on the intersection of the $y = 0$ plane with the stellar surface. The inner boundary of the simulation domain (the stellar surface) is represented by the *red sphere*. Magnetic field lines that connect to the planetary magnetosphere are drawn in *pink* (taken from [18], courtesy The Astrophysical Journal)

in binary stars and interaction stellar winds-planetary magnetospheres in planetary systems).

4.4.3 Couplings with Magnetic Star–Planet Interactions

In previous sections, we have thus shown that tidal interactions generate torques, which apply both on planets and on their host star in planetary systems. However, this latter is also submitted to the torque due to its own wind. For example, in the case of solar-type stars, the pressure-driven magnetic wind induces mass and angular momentum losses that are strongly correlated to their magnetic activity and will have strong interactions with planetary magnetospheres (Fig. 4.22). In this context, [3] have shown that it is necessary to take into account both magnetic and tidal torques to predict the correct final states. In this way, the potential modification of stellar activity by planetary companions (see the previous section) has thus to be constrained while robust prescription for the torque exerted on stars because of their winds have to be obtained. Recent works have been aimed to reach this objective [77, 91]. In a near future, coupling physical descriptions of tidal and stellar wind torques will thus be necessary to obtain a consistent modelling of the host-star rotation evolution.

4.5 Conclusion

In this review, we have tried to give the must complete picture of tidal interactions in planetary and in multiple stellar systems. In the first section, we have recalled the properties of the tidal potential, and we have shown the crucial dependence of systems' evolution on dissipative mechanisms that convert the kinetic energy of tidal flows into heat. Next, we have shown the important diversity of mechanisms that take place in stellar and planetary interiors. We have given their main properties, namely their relative amplitude and their dependence on the internal structure and on the tidal frequency that can be very complex. This shows how it is now necessary to study tidal interactions with a good description of the bodies' internal structure and to go beyond the rough approximations that are often adopted to describe systems' evolution due to tidal interactions. Finally, we have pointed the possible important impact of tidal interactions on stellar evolution.

Acknowledgements S. Mathis thanks warmly the organisers of the school, C. Neiner and J.-P. Rozelot, for their invitation to give this lecture. This work was supported in part by the Programme National de Planétologie (CNRS/INSU), the Programme National de Physique Stellaire (CNRS/INSU), the EMERGENCE-UPMC project EME0911, and the CNRS *Physique théorique et ses interfaces* program.

References

1. Alexander, M.E.: Astrophys. Space Sci. **23**, 459 (1973)
2. Baraffe, I.: Space Sci. Rev. **116**, 67 (2005)
3. Barker, A.J., Ogilvie, G.I.: Mon. Not. R. Astron. Soc. **395**, 2268 (2009)
4. Barker, A.J., Ogilvie, G.I.: Mon. Not. R. Astron. Soc. **404**, 1849 (2010)
5. Barker, A.J., Ogilvie, G.I.: Mon. Not. R. Astron. Soc. **417**, 745 (2011)
6. Biot, M.A.: J. Appl. Phys. **25**, 1385 (1954)
7. Bodenheimer, P., Lin, D.N.C., Mardling, R.A.: Astrophys. J. **548**, 466 (2001)
8. Brun, A.-S., Toomre, J.: Astrophys. J. **570**, 865 (2002)
9. Brun, A.-S., Miesch, M.S., Toomre, J.: Astrophys. J. **614**, 1073 (2004)
10. Bryan, G.H.: Philos. Trans. R. Soc. Lond. A **180**, 187 (1889)
11. Catala, C., et al.: Mon. Not. R. Astron. Soc. **374**, L42 (2007)
12. Cébron, D.: Ph.D. Thesis, Université de Provence, Aix Marseille I (2011)
13. Cébron, D., et al.: Phys. Earth Planet. Inter. **182**, 119 (2010)
14. Cébron, D., Maubert, P., Le Bars, M.: Geophys. J. Int. **182**, 1311 (2010)
15. Cébron, D., et al.: EPJ Web Conf. **11**, 03003 (2011)
16. Chapman, S., Lindzen, R.: Atmospheric Tides. Reidel, Dordrecht (1970)
17. Charnoz, S., et al.: Icarus **216**, 535 (2011)
18. Cohen, O., et al.: Astrophys. J. Lett. **704**, L85 (2009)
19. Correia, A.C.M., Laskar, J.: J. Geophys. Res. **108**(E11), 9-1 (2003)
20. Correia, A.C.M., Laskar, J.: Icarus **163**, 24 (2003)
21. Correia, A.C.M., Laskar, J., de Surgy, O.N.: Icarus **163**, 1 (2003)
22. Correia, A.C.M., Levrard, B., Laskar, J.: Astron. Astrophys. **488**, L63 (2008)
23. Darwin, G.H.: Philos. Trans. R. Soc. Lond. **171**, 713 (1880)
24. Darwin, G.H.: Philos. Trans. R. Soc. Lond. **172**, 491 (1881)
25. Decressin, T., et al.: Astron. Astrophys. **495**, 271 (2009)
26. Dermott, S.F.: Icarus **37**, 310 (1979)

27. Donati, J.-F., et al.: Mon. Not. R. Astron. Soc. **385**, 1179 (2008)
28. Efroimsky, M.: Astrophys. J. (2011, submitted). arXiv:1105.3936
29. Efroimsky, M., Lainey, V.: J. Geophys. Res. **112**, E12003 (2007)
30. Efroimsky, M., Williams, J.G.: Celest. Mech. Dyn. Astron. **104**, 257 (2009)
31. Eggleton, P.P., Kiseleva, L.G., Hut, P.: Astrophys. J. **499**, 853 (1998)
32. Farès, R., et al.: Mon. Not. R. Astron. Soc. **398**, 1383 (2009)
33. Ferraz-Mello, S., Rodrìguez, A., Hussmann, H.: Celest. Mech. Dyn. Astron. **101**, 171 (2008)
34. Fortney, J.J., Nettelmann, N.: Space Sci. Rev. **152**, 423 (2009)
35. Goldreich, P., Keeley, D.A.: Astrophys. J. **211**, 934 (1977)
36. Goldreich, P., Nicholson, P.D.: Astrophys. J. **342**, 1075 (1989)
37. Goldreich, P., Nicholson, P.D.: Astrophys. J. **342**, 1079 (1989)
38. Goldreich, P., Soter, S.: Icarus **5**, 375 (1966)
39. Goodman, J., Lackner, C.: Astrophys. J. **696**, 2054 (2009)
40. Gu, P.G., Ogilvie, G.I.: Mon. Not. R. Astron. Soc. **395**, 422 (2009)
41. Guillot, T.: Planet. Space Sci. **47**, 1183 (1999)
42. Guillot, T.: Annu. Rev. Earth Planet. Sci. **33**, 493 (2005)
43. Guimbard, D., et al.: J. Fluid Mech. **660**, 240 (2010)
44. Henning, W.G., O'Connell, R.J., Sasselov, D.D.: Astrophys. J. **707**, 1000 (2009)
45. Herreman, W., Le Bars, M., Le Gal, P.: Phys. Fluids **21**, 046602 (2009)
46. Hut, P.: Astron. Astrophys. **99**, 126 (1981)
47. Ivanov, P.B., Papaloizou, J.C.B.: Mon. Not. R. Astron. Soc. **353**, 1161 (2004)
48. Ivanov, P.B., Papaloizou, J.C.B.: Mon. Not. R. Astron. Soc. **407**, 1609 (2010)
49. Kaula, W.M.: Astron. J. **67**, 300 (1962)
50. Kaula, W.M.: Rev. Geophys. Space Phys. **2**, 661 (1964)
51. Kerswell, R.: Annu. Rev. Fluid Mech. **34**, 83 (2002)
52. Kippenhahn, R., Weigert, A.: Stellar Structure and Evolution. Springer, Berlin (1990)
53. Lacaze, L., Le Gal, P., le Dizès, S.: J. Fluid Mech. **505**, 22 (2004)
54. Lacaze, L., Le Gal, P., le Dizès, S.: Phys. Earth Planet. Inter. **151**, 194 (2005)
55. Lacaze, L., et al.: Geophys. Astrophys. Fluid Dyn. **100**, 299 (2006)
56. Lainey, V., Arlot, J.-E., Karatekin, Ö., van Hoolst, T.: Nature **459**, 957 (2009)
57. Lainey, V., et al.: Astrophys. J. **752**, 14 (2012)
58. Lambeck, K.: The Earth's Variable Rotation: Geophysical Causes and Consequences. Cambridge University Press, Cambridge (1980)
59. Lavorel, G., Le Bars, M.: Phys. Fluids **22**, 114101 (2010)
60. Le Bars, M., Le Dizès, S.: J. Fluid Mech. **563**, 189 (2006)
61. Le Bars, M., Le Dizès, S., Le Gal, P.: J. Fluid Mech. **585**, 323 (2007)
62. Le Bars, M., et al.: Phys. Earth Planet. Inter. **178**, 48 (2010)
63. Leconte, J., Chabrier, G., Baraffe, I., Levrard, B.: Astron. Astrophys. **516**, A64 (2010)
64. Levrard, B., Winisdoerffer, C., Chabrier, G.: Astrophys. J. **692**, L9 (2009)
65. Lord Kelvin: The tides. In: Evening Lecture to the British Association at the Southampton Meeting, Friday, 25 August 1882, Scientific Papers, The Harvard Classics (1882)
66. Love, A.E.H.: Some Problems of Geodynamics. Cambridge University Press, Cambridge (1911)
67. MacDonald, G.J.F.: Rev. Geophys. Space Phys. **2**, 467 (1964)
68. Mardling, R.A., Lin, D.N.C.: Astrophys. J. **573**, 829 (2002)
69. Mathis, S.: Astron. Astrophys. **506**, 811 (2009)
70. Mathis, S.: Astron. Nachr. **331**, 883 (2010)
71. Mathis, S., de Brye, N.: Astron. Astrophys. **526**, A65 (2011)
72. Mathis, S., de Brye, N.: Astron. Astrophys. **540**, A37 (2012)
73. Mathis, S., Le Poncin-Lafitte, C.: Astron. Astrophys. **497**, 889 (2009)
74. Mathis, S., Zahn, J.-P.: Astron. Astrophys. **425**, 229 (2004)
75. Mathis, S., Palacios, A., Zahn, J.-P.: Astron. Astrophys. **425**, 243 (2004)
76. Mathis, S., Talon, S., Pantillon, F.-P., Zahn, J.-P.: Sol. Phys. **251**, 101 (2008)
77. Matt, S., Pudritz, R.E.: Astrophys. J. **678**, 1109 (2008)

78. Melchior, P.: The Eart Tides. Pergamon, New York (1966)
79. Neron de Surgy, O., Laskar, J.: Astron. Astrophys. **318**, 975 (1997)
80. Ogilvie, G.I.: J. Fluid Mech. **543**, 19 (2005)
81. Ogilvie, G.I.: Mon. Not. R. Astron. Soc. **396**, 794 (2009)
82. Ogilvie, G.I., Lin, D.N.C.: Astrophys. J. **610**, 477 (2004)
83. Ogilvie, G.I., Lin, D.N.C.: Astrophys. J. **661**, 1180 (2007)
84. Papaloizou, J.C.B., Ivanov, P.B.: Mon. Not. R. Astron. Soc. **364**, L66 (2005)
85. Papaloizou, J.C.B., Ivanov, P.B.: Mon. Not. R. Astron. Soc. **407**, 1631 (2010)
86. Papaloizou, J.C.B., Savonije, G.J.: Mon. Not. R. Astron. Soc. **291**, 651 (1997)
87. Peale, S.J., Cassen, P.: Icarus **36**, 245 (1978)
88. Penev, K., Sasselov, D.: Astrophys. J. **731**, id. 67 (2011)
89. Penev, K., Barranco, J., Sasselov, D.: Astrophys. J. **705**, 285 (2009)
90. Penev, K., Sasselov, D., Robinson, F., Demarque, P.: Astrophys. J. **704**, 230 (2009)
91. Pinto, R.F., Brun, A.-S., Jouve, L., Grappin, R.: Astrophys. J. **732**, id. 72 (2011)
92. Remus, F., Mathis, S., Zahn, J.-P.: Astron. Astrophys. (2012, in press). arXiv:1205.3536
93. Remus, F., Mathis, S., Zahn, J.-P., Lainey, V.: Astron. Astrophys. **541**, A165 (2012)
94. Rieutord, M.: Astron. Astrophys. **259**, 581 (1992)
95. Rieutord, M., Bonazzola, S.: Mon. Not. R. Astron. Soc. **227**, 295 (1987)
96. Rieutord, M., Valdettaro, L.: J. Fluid Mech. **341**, 77 (1997)
97. Rieutord, M., Valdettaro, L.: J. Fluid Mech. **643**, 363 (2010)
98. Rieutord, M., Zahn, J.-P.: Astrophys. J. **474**, 760 (1997)
99. Rieutord, M., Valdettaro, L., Georgeot, B.: J. Fluid Mech. **435**, 103 (2001)
100. Rieutord, M., Valdettaro, L., Georgeot, B.: J. Fluid Mech. **463**, 345 (2002)
101. Rocca, A.: Astron. Astrophys. **111**, 252 (1982)
102. Rocca, A.: Astron. Astrophys. **175**, 81 (1987)
103. Rocca, A.: Astron. Astrophys. **213**, 114 (1989)
104. Santos, N.C., et al.: Our non-stable universe. In: JENAM-2007 (2007)
105. Savonije, G.-J.: EAS Publ. Ser. **29**, 91 (2008)
106. Savonije, G.J., Papaloizou, J.C.B.: Mon. Not. R. Astron. Soc. **291**, 633 (1997)
107. Savonije, G.J., Witte, M.G.: Astron. Astrophys. **386**, 211 (2002)
108. Savonije, G.J., Papaloizou, J.C.B., Alberts, F.: Mon. Not. R. Astron. Soc. **277**, 471 (1995)
109. Scharlemann, E.T.: Astrophys. J. **246**, 292 (1981)
110. Schneider, J., et al.: Astron. Astrophys. **532**, A79 (2011)
111. Talon, S.: EAS Publ. Ser. **32**, 81 (2008)
112. Talon, S., Charbonnel, C.: Astron. Astrophys. **440**, 981 (2005)
113. Talon, S., Kumar, P.: Astrophys. J. **503**, 387 (1998)
114. Talon, S., Zahn, J.-P.: Astron. Astrophys. **317**, 749 (1997)
115. Tassoul, M., Tassoul, J.-L.: Astrophys. J. **395**, 259 (1992)
116. Tassoul, M., Tassoul, J.-L.: Astrophys. J. **395**, 604 (1992)
117. Tassoul, M., Tassoul, J.-L.: Astrophys. J. **481**, 363 (1997)
118. Terquem, C., Papaloizou, J.C.B., Nelson, R.P., Lin, D.N.C.: Astrophys. J. **502**, 788 (1999)
119. Tobie, G.: Impact du chauffage de marée sur l'évolution géodynamique d'Europe et de Titan. PhD thesis, Université Paris 7, Denis Diderot (2003)
120. Tobie, G., Mocquet, A., Sotin, C.: Icarus **177**, 534 (2005)
121. Waleffe, F.A.: Phys. Fluids **2**, 76 (1990)
122. Weinberg, N.N., Arras, P., Quataert, E., Burkart, J.: Astrophys. J. **751**, 136 (2011)
123. Witte, M.G., Savonije, G.J.: Astron. Astrophys. **341**, 842 (1999)
124. Witte, M.G., Savonije, G.J.: Astron. Astrophys. **350**, 129 (1999)
125. Witte, M.G., Savonije, G.J.: Astron. Astrophys. **366**, 840 (2001)
126. Witte, M.G., Savonije, G.J.: Astron. Astrophys. **386**, 222 (2002)
127. Wu, Y.: Astrophys. J. **635**, 674 (2005)
128. Wu, Y.: Astrophys. J. **635**, 688 (2005)
129. Zahn, J.-P.: Ann. Astrophys. **29**, 313 (1966)
130. Zahn, J.-P.: Ann. Astrophys. **29**, 489 (1966)

131. Zahn, J.-P.: Ann. Astrophys. **29**, 565 (1966)
132. Zahn, J.-P.: Astron. Astrophys. **41**, 329 (1975)
133. Zahn, J.-P.: Astron. Astrophys. **57**, 383 (1977)
134. Zahn, J.-P.: Astron. Astrophys. **220**, 112 (1989)
135. Zahn, J.-P.: Astron. Astrophys. **265**, 115 (1992)
136. Zahn, J.-P.: Astron. Astrophys. **288**, 829 (1994)
137. Zahn, J.-P.: EAS Publ. Ser. **26**, 49 (2007)
138. Zahn, J.-P.: EAS Publ. Ser. **29**, 67 (2008)
139. Zahn, J.-P., Talon, S., Matias, J.: Astron. Astrophys. **322**, 320 (1997)

Chapter 5
Interactions in Massive Binary Stars as Seen by Interferometry

F. Millour, A. Meilland, P. Stee, and O. Chesneau

Abstract With the advent of large-collecting-area instruments, the number of objects that can be reached by optical long-baseline interferometry is steadily increasing. We present here a few results on massive binary stars, showing the interest of using this technique for studying the insight of interactions in these systems. Indeed, many massive stars with extended environments host, or are suspected to host, companion stars. These companions could have an important role in shaping the circumstellar environment of the system. These examples provide a view in which binarity could be an ingredient, among many others, for the activity of these stars.

5.1 Introduction

The most massive stars are still a puzzle, especially regarding their evolution. They are the best candidates to be progenitor of type II supernovas, which are one source of metallic enrichment of the interstellar medium, but they also input kinetic energy to their vicinity, triggering densification and collapse of neighboring interstellar clouds.

In the past years, many new observing techniques have appeared that allow one to access unprecedented angular resolution. Therefore, the detection and characterization of circumstellar environments (hereafter CSEs) around massive stars has become a reality. A byproduct is that several new companion stars have been discovered using high-angular resolution techniques, being adaptive optics, lucky imaging, or long-baseline interferometry.

We try here to give an insight of the influence of stellar companions to the structure of CSEs by showing a few examples of massive stars, ranging from the classical Be stars up to the most massive blue supergiant stars.

F. Millour (✉) · A. Meilland · P. Stee · O. Chesneau
OCA, Bd de l'Observatoire, 06304 Nice, France
e-mail: fmillour@oca.eu

A. Meilland
e-mail: ame@oca.eu

J.-P. Rozelot, C. Neiner (eds.), *The Environments of the Sun and the Stars*,
Lecture Notes in Physics 857,
DOI 10.1007/978-3-642-30648-8_5, © Springer-Verlag Berlin Heidelberg 2013

5.2 Interactions in Intermediate-Mass Stars: Classical Be Stars

Binarity may play a nonnegligible role in the formation of circumstellar disks around classical Be stars. These stars close to main sequence are known to be fast rotators; however, the most recent studies [8, 15] indicate that they are not all critical-rotators, even though a very recent work shed new light on this fact [31]. Still, rotation cannot be the only mechanism involved in the mass-ejection. Recent studies like [37] try to investigate the role of the gravitational influence of a companion star in a few extreme cases. However, such influence is not the only physical process proposed as an additional momentum source to cancel the stellar gravity. Radiative pressure [1], nonradial pulsations [44], or even magnetism [25] have also been suggested to play such a role.

To progress in the understanding of the physical processes responsible for the Be phenomenon, one needs to resolve their CSE both spatially to obtain information on the distribution of matter surrounding the star and also spectrally to get access to the kinematics of the ejected matter. Consequently, spectro-interferometry is the most suitable technique to study these objects. First VLTI/AMBER [42] and CHARA/VEGA [38] observations of Be stars have evidenced that the matter is mostly concentrated into the equatorial plane and that this geometrically thin disk is dominated by rotation with a rotational law close to the Keplerian one [4, 10, 23, 26, 27, 31].

An example of spectro-interferometric observations with the VLTI/AMBER well-fitted by a simple geometrically thin rotating disk model is plotted in Fig. 5.1. As seen in this figure, a rotating disk provides typical "W"-shaped visibilities and "S"-shaped phases as long as the disk is not fully resolved by the interferometer (top plots), whereas it gives more complex phases shape (double-"S") when it is fully resolved (bottom plots).

Few interferometric observations also suggest the presence of a nonnegligible polar wind [20, 27]. Moreover, thanks to interferometric measurements of the projected disk flattening, the star inclination angle can be inferred without ambiguities, and consequently, the rotational velocity of the central star can be determined. In the first spectro-interferometric survey of Be star, Meilland et al. [31] show that the stars are rotating in average at an $\overline{\Omega/\Omega_c} = 0.95 \pm 0.02$ rate, a result not quite compatible with estimation from [15, see Fig. 5.2]. Though this rotation rate is large, the star still needs a perturbation to expel its disk. Possible perturbations are nonradial pulsations or the gravitational perturbation of a companion star. Finally, interferometry can also help to discover new companions as evidenced in the case of δ Cen [28].

Some Be stars show large variability that may be related to binarity. For instance, this is the case of two well-studied objects, Achernar and δ Sco. Achernar is a quasi-cyclic Be star with a period of formation and dissipation of the equatorial disk of about 12 years. It was observed at a minimum of activity by Domiciano de Souza et al. [11], and the authors found that the star was strongly flattened by close-to-critical rotation. Using these interferometric observations and spectroscopic follow-up of a full cycle of activity from Vinicius et al. [47]; Kanaan et al. [19] managed to fully model the environment of this object, which consists in a steady polar wind

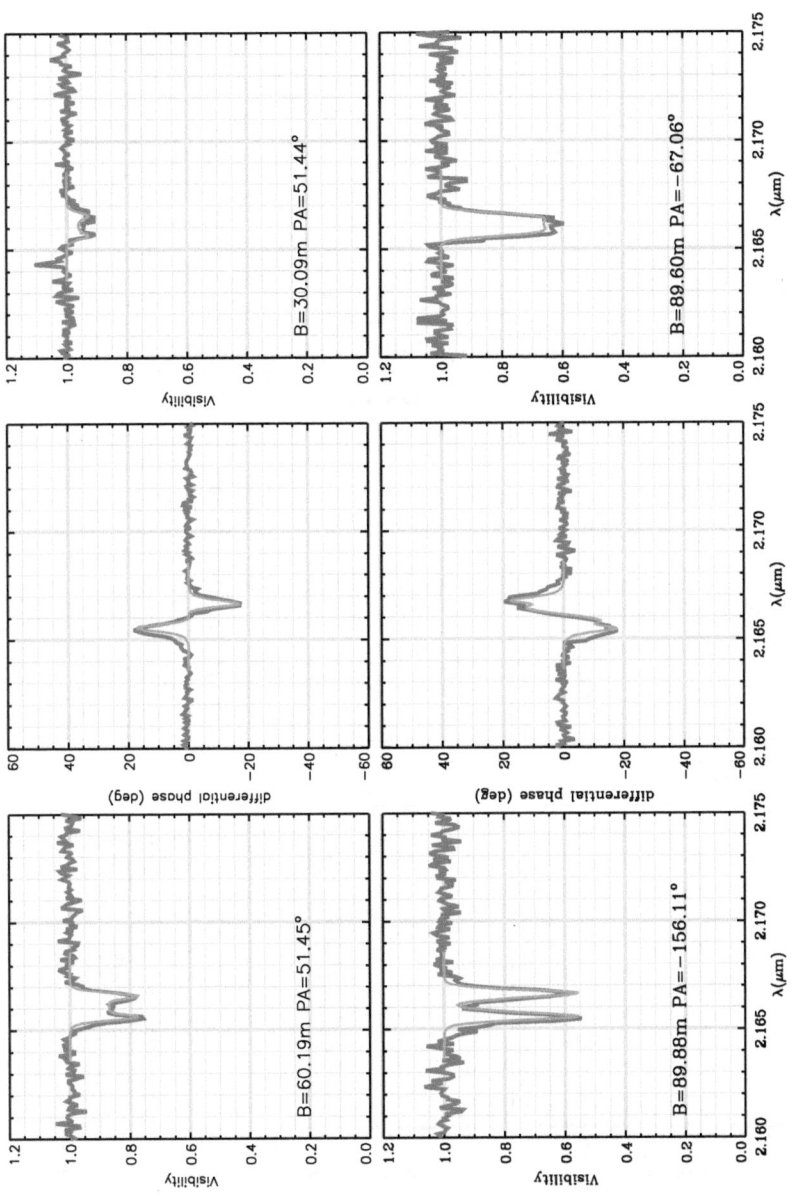

Fig. 5.1 α Col selected differential visibilities and phases from 2 VLTI/AMBER HR measurements (*red line*). Each row corresponds to one VLTI/AMBER measurement (three different baselines). The *top row* shows short baselines (barely resolved disk), while the *bottom row* shows long baselines (fully resolved disk). The visibilities and phases of the best-fit geometrically-thin-rotating-disk model is over-plotted in *green*. From [31] (Courtesy Astronomy and Astrophysics)

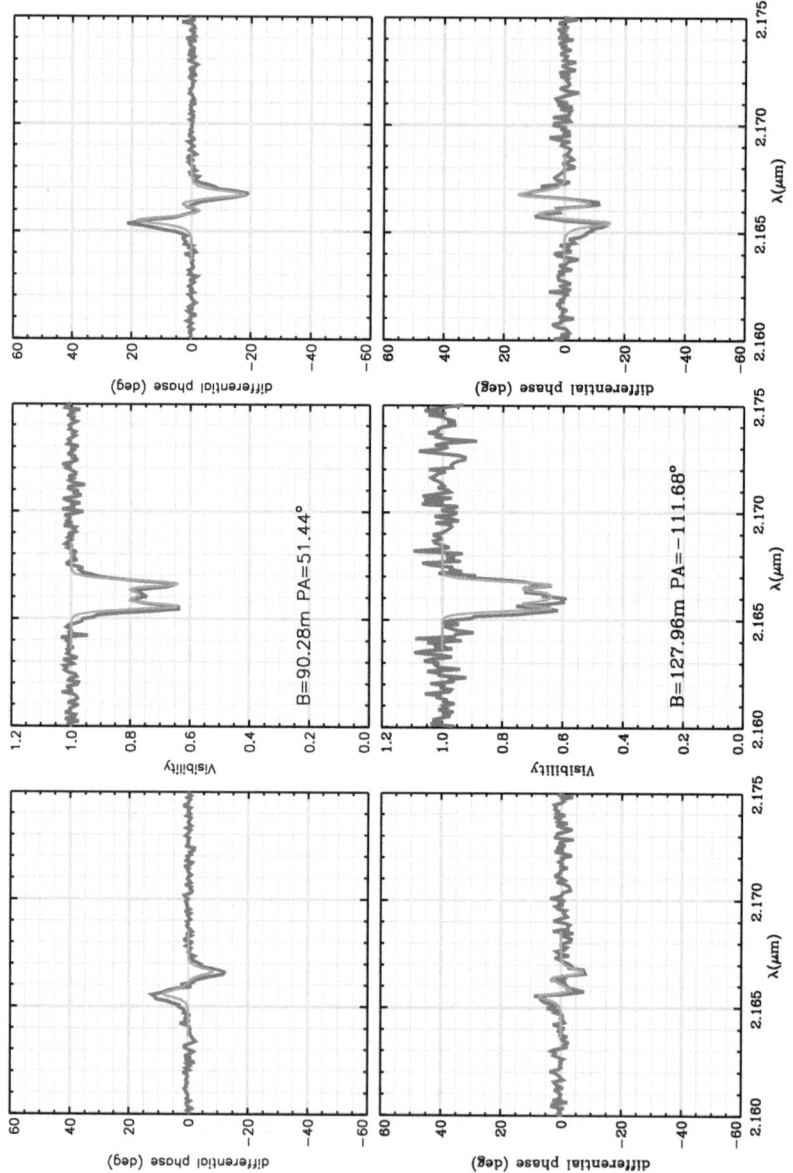

Fig. 5.1 (Continued)

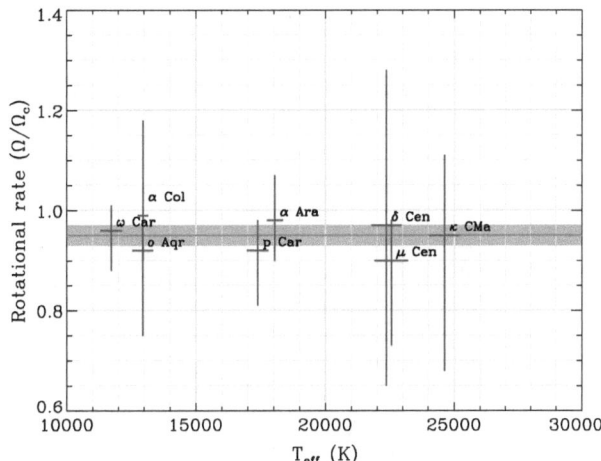

Fig. 5.2 A set of eight Be stars observed with long-baseline interferometry, showing for the first time the rotation rate of the central star free of any inclination effect. We show here that the rotation rate is roughly constant whatever the star temperature, at $\overline{\Omega}/\Omega_c = 0.95 \pm 0.02$ of the critical rotation. From [31]

driven by radiative pressure and a transient equatorial disk which is produced during a brief outburst. The ejected matter then propagates into the CSE with an expansion velocity of the order of 0.2 km s^{-1}. The cause of the ejection remained mysterious until the discovery of a companion star using the VLT/NACO instrument [21]. New observations will soon be executed to constrain the companion orbit and determine whether or not it could be the cause of the cyclic variations of Achernar.

On the other hand, δ Sco is a well-known binary system. First evidence of its multiplicity was reported by Innes [18] using the lunar occultation technique. However, this work was forgotten for a long time, and the binary nature of δ Sco was rediscovered with three different techniques in 1974: by speckle-interferometry [24], lunar occultation [13], and intensity interferometry [16]. The extremely eccentric orbit ($e \simeq 0.94$) was then constrained by many authors using speckle and long-baseline interferometry [2, 17, 35, 45, 46]. However, the δ Sco system did not show clear evidence of the Be phenomenon until June 2000, close to periastron (September 2000). At this epoch, Otero et al. [41] found a 0.4 mag brightening of the object. Simultaneous spectroscopic observations published in [14] showed evidences of strong Hα emission lines. Using spectroscopic measurements obtained at different epochs and assuming that the matter was concentrated in a Keplerian rotating disk, Miroshnichenko et al. [36] suggested that the ejected matter was expanding at about 0.4 km s^{-1}, a velocity compatible with the one found for Achernar. Thanks to new spectrally resolved VLTI/AMBER and CHARA/VEGA observations, Meilland et al. [30] managed to probe the CSE structure and confirmed the Miroshnichenko hypothesis that the ejected matter is concentrated in a Keplerian rotating disk. They also constrained the stellar rotational velocity and found that this star was likely to rotate far below its critical limit, at about $0.7V_c$. Finally, they detected asymmetries in their data that can hardly be modeled under Okazaki [40] one-armed oscillation framework and that could be due to a tidal warping of the disk by the companion periastron passage (see Fig. 5.3), as is described in [37].

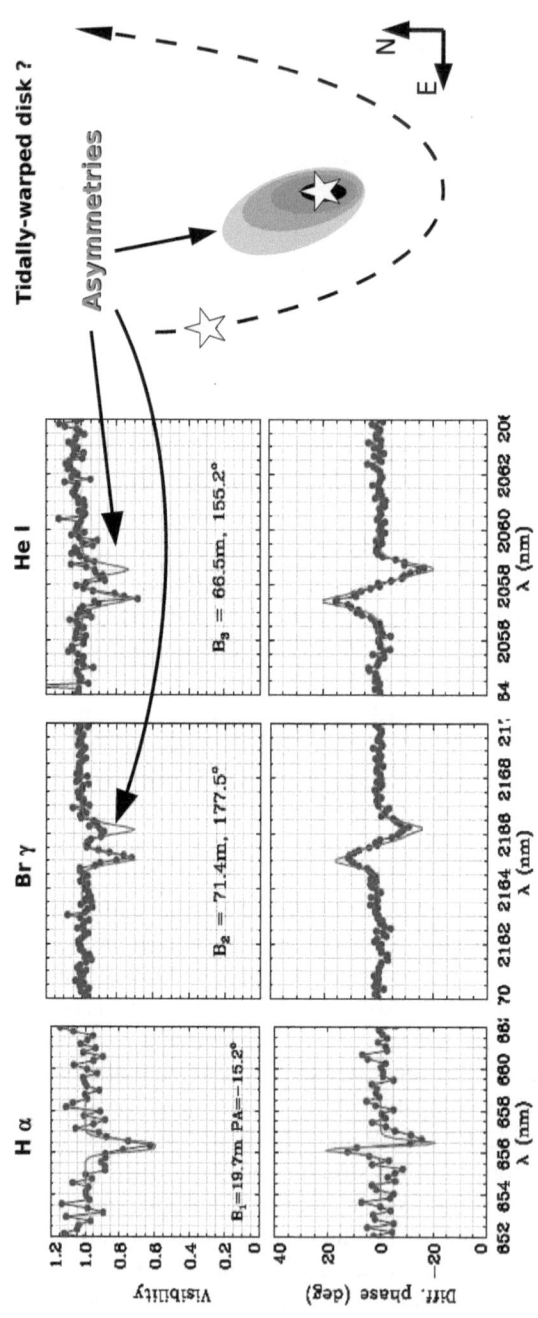

Fig. 5.3 Illustration of the δ Sco system close to the periastron passage. The three left figures are from [30], showing the AMBER + VEGA data (*blue*) together with a kinematic model of a rotating disk (*red*). Meilland et al. [30] proposed that the differences between the model and the data (*arrows*) could be the signature of a tidally warped disk around the primary star (right sketch, not to scale—the disk appears here approx. 10 times larger for clarity)

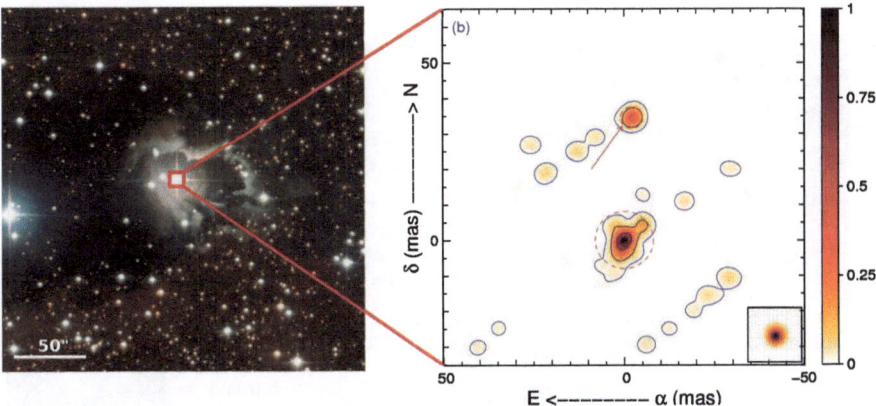

Fig. 5.4 Images of HD87643 at two highly (1:1000) different scales. Left is the large-scale nebula, exhibiting its complex shape and its inner arc-like structures; right is the interferometric small-scale image, showing the companion star (*arrow*) and the circumstellar disk (*dotted circle*). Figures from [32]

In the case of these two stars, the binarity clearly plays a important role in the mass-ejection process. However, whether the companion action is direct, i.e., by canceling the residual stellar gravity at the surface of the central star, or indirect, i.e., by impacting on other physical processes such as nonradial pulsations, remains an open question. Moreover, the influence of binarity on the Be phenomenon is not limited to the mass-ejection; it can also affect the reorganization of the gas in the CSE.

5.3 Interactions in the Most Massive Stars

As seen in the previous section, disks encountered around Be stars are more and more found to be rotating close to Keplerian velocities. This applies also to some B[e] candidate supergiant stars.

The situation for Be stars is somewhat complicated, as several hypotheses exist to explain the disk formation and steadiness. For B[e] stars, which are surrounded by dense disks of plasma *and* dust, the situation is even more complex. The dust survives much closer to these hot stars than expected so far [12, 29, 32], meaning that complex radiative transfer processes such as line-blanketing could occur in the gas disk of these stars. The B[e] supergiant stars critical rotation rate is strongly decreased by the increase of their radius while leaving the main sequence. Therefore, rotation alone is certainly not sufficient to explain the creation of a circumstellar disk without invoking the influence of a close companion [34].

Such influence seems clear in the few cases where companion stars were detected, like in the binary system HD 87643 [32, see Fig. 5.4]. In this B[e] system, enshrouded inside a complex nebula reminiscent of the ones found around LBVs,

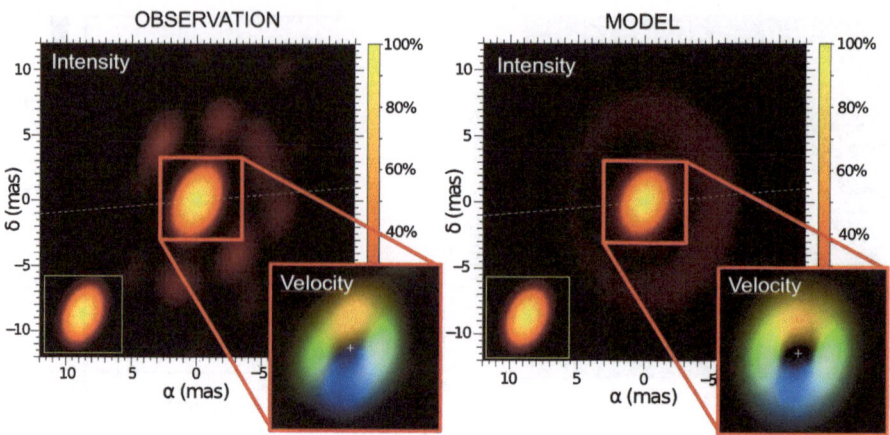

Fig. 5.5 The VLTI observations of the A[e] system HD62623 (here, the figure of the press release) showed that the dense circumbinary disk [29], outer red ring in both the image (*left*) and the model (*right*), seems constantly fed by an inner plasma disk (*central yellow dot*), probably in Keplerian rotation (*insets*), whose angular momentum originates from a solar mass companion [33]

the presence of the companion provides a key to understand the whole range of features observed in the system, if it has an eccentric orbit:

- the extended nebula that could come from a past outburst of the system at one periastron passage,
- the series of arc-like structures found in the same nebula, tracing more recent periastron passages
- the main star disk, formed by the direct interaction with the companion star

The supergiant A[e] star HD 62623 is also informative in this context [29, 33]. It is an A type supergiant showing the "B[e] phenomenon," namely a spectrum dominated by strong emission lines (allowed and forbidden), and a large infrared excess. Spectrally and spatially resolved AMBER/VLTI observations in the Brγ line have shown that the supergiant star lies in a cavity and is surrounded by a rotating disk of plasma. The Brγ line at the location of the central star is *in absorption* showing that it is a normal A-type star, albeit with a significantly large *vsini* of about 50 km s^{-1}. By contrast, the Balmer and Bracket lines are wider (*vsini* of about 120 km s^{-1}), and the AMBER observations demonstrated that they originate from a disk of plasma, most probably in Keplerian rotation (see Fig. 5.5). In absence of any proof of binarity, it is often difficult to understand how such a dense equatorial disk could have been generated. However, HD62623 is a known binary with a stellar companion that orbits close to the supergiant with a period of about 136 days [43]. The mass ratio inferred is very large, and the companion is probably a solar-mass star, hence unseen in the AMBER images. Plets et al. [43] proposed that an efficient angular momentum transfer occurs near the L2 Lagrangian point of the system, propelling the mass lost from the supergiant star by its radiative wind and probably also by strong tides into a stable dense circumbinary disk.

Similarly, the evolved system υ Sgr was recently investigated, again using spectroscopy and optical interferometry in the near-IR and the visible [3, 22, 39], evidencing a dense circumbinary disk, apparently long-lived.

5.4 Novas, Bipolar Nebulae, and the Underlying Binary System

Novas are formed of a white dwarf (WD) and a red giant star, whose material is accreted on the WD surface. When enough material is accreted, thermonuclear reactions can light up, expelling a fireball, which we directly observe, usually using spectroscopy, or imaging a few years after. Novas are suspected to be the progenitors of type I supernovas, as the accreted material piles-up, nova after nova, providing the good conditions for core-collapse.

Such fireballs are seen as bipolar nebulae years after the explosion, but an open question remained to know if that was an intrinsic shape of the explosion or if it was shaped after, by the CSE.

A new field of research was open by the first spectro-interferometric observations of a nova in [5]. This letter evidenced an elongation of the nova fireball, just a gasp after the outburst (5.5 days). This first clue of bipolarity for one nova could not be repeatedly achieved when observing a second nova with the VLTI [6], due to a different observing configuration. Nevertheless, more recently, Chesneau et al. [7] provided firm evidence of bipolarity on a third nova, observed in 2011.

The presence of bipolar nebulae at the very first moments of outbursts in massive stars (e.g., LBVs like η Car) or less massive systems (e.g., planetary nebulae, novas) is still a broad and controversial subject. Indeed, bipolarity in the nebula is very often associated with binarity in the core [9]. But the inverse is not true, namely that binarity will essentially imply at a moment of the life of the system a bipolar ejection of material.

The key question is therefore to link the initial parameters of the system to:

- its evolution,
- the formation of a circumstellar disk (such as the ones encountered around B[e] stars or interacting systems such as υ Sgr or δ Sco),
- the occurrence of outburst events, forming rapidly a dense bipolar nebula,

but it still has to be answered.

5.5 Discussion, Concluding Remarks

With this paper, we tried to link the presence of a companion star to the presence of CSEs, by providing recent examples of high-angular resolution observations.

Interferometry has brought the view of highly structured CSEs around several massive stars, in the form of thin or thick disks, plus sometimes the presence of a companion star. The formation scenario of these CSEs remains to be determined, but the ingredients seem to count, among others, a binary system, including a hot, massive star, a dense disk, and perhaps also a fast-rotating star.

Acknowledgements The authors are thankful to the organizers of this very nice school. We also thank the VLTI team which is improving this wonderful instrument at every observation.

References

1. Abbott, D.C.: In: Conti, P.S., De Loore, C.W.H. (eds.) IAU Symposium, Mass Loss and Evolution of O-Type Stars, vol. 83, pp. 237–239 (1979)
2. Bedding, T.R.: Astron. J. **106**, 768 (1993)
3. Bonneau, D., Chesneau, O., Mourard, D., et al.: Astron. Astrophys. **532**, A148 (2011)
4. Carciofi, A.C., Okazaki, A.T., Le Bouquin, J.-B., et al.: Astron. Astrophys. **504**, 915 (2009)
5. Chesneau, O., Lykou, F., Balick, B., et al.: Astron. Astrophys. **473**, L29 (2007)
6. Chesneau, O., Banerjee, D.P.K., Millour, F., et al.: Astron. Astrophys. **487**, 223 (2008)
7. Chesneau, O., Meilland, A., Banerjee, D.P.K., et al.: Astron. Astrophys. **534**, 1 (2011)
8. Cranmer, S.R.: Astrophys. J. **634**, 585 (2005)
9. de Marco, O.: Publ. Astron. Soc. Pac. **121**, 316 (2009)
10. Delaa, O., Stee, P., Meilland, A., et al.: Astron. Astrophys. **529**, A87 (2011)
11. Domiciano de Souza, A., Kervella, P., Jankov, S., et al.: Astron. Astrophys. **407**, L47 (2003)
12. Domiciano de Souza, A., Driebe, T., Chesneau, O., et al.: Astron. Astrophys. **464**, 81 (2007)
13. Dunham, D.W.: In: I.O.T. Association (ed.) Occultation Newsletter, vol. 1, p. 4 (1974)
14. Fabregat, J., Reig, P., Otero, S.: IAU Circ. **7461**, 1 (2000)
15. Frémat, Y., Zorec, J., Hubert, A.-M., Floquet, M.: Astron. Astrophys. **440**, 305 (2005)
16. Hanbury Brown, R., Davis, J., Allen, L.R.: Mon. Not. R. Astron. Soc. **167**, 121 (1974)
17. Hartkopf, W.I., Mason, B.D., McAlister, H.A.: Astron. J. **111**, 370 (1996)
18. Innes, R.T.A.: Mon. Not. R. Astron. Soc. **61**, 358 (1901)
19. Kanaan, S., Meilland, A., Stee, P., et al.: Astron. Astrophys. **486**, 785 (2008)
20. Kervella, P., Domiciano de Souza, A.: Astron. Astrophys. **453**, 1059 (2006)
21. Kervella, P., Domiciano de Souza, A.,, Bendjoya, P.: Astron. Astrophys. **484**, L13 (2008)
22. Koubský, P., Harmanec, P., Yang, S., et al.: Astron. Astrophys. **459**, 849 (2006)
23. Kraus, S., Monnier, J.D., Che, X., et al.: Astrophys. J. **744**, 19 (2012)
24. Labeyrie, A., Bonneau, D., Stachnik, R.V., Gezari, D.Y.: Astrophys. J. Lett. **194**, L147 (1974)
25. Li, Q., Cassinelli, J.P., Brown, J.C., Waldron, W.L., Miller, N.A.: Astrophys. J. **672**, 1174 (2008)
26. Meilland, A., Millour, F., Stee, P., et al.: Astron. Astrophys. **464**, 73 (2007)
27. Meilland, A., Stee, P., Vannier, M., et al.: Astron. Astrophys. **464**, 59 (2007)
28. Meilland, A., Millour, F., Stee, P., et al.: Astron. Astrophys. **488**, L67 (2008)
29. Meilland, A., Kanaan, S., Borges Fernandes, M., et al.: Astron. Astrophys. **512**, A73 (2010)
30. Meilland, A., Delaa, O., Stee, P., et al.: Astron. Astrophys. **532**, A80 (2011)
31. Meilland, A., Millour, F., Stee, P., et al.: Astron. Astrophys. **538**, 110 (2012)
32. Millour, F., Chesneau, O., Borges Fernandes, M., et al.: Astron. Astrophys. **507**, 317 (2009)
33. Millour, F., Meilland, A., Chesneau, O., et al.: Astron. Astrophys. **526**, A107 (2011)
34. Miroshnichenko, A.S.: Astrophys. J. **667**, 497 (2007)
35. Miroshnichenko, A.S., Fabregat, J., Bjorkman, K.S., et al.: Astron. Astrophys. **377**, 485 (2001)
36. Miroshnichenko, A.S., Bjorkman, K.S., Morrison, N.D., et al.: Astron. Astrophys. **408**, 305 (2003)
37. Moreno, E., Koenigsberger, G., Harrington, D.M.: Astron. Astrophys. **528**, A48 (2011)
38. Mourard, D., Clausse, J.M., Marcotto, A., et al.: Astron. Astrophys. **508**, 1073 (2009)
39. Netolický, M., Bonneau, D., Chesneau, O., et al.: Astron. Astrophys. **499**, 827 (2009)

40. Okazaki, A.T.: Astron. Astrophys. **318**, 548 (1997)
41. Otero, S., Fraser, B., Lloyd, C.: Inf. Bull. Var. Stars **5026**, 1 (2001)
42. Petrov, R.G., Malbet, F., Weigelt, G., et al.: Astron. Astrophys. **464**, 1 (2007)
43. Plets, H., Waelkens, C., Trams, N.R.: Astron. Astrophys. **293**, 363 (1995)
44. Rivinius, T., Baade, D., Stefl, S., et al.: Astron. Astrophys. **336**, 177 (1998)
45. Tango, W.J., Davis, J., Jacob, A.P., et al.: Mon. Not. R. Astron. Soc. **396**, 842 (2009)
46. Tycner, C., Ames, A., Zavala, R.T., et al.: Astrophys. J. Lett. **729**, L5 (2011)
47. Vinicius, M.M.F., Zorec, J., Leister, N.V., Levenhagen, R.S.: Astron. Astrophys. **446**, 643 (2006)

Part III
Interferometric and Other Techniques to Peer Inside Stellar Environments

Chapter 6
Very Close Environments of Young Stars

F. Malbet

Abstract Long-baseline interferometry at infrared wavelengths allows the inner-most regions around young stars to be observed. These observations directly probe the location of the dust and gas in the disks. The characteristic sizes of these regions found are larger than previously thought. These results have motivated in part a new class of models of the inner disk structure, but the precise understanding of the origin of these low visibilities is still in debate. Mid-infrared observations probe disk emission over a larger range of scales revealing mineralogy gradients in the disk. Spectrally resolved observations allow the dust and gas to be studied separately showing that the Brackett gamma emission can find its origin either in a wind or in a magnetosphere and that there is probably no correlation between the location of the Brackett gamma emission and accretion. In a certain number of cases, the very high spatial resolution reveals very close companions and can determine their masses. Overall, these results provide essential information on the structure and the physical properties of close regions surrounding young stars especially where planet formation is suspected to occur.

6.1 Introduction

Many physical phenomena occur in the inner regions of the disks which surround young stars. The matter which eventually falls onto the stellar surface works its way through an accreting circumstellar disk which is subject to turbulence, convection, external and internal irradiation. The disks, which are rotating in a quasi-Keplerian motion, are probably the birth location of future planetary systems. Strong outflows, winds, and even jets find their origin in the innermost regions of many young stellar systems. The mechanisms of these ejection processes are not well understood, but they are probably connected to accretion. Most stars are born in multiple systems which can be very tightly bound and therefore have a strong impact on the physics of the disk inner regions.

F. Malbet (✉)
Institut de Planétologie et d'Astrophysique de Grenoble, Université J. Fourier/CNRS, BP 53, 38041 Grenoble cedex 9, France
e-mail: Fabien.Malbet@obs.ujf-grenoble.fr

J.-P. Rozelot, C. Neiner (eds.), *The Environments of the Sun and the Stars*, 163
Lecture Notes in Physics 857,
DOI 10.1007/978-3-642-30648-8_6, © Springer-Verlag Berlin Heidelberg 2013

The details of all these physical processes are not well understood yet because of lack of data to constrain them. The range of physical parameters which define best the inner disk regions in young stellar objects are:

- radius ranging from 0.1 AU to 10 AU
- temperature ranging from 150 K to 4000 K
- velocity ranging from 10 km/s to a few 100 km/s

The instrumental requirements to investigate the physical conditions in such regions are therefore driven by the spectral coverage which must encompass the near- and mid-infrared from 1 to 20 μm. Depending on the distance of the object (typically between 75 pc and 450 pc), the spatial resolution required to probe the inner parts of disks ranges between fractions and a few tens of milli-arcseconds. Since the angular resolution of astronomical instruments depends linearly on the wavelength and inversely on the telescope diameter, observing in the near- and mid-infrared wavelength domain points toward telescopes of sizes ranging from ten to several hundreds of meters. The only technique that allows such spatial resolution is therefore infrared interferometry.

Millan-Gabet et al. [35], published in *Protostars and Planets V*, reviewed the main results obtained with infrared interferometry in the domain of young stars between 1998 and 2005. I presented a review talk on this subject at the IAU Symposium 243 on *Star-disk interaction in young stars* [27] as well as R. Akeson in several schools in 2008 [1, 2]. This contribution to the school is an update of these results.

Section 6.2 briefly explains the principles of infrared interferometry and lists the literature on the observations carried out with this technique. Section 6.3 focuses on the main results obtained on disk physics (sizes, structures, dust and gas components, etc.), and Sect. 6.4 presents results on other phenomena constrained by interferometry (winds, magnetospheric accretion, multiple systems, etc.). In Sect. 6.5, I finish the review with the type of results that can be expected in the future.

6.2 Infrared Interferometry

6.2.1 Principle and Observations

Long baseline optical interferometry consists in mixing the light received from an astronomical source and collected by several independent telescopes separated from each other by tens or even hundreds of meters. The light beams are then overlapped and form an interference pattern if the optical path difference between the different arms of the interferometer—taking into account paths from the source up to the detector—is smaller than the coherence length of the incident wave (typically of the order of several microns). This interference figure is composed of fringes, i.e. a succession of stripes of faint (destructive interferences) and bright (constructive

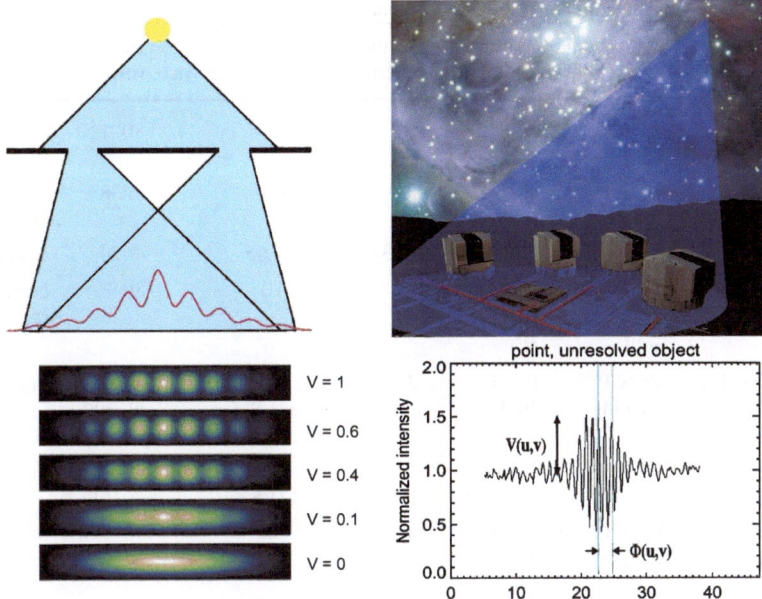

Fig. 6.1 Principle of interferometry. *Upper panels*: Young's slit experiment (*left*) compared to optical interferometry (*right*): in both cases light travels from a source to a plane where the incoming wavefront is split. The telescope apertures play the same role as the Young's slits. The difference lies in the propagation of light after the plane. In the case of optical interferometry, the instrument controls the propagation of light down to the detectors. At the detector plane, the light beams coming from the two apertures are overlapped. *Lower panels*: interference fringes whose contrast changes with the morphology of the source. *Left panel* shows fringes whose contrast varies from 0 to 1. *Right panel* displays actual stellar fringes but scanned along the optical path. The measure of the complex visibilities corresponds to the amplitude of the fringes for the visibility amplitude and the position of the fringes in wavelength units for the visibility phase

interferences) intensity. By measuring the contrast of these fringes, i.e. the normalized flux difference between the darkest and brightest regions, information about the morphology of the observed astronomical source can be recovered. Figure 6.1 illustrates this principle.

6.2.2 Instruments Available for Inner Regions Studies

Interferometric observations of young stellar objects were and are still performed at six different facilities on seven different instruments (see Table 6.1). We can classify these observations into three different categories:

- *Small-aperture interferometers*: PTI, IOTA, and ISI were the first facilities to be operational for YSO observations in the late 1990s (see Fig. 6.2). They have provided mainly the capability of measuring visibility amplitudes and lately closure

Table 6.1 Interferometers involved in YSO science

Facility	Instrument observable	Wavelength (microns)	Numbers of apertures	Aperture diameter (m)	Baseline (m)	Status
PTI	V^2	H, K	3	0.4	80–110	closed in 01/2009
IOTA	V^2, CP	H, K	3	0.4	5–38	closed in 06/2007
ISI	heterodyne	11	2 (3)	1.65	4–70	in operation
KI	V^2,	K spectral, L	2	10	80	closed in 08/2012
KI	nulling	N				
VLTI/AMBER	V^2, CP imaging	1–2.5 /high spectral	3	8.2 /1.8	40–130 /8–200	in operation
VLTI/MIDI	V^2	8–13 /spectral	2	8.2 /1.8	40–130 /8–200	in operation
VLTI/PIONIER	V^2, CP, imaging	$H, (K)$ /low spectral	4	8.2 /1.8	40–130	in operation until 2014
CHARA/MIRC, /CLIMB	V^2, CP, imaging	1–2.5	2/4 (6)	1	50–350	in operation
CHARA/VEGA	V^2, CP	V/spectral	3 (6)	1	50–350	in operation
LBT	imaging, nulling	1–10	2	8.4	6–23	in construction
MROI	imaging	1–2.5	4 (10)	1	6–23	in construction

V^2: visibility measurement; CP: closure phase

Acronyms. PTI: *Palomar Testbed Interferometer*; IOTA: *Infrared and Optical Telescope Array* (closed since 2006); ISI: *Infrared Spatial Interferometer*; KI: *Keck Interferometer*; VLTI: *Very Large Telescope Interferometer*; CHARA: *Center for High Angular Resolution Array*; MROI: *Magdalena Ridge Optical Interferometer*; LBT: *Large Binocular Telescope* (not yet operational)

phases. The latest one, CHARA, has an aperture diameter of 1 m. The instruments are mainly accessible through team collaboration. IOTA was shut down in 2007 and PTI in January 2009. MROI is in construction.

- *Large-aperture interferometers*: KI, VLTI, and soon LBT are facilities with apertures larger than 8 m. The instruments are widely open to the astronomical community through general calls for proposals. Lately, these facilities have significantly increased the number of young objects observed. KI has announced its closure for the second semester of 2012.

- *Instruments with spectral resolution*: CHARA, MIDI, and AMBER, KI/ASTRA provide spectral resolution from a few hundred up to 10,000, whereas other instruments mainly provided broadband observations. The spectral resolution al-

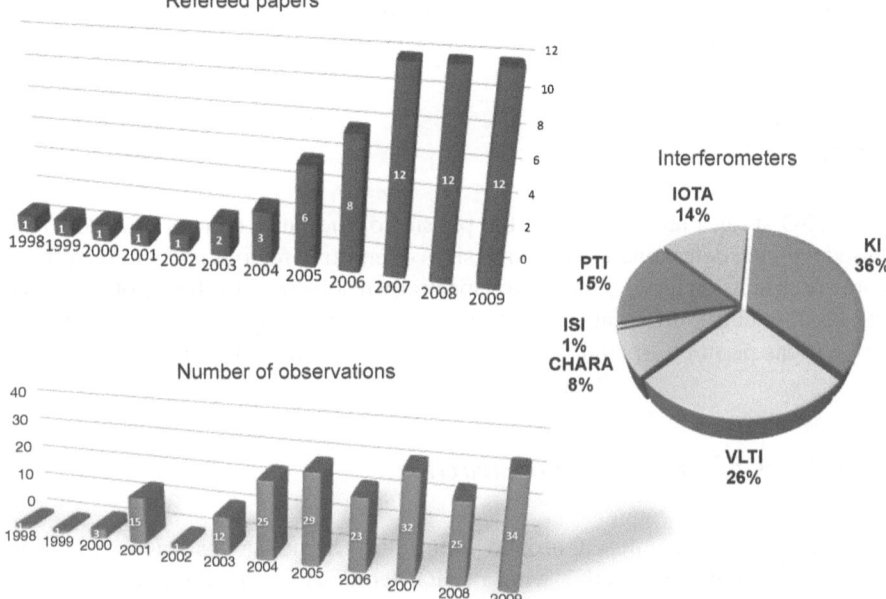

Fig. 6.2 Young stellar objects observed by interferometry and number of refereed papers published in the period 1998–2009 (*top graph*). The *bottom left graph* presents the number of YSO targets observed in the same period, and the *right bottom* one gives the distribution by interferometer

lows the various phenomena occurring in the environment of young stars to be separated.

- *Instruments with imaging*: CHARA/MIRC, VLTI/AMBER, VLTI/PIONIER, and soon MROI provide imaging capabilities.
- *Instruments with nulling*: KI in *N* band and soon LBT provide nulling capabilities that allow one to reach high dynamic observations.

6.2.3 Elements of Bibliography

Figure 6.2 displays the number of published results until the end of 2009 and show that it is increasing with time and improved facilities. At the date of the conference there were 75 refereed articles published in the field of young stars corresponding to up to 100 young stellar objects observed.

The distribution of observed object is rather well distributed among the various facilities. Several categories of young stellar systems have been observed at milliarcsecond scales mainly in the near-infrared wavelength domain, but also in the mid-infrared one. They include the brightest Herbig Ae/Be stars, the fainter T Tauri stars, and the few FU Orionis. Finally most observations were carried out in broad

band, but the advent of large aperture interferometers like the VLTI and KI allowed
higher spectral resolution to be obtained.

6.3 Inner Disk Physics

Most YSO studies are focused on the physics of the inner regions of disks. They
started with the determination of rough sizes of emitting regions and naturally led to
more constraints on the disk structure. Spectrally resolved mid-infrared observations
are able to identify different types of dust grains. Spectrally resolved near-infrared
observations permit to spatially discriminate between gas and dust.

6.3.1 Sizes of Circumstellar Structures

Disks are known to be present around young stars. Some ten years ago, disks were
believed to behave "normally" with a radial temperature distribution following a
power-law $T \propto r^{-q}$ with q ranging between 0.5 and 0.75. The value of q depends
on the relative effect of irradiation from the central star in comparison with heat
dissipation due to accretion. This model was successful to reproduce ultraviolet
and infrared excesses in spectral energy distributions (SEDs). Malbet and Bertout
[28] investigated the potential of optical long baseline interferometry to study the
disks of T Tauri stars and FU Orionis stars. They found that the structure would be
marginally resolved but observations would be possible with baselines of the order
of 100 m with a visibility amplitude remaining high.

First observations of the brighter Herbig Ae/Be stars showed that the observed
visibilities were much smaller than expected especially for those objects where the
accretion plays a little role. Monnier and Millan-Gabet [36] pointed out that the
interferometric sizes of these objects were much larger than expected from the standard disk model. They plotted the sizes obtained as functions of the stellar luminosity and found that there was a strong correlation following an $L^{0.5}$ law over two
decades. This behavior is consistent with the variation of radius of dust sublimation with respect to the central star luminosity: the dust distribution radius shifts to
larger radii for more luminous objects because the temperature is larger than the
sublimation limit ($\sim$1000–1500 K). Only the most massive Herbig Be stars seem to
be compliant with the standard accretion disk model.

In the meantime, in order to account for the near-infrared characteristics of SEDs
and in particular a flux excess around $\lambda = 3$ μm, Natta et al. [38] proposed that
disks around Herbig Ae/Be stars have an optically thin inner cavity and create a
puffed-up inner wall of optically thick dust at the dust sublimation radius. More
realistic models were developed afterward which take more physical properties into
account [12, 21, 37]. However as pointed out by Vinković et al. [51] and recently by
Schegerer et al. [42] on the specific and actual case of RY Tau, the disk models are

not the only ones that can reproduce the measurements: models with a disk halo or envelope can also match the data.

Observations at KI [3, 11, 15] found also large NIR sizes for lower-luminosity T Tauri stars, in many cases even larger than would be expected from extrapolation of the HAe relation. It is interpreted by the fact that the accretion disk contributes significantly to the luminosity emitted by the central region, and therefore this additional luminosity must be taken into account in the relationship. However in these systems uncertainties are still large, and very few measurements per object have been obtained. In order to interpret all the T Tauri measurements, Akeson et al. [3] need to introduce a new physical component like optically thick gas emission in the inner hole and extended structure around the objects.

Characteristic dimensions of the emitting regions at 10 μm were found by Leinert et al. [25] to be ranging from 1 to 10 AU. The sizes of their sample stars correlate with the slope of the 10–25 μm infrared spectrum: the reddest objects are the largest ones. Such a correlation is consistent with a different geometry in terms of flaring or flat (self-shadowed) disks for sources with strong or moderate mid-infrared excess, respectively, demonstrating the power of interferometry not only to probe characteristic sizes of disks but also to derive information on the vertical disk structure.

6.3.2 Constraints on Disk Structure

Theoreticians start discussing slightly different scenarios of the inner regions around young stars. For example, the inner puffed-up wall is modeled with a curved shape by Isella and Natta [21] due to the very large vertical density gradient and the dependence of grain evaporation temperature on gas density as expected when a constant evaporation temperature is assumed. Tannirkulam et al. [45] proposed that the geometry of the rim depends on the composition and spatial distribution of dust due to grain growth and settling.

Vinković and Jurkić [50] presented a model-independent method of comparison of NIR visibility data of YSOs. The method based on scaling the measured baseline with the YSO distance and luminosity removes the dependence of visibility on these two variables. They found that low luminosity Herbig Ae stars are best explained by the uniform brightness ring and the halo model, T Tauri stars with the halo model, and high luminosity Herbig Be stars with the accretion disk model, but they admit that the validity of each model is not well established.

At the moment, only few objects have been thoroughly studied: FU Orionis [29, 30, 39] has been observed on 42 nights over a period of six years from 1998 to 2003 with 287 independent measurements of the fringe visibility at six different baselines ranging from 20 to 110 m in length, in the H and K bands. The data not only resolves FU Ori at the AU scale, but also allows the accretion disk scenario to be tested. The most probable interpretation is that FU Ori hosts an active accretion disk whose temperature law is consistent with the standard model. In the mid-infrared, Quanz et al. [39] resolved structures that are also best explained with an optically thick accretion disk. A simple accretion disk model fits the observed SED and visibilities reasonably well and does not require the presence of any additional structure such as a dusty envelope. However Zhu et al. [53] revisited this issue using detailed radiative transfer calculations to model the recent, high signal-to-noise ratio data obtained on FU Ori from the IRS instrument on the Spitzer Space Telescope. They found that a physically plausible flared disk irradiated by the central accretion disk matches the observations. Their accretion disk model with outer disk irradiation by the inner disk reproduces the spectral energy distribution, and their model is consistent with near-infrared interferometry, but there remains some inconsistencies with mid-infrared interferometric results. This is why one should remain careful with results coming from surveys having only few measurements per object.

Millan-Gabet et al. [33] obtained K-band observations of three other FU Orionis objects, V1057 Cyg, V151 Cyg, and Z CMa-SE and found that all three objects appear significantly more resolved than expected from simple models of accretion disks tuned to fit the SEDs. They believe that emission at the scale of tens of AU in the interferometer field of view is responsible for the low visibilities, originating in scattering by large envelopes surrounding these objects. Li Causi et al. [26] have measured again interferometric visibilities of Z CMa with VLTI/AMBER.

Kraus et al. [23] have shown that they are able to derive the temperature radial distribution of the disk around MWC 147 from the interferometric measurements using the spectral variation of the visibilities in low resolution with AMBER and MIDI (see Fig. 6.3). A similar work has been attempted at PTI with larger uncertainties [16].

On the pure theoretical side, very few physical models achieved to fit interferometric data simultaneously with SEDs. Using a two-layer accretion disk model, Lachaume et al. [24] found satisfactory fits for SU Aur, in solutions that are characterized by the midplane temperature being dominated by accretion, while the emerging flux is dominated by reprocessed stellar photons. Since the midplane temperature drives the vertical structure of the disk, there is a direct impact on the measured visibilities that are not necessarily taken into account by other models. In the MIR range, Schegerer et al. [42] was able to fit both the SED and the visibilities of RY Tau as mentioned earlier with two different models, although several other models have been dismissed with MIDI data. The region where the measurements would allow us to choose between the two remaining models is located inward and can be observed at shorter wavelengths.

6.3.3 Dust Mineralogy

The mid-infrared wavelength region contains strong resonances of abundant dust species, both oxygen-rich (amorphous or crystalline silicates) and carbon-rich (polycyclic aromatic hydrocarbons, or PAHs). Therefore, spectroscopy of optically thick protoplanetary disks offers a diagnostic of the chemical composition and grain size of dust in disk atmospheres.

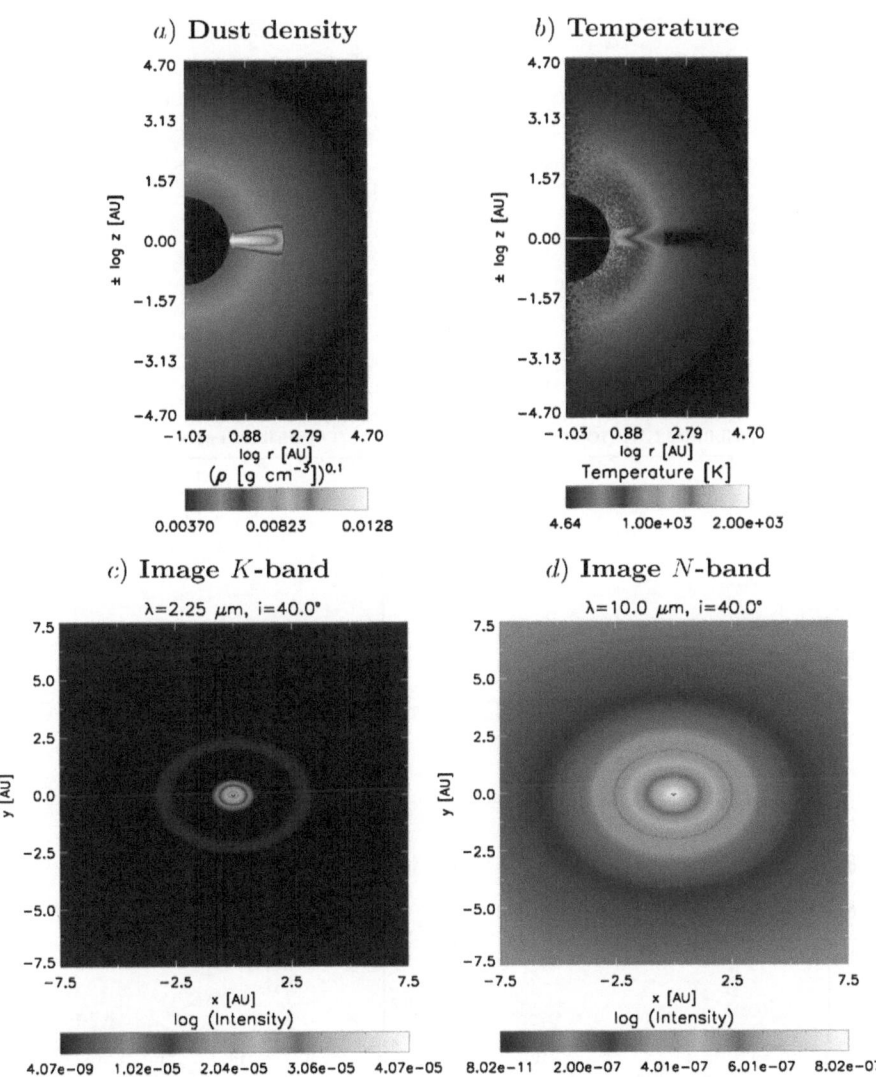

Fig. 6.3 Radiative transfer model constrained by interferometric measurements on MWC 147 by Kraus et al. [23]. The model computed for MWC 147 assumes a spherical shell geometry. Panels (**a**) and (**b**) show the dust density and the temperature distribution. Panels (**c**) and (**d**) show the ray-traced images for two representative NIR (2.25 μm) and MIR (10.0 μm) wavelengths. Panel (**e**) shows the SED for various inclination angles, whereas panel (**f**) gives the SED for the best-fit inclination angle and separates the flux which originates in stellar photospheric emission, thermal emission, dust irradiation, and accretion luminosity. Finally, panels (**g**) to (**j**) depict the NIR and MIR visibilities computed from their radiative transfer models

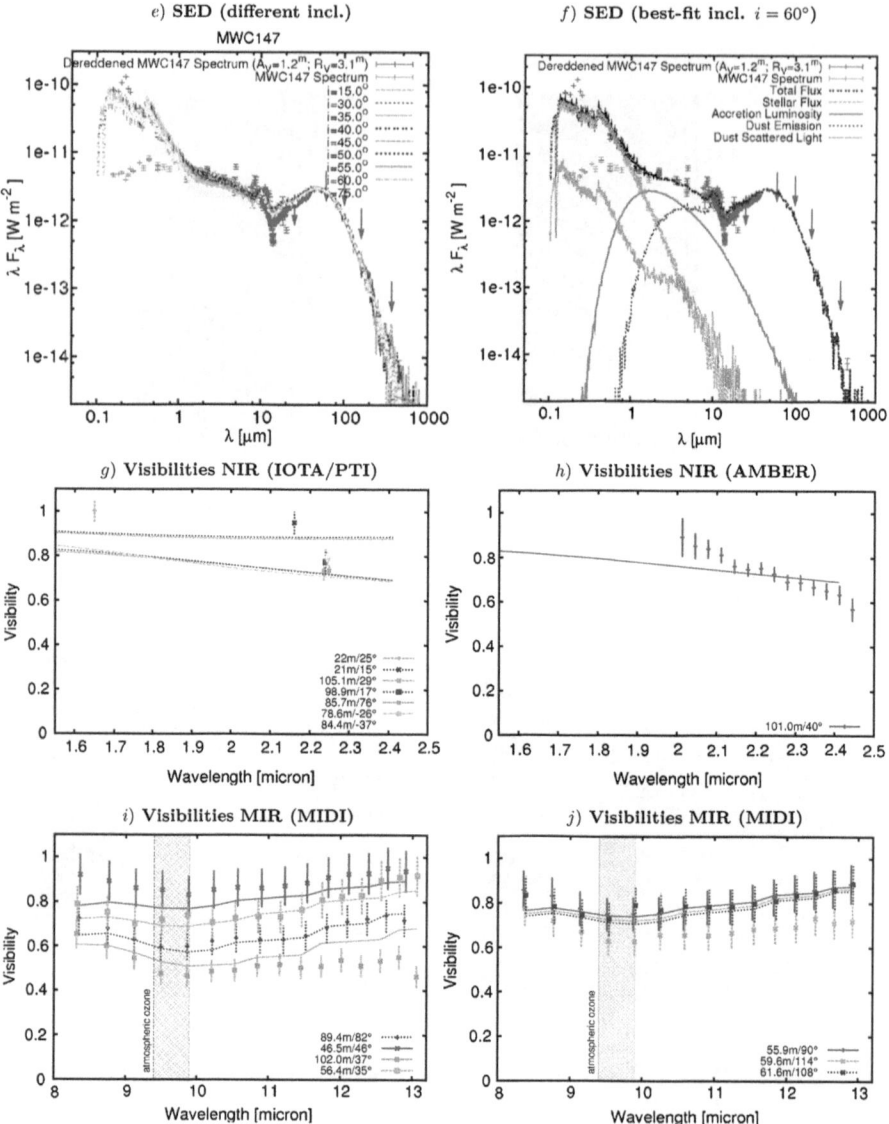

Fig. 6.3 (Continued)

Van Boekel et al. [49] spatially resolved three protoplanetary disks surrounding Herbig Ae/Be stars across the *N* band. The correlated spectra measured by MIDI at the VLTI correspond to disk regions ranging from 1 to 2 AUs. By combining these measurements with unresolved spectra, the spectrum corresponding to outer disk regions at 220 AU can also be derived. These observations have revealed that the dust in these regions was highly crystallized (40 to 100 %), more than any other

Fig. 6.4 Spectrally dispersed visibility amplitudes of 51 Oph in the CO bandhead spectral region. Overimposed is the spectrum as measured by VLTI/AMBER (*black line*). The *blue curve* corresponds to the addition of a simple uniform disk model for the excess emission in the line with a typical diameter of 0.2 AU. From Tatulli et al. [47]

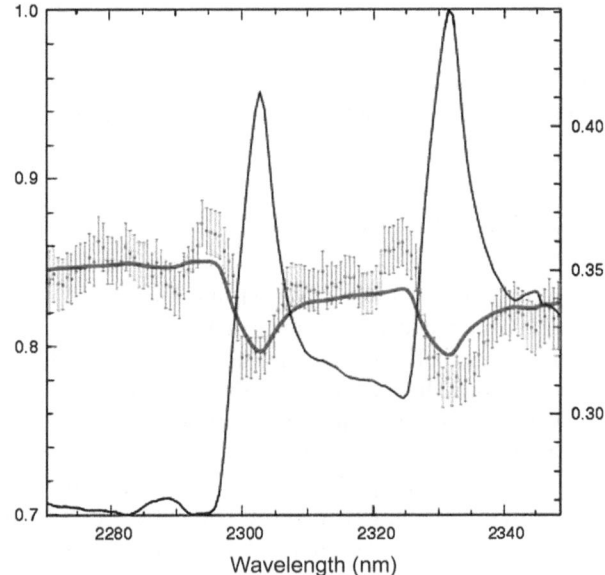

dust observed in young stars until now. The spectral shape of the inner-disk spectra shows surprising similarity with Solar System comets. Their observations imply that silicates crystallize before terrestrial planets are formed, consistent with the composition of meteorites in the Solar System. Similar measurements were also carried out by Ratzka et al. [40] on the T Tauri system, TW Hya. According to the correlated flux measured with MIDI, most of the crystalline material is located in the inner, unresolved part of the disk, about 1 AU in radius.

6.3.4 Gas/Dust Connection

Gil et al. [19] observed the young stellar system 51 Oph confirming the interpretation of Thi et al. [48] and more recently Berthoud et al. [7] of a disk seen nearly edge-on: the radial distribution of excitation temperatures for the vibrational levels of CO overtone ($\Delta v = 2$) emission from hot gas is consistent with the gas being in radiative thermal equilibrium except at the inner edge, where low vibrational bands have higher excitation temperatures. Tatulli et al. [47] confirmed the high inclination of the disk but also detect the CO bandheads allowing the dust responsible for the continuum to be separated from the gas. As a matter of fact, the visibilities in the CO bands is lower than the ones measured in the continuum implying that the region responsible for this gas emission is smaller than the region responsible for the dust emission. Figure 6.4 illustrates this result and shows that the combination of very high spatial information with spectral resolution opens brand new perspectives in the studies of the inner disk properties by discriminating between components.

6.4 Other AU-Scale Phenomena

Several other physical phenomena have been investigated in the innermost region of disks: wind, magnetosphere, and close companions.

6.4.1 Outflows and Winds

The power of spectrally resolved interferometric measurements provides detailed wavelength dependence of inner disk continuum emission (see end of Sect. 6.3.2). These new capabilities enable also detailed studies of hot winds and outflows, and therefore the physical conditions and kinematics of the gaseous components in which emission and absorption lines arise like $Br\gamma$ and H_2 ones. With VLTI/AMBER, Malbet et al. [31] spatially resolved the luminous Herbig Be object MWC 297, measuring visibility amplitudes as a function of wavelength at intermediate spectral resolution $R = 1500$ across a 2.0–2.2 µm band and in particular the $Br\gamma$ emission line. The interferometer visibilities in the $Br\gamma$ line are about 30 % lower than those of the nearby continuum, showing that the $Br\gamma$ emitting region is significantly larger than the NIR continuum region. Known to be an outflow source, a preliminary model has been constructed in which a gas envelope, responsible for the $Br\gamma$ emission, surrounds an optically thick circumstellar disk. The characteristic size of the line-emitting region being 40 % larger than that of the NIR disk. This model is successful at reproducing the VLTI/AMBER measurements as well as previous continuum interferometric measurements at shorter and longer baselines [14, 32], the SED, and the shapes of the Hα, Hβ, and $Br\gamma$ emission lines. The precise nature of the MWC 297 wind, however, remains unclear; the limited amount of data obtained in these first observations cannot, for example, discriminate between a stellar or disk origin for the wind, or between competing models of disk winds (e.g., Ferreira et al. [18], Shu et al. [44]). Recently, Weigelt et al. [52] have provided new measurements with a comprehensive modeling of the environment.

Benisty et al. [4] has nicely shown the wind effect at milli-arcsecond resolution in one of the component of the Z CMa binary during an outburst event.

6.4.2 Magnetosphere

The origin of the hydrogen line emission in Herbig Ae/Be stars is still unclear. The lines may originate either in the gas which accretes onto the star from the disk, as in magnetospheric accretion models [20], or in winds and jets, driven by the interaction of the accreting disk with a stellar [43] or disk [10] magnetic field. For all models, emission in the hydrogen lines is predicted to occur over very small spatial scales, a few AUs at most. To understand the physical processes that happen at these scales, one needs to combine very high spatial resolution with enough spectral resolution to resolve the line profile.

Fig. 6.5 A systematic study of the origin of the Brγ emission in Herbig Ae/Be stars from Kraus et al. [22]. *Top*: the fitted ring radii for the continuum (*black points*) and Brγ line (*red points*) plotted as functions of stellar luminosity. The spatial extension of the stellar surface is represented in grey and the dust sublimation radius corresponding to the dust sublimation temperatures of 2000, 1500, and 1000 K. *Bottom*: the ratio of normalized size of the Brγ region by the size of the continuum-emitting region. *Circles in blue* emphasize the system for which a magnetospheric accretion scenario is a compatible model, in green the ones with probable disk winds, and in brown the ones with X wind or disk wind

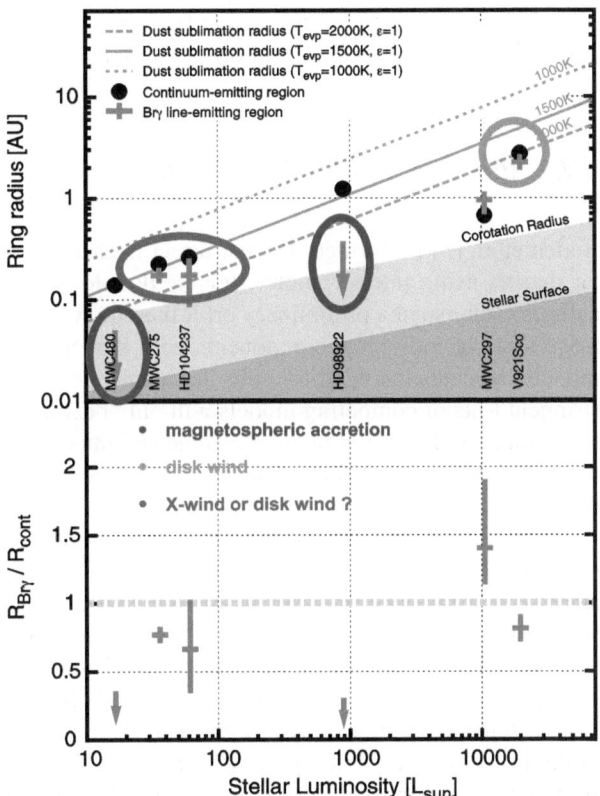

On the one hand, Tatulli et al. [46] performed interferometric observations of the Herbig Ae star HD 104237, obtained with the VLTI/AMBER instrument with $R = 1500$ high spectral resolution. The observed visibility was identical in the Brγ line and in the continuum, even though the line represents 35 % of the continuum flux. This immediately implies that the line and continuum emission regions have the same apparent size. Using simple models to describe the Brγ emission, they showed that the line emission is unlikely to originate in either magnetospheric accreting columns of gas or in the gaseous disk but more likely in a compact outflowing disk wind launched in the vicinity of the rim, about 0.5 AU from the star. The main part of the Brγ emission in HD 104237 is unlikely to originate in magnetospheric accreting matter.

On the other hand, Eisner [13] measured an increase of the Brγ visibility in MWC 480 implying that the region of emission of the hydrogen line is very compact, less than 0.1 mas in radius which could be interpreted as emission originating in the magnetosphere of the system.

At the present time, given the limited number of samples, it is difficult to derive a general tendency, but it seems that all possible scenario can be found like Kraus et al. [22] show in a mini-survey of five Herbig Ae/Be stars (Fig. 6.5). Eisner et al. [17] support the hypothesis that Brγ emission generally traces magnetospherically driven accretion and/or outflows in young star/disk systems.

6.4.3 Binaries and Multiple Systems

Boden et al. [8] performed the first direct measurement of pre-main sequence stellar masses using interferometry, for the double-lined system HD 98800-B. These authors established a preliminary orbit that allowed determination of the (subsolar) masses of the individual components with 8 % accuracy. Comparison with stellar models indicates the need for subsolar abundances for both components, although stringent tests of competing models will only become possible when more observations improve the orbital phase coverage and thus the accuracy of the stellar masses derived. Boden et al. [9] published another determination of dynamical masses for the pre-main sequence system V773 Tau A.

As another example, based on a low-level oscillation in the visibility amplitude signature in the PTI data of FU Ori, Malbet et al. [30] claimed the detection of an off-centered spot embedded in the disk that could be physically interpreted as a young stellar or protoplanetary companion located at $\sim$10 AUs and could possibly be at the origin of the FU Ori outburst itself. Using another technique, Millan-Gabet et al. [34] reported on the detection of localized off-center emission at 1–4 AU in the circumstellar environment of AB Aurigae. They used closure-phase measurements in the near-infrared. When probing sub-AU scales, all closure phases are close to zero degrees, as expected given the previously determined size of the AB Aurigae inner-dust disk. However, a clear closure-phase signal of $-3.5° \pm 0.5°$ is detected on one triangle containing relatively short baselines, requiring a high degree of asymmetry from emission at larger AU scales in the disk. They interpret such detected asymmetric near-infrared emission as a result of localized viscous heating due to a gravitational instability in the AB Aurigae disk, or to the presence of a close stellar companion or accreting substellar object.

Berger et al. [6] present the first image of a triple system in GW Ori. They show that accretion disk models of GW Ori will need to be completely reconsidered because of this outer companion C and the unexpected brightness of companion B.

6.5 Future Prospects

As emphasized in this review, more interferometric data is required with better accuracy and also wider coverage of the baselines in order to constrain better the models that have been proposed. Like for radio astronomy, these supplementary data will allow image reconstruction without any prior knowledge of the observed structure. Several facilities are already ready to obtain interferometric images although with few pixels across the field: MIRC at CHARA and AMBER and PIONIER at the

VLTI in the near-infrared. However at the moment MIRC is limited in sensitivity and AMBER in number of telescopes (3), which makes it difficult to routinely image although images has been obtained on two YSOs [5, 41].

In the mid-infrared the MATISSE instrument has been proposed to ESO to provide imaging with four telecopes at the VLTI. GRAVITY is also a proposed second generation VLTI instrument which will be able to combine four beams with a better sensitivity than AMBER and PIONIER. The LBT will also provide imaging capability.

All these instruments provide spectral resolution that make them indeed spectro-imagers. Therefore in the future, one should be able to obtain a wealth of information from the innermost regions of disks around young stars. However in the meantime, observations are already mature enough to allow detailed modeling of the phenomena occurring in these inner regions of young stellar objects.

References

1. Akeson, R.: Observations of circumstellar disks with infrared interferometry. New Astron. Rev. **52** 94–104 (2008). doi:10.1016/j.newar.2008.04.006
2. Akeson, R.L.: Observations of circumstellar disks with infrared interferometry. J. Phys. Conf. Ser. **131**(1), 012019 (2008). doi:10.1088/1742-6596/131/1/012019
3. Akeson, R.L., Walker, C.H., Wood, K., Eisner, J.A., Scire, E., Penprase, B, Ciardi, D.R., van Belle, G.T., Whitney, B., Bjorkman, J.E.: Observations and modeling of the inner disk region of T Tauri stars. Astrophys. J. **622**, 440–450 (2005). doi:10.1086/427770. arXiv:astro-ph/0412438
4. Benisty, M., Malbet, F., Dougados, C., Natta, A., Le Bouquin, J.B., Massi, F., Bonnefoy, M., Bouvier, J., Chauvin, G., Chesneau, O., Garcia, P.J.V., Grankin, K., Isella, A., Ratzka, T., Tatulli, E., Testi, L., Weigelt, G., Whelan, E.T.: The 2008 outburst in the young stellar system Z CMa. I. Evidence of an enhanced bipolar wind on the AU-scale. Astron. Astrophys. **517**, L3 (2010). doi:10.1051/0004-6361/201014776. arXiv:1007.0682
5. Benisty, M., Renard, S., Natta, A., Berger, J.P., Massi, F., Malbet, F., Garcia, P.J.V., Isella, A., Mérand, A., Monin, J.L., Testi, L., Thiébaut, E., Vannier, M., Weigelt, G.: A low optical depth region in the inner disk of the Herbig Ae star HR 5999. Astron. Astrophys. **531**, A84 (2011). doi:10.1051/0004-6361/201016091. arXiv:1106.4150
6. Berger, J.P., Monnier, J.D., Millan-Gabet, R., Renard, S., Pedretti, E., Traub, W., Bechet, C., Benisty, M., Carleton, N., Haguenauer, P., Kern, P., Labeye, P., Longa, F., Lacasse, M., Malbet, F., Perraut, K., Ragland, S., Schloerb, P., Schuller, P.A., Thiébaut, E.: First astronomical unit scale image of the GW Orionis triple system. Direct detection of a new stellar companion. Astron. Astrophys. **529**, L1 (2011). doi:10.1051/0004-6361/201016219. arXiv:1103.3888
7. Berthoud, M.G., Keller, L.D., Herter, T.L., Richter, M.J., Whelan, D.G.: Near-IR CO overtone emission in 51 Ophiuchi. Astrophys. J. **660**, 461–468 (2007). doi:10.1086/512056
8. Boden, A.F., Sargent, A.I., Akeson, R.L., Carpenter, J.M., Torres, G., Latham, D.W., Soderblom, D.R., Nelan, E., Franz, O.G., Wasserman, L.H.: Dynamical masses for low-mass pre-main-sequence stars: a preliminary physical orbit for HD 98800 B. Astrophys. J. **635**, 442–451 (2005). doi:10.1086/497328. arXiv:astro-ph/0508331
9. Boden, A.F., Torres, G., Sargent, A.I., Akeson, R.L., Carpenter, J.M., Boboltz, D.A., Massi, M., Ghez, A.M., Latham, D.W., Johnston, K.J., Menten, K.M., Ros, E.: Dynamical masses for pre-main-sequence stars: a preliminary physical orbit for V773 Tau A. Astrophys. J. **670**, 1214–1224 (2007). doi:10.1086/521296. arXiv:0706.2376

10. Casse, F., Ferreira, J.: Magnetized accretion-ejection structures, IV: magnetically-driven jets from resistive, viscous, Keplerian discs. Astron. Astrophys. **353**, 1115–1128 (2000). arXiv:astro-ph/9911471

11. Colavita, M., Akeson, R., Wizinowich, P., Shao, M., Acton, S., Beletic, J., Bell, J., Berlin, J., Boden, A., Booth, A., Boutell, R., Chaffee, F., Chan, D., Chock, J., Cohen, R., Crawford, S., Creech-Eakman, M., Eychaner, G., Felizardo, C., Gathright, J., Hardy, G., Henderson, H., Herstein, J., Hess, M., Hovland, E., Hrynevych, M., Johnson, R., Kelley, J., Kendrick, R., Koresko, C., Kurpis, P., Le Mignant, D., Lewis, H., Ligon, E., Lupton, W., McBride, D., Mennesson, B., Millan-Gabet, R., Monnier, J., Moore, J., Nance, C., Neyman, C., Niessner, A., Palmer, D., Reder, L., Rudeen, A., Saloga, T., Sargent, A., Serabyn, E., Smythe, R., Stomski, P., Summers, K., Swain, M., Swanson, P., Thompson, R., Tsubota, K., Tumminello, A., van Belle, G., Vasisht, G., Vause, J., Walker, J., Wallace, K., Wehmeier, U.: Observations of DG Tauri with the Keck interferometer. Astrophys. J. Lett. **592**, L83–L86 (2003). doi:10.1086/377704. arXiv:astro-ph/0307051

12. Dullemond, C.P., Dominik, C., Natta, A.: Passive irradiated circumstellar disks with an inner hole. Astrophys. J. **560**, 957–969 (2001). doi:10.1086/323057. arXiv:astro-ph/0106470

13. Eisner, J.A.: Water vapour and hydrogen in the terrestrial-planet-forming region of a protoplanetary disk. Nature **447**, 562–564 (2007). doi:10.1038/nature05867. arXiv:0706.1239

14. Eisner, J.A., Lane, B.F., Hillenbrand, L.A., Akeson, R.L., Sargent, A.I.: Resolved inner disks around Herbig Ae/Be stars. Astrophys. J. **613**, 1049–1071 (2004). doi:10.1086/423314. arXiv:astro-ph/0406356

15. Eisner, J.A., Hillenbrand, L.A., White, R.J., Akeson, R.L., Sargent, A.I.: Observations of T Tauri disks at Sub-AU radii: implications for magnetospheric accretion and planet formation. Astrophys. J. **623**, 952–966 (2005). doi:10.1086/428828. arXiv:astro-ph/0501308

16. Eisner, J.A., Chiang, E.I., Lane, B.F., Akeson, R.L.: Spectrally dispersed K-band interferometric observations of Herbig Ae/Be sources: inner disk temperature profiles. Astrophys. J. **657**, 347–358 (2007). doi:10.1086/510833. arXiv:astro-ph/0611447

17. Eisner, J.A., Monnier, J.D., Woillez, J., Akeson, R.L., Millan-Gabet, R., Graham, J.R., Hillenbrand, L.A., Pott, J.U., Ragland, S., Wizinowich, P.: Spatially and spectrally resolved hydrogen gas within 0.1 AU of T Tauri and Herbig Ae/Be stars. Astrophys. J. **718**, 774–794 (2010). doi:10.1088/0004-637X/718/2/774. arXiv:1006.1651

18. Ferreira, J., Bessolaz, N., Zanni, C., Combet, C.: Large scale magnetic fields in discs: jets and reconnection X-winds. In: Bouvier, J., Appenzeller, I. (eds.) IAU Symposium, vol. 243, pp. 307–314 (2007). doi:10.1017/S1743921307009660

19. Gil, C., Malbet, F., Schöller, M., Chesneau, O., Leinert, C.: Observations of 51 Ophiuchi with MIDI at the VLTI. In: Richichi, A., Delplancke, F., Paresce, F., Chelli, A. (eds.) The Power of Optical/IR Interferometry: Recent Scientific Results and 2nd Generation, p. 187 (2008)

20. Hartmann, L., Hewett, R., Calvet, N.: Magnetospheric accretion models for T Tauri stars, I: Balmer line profiles without rotation. Astrophys. J. **426**, 669–687 (1994). doi:10.1086/174104

21. Isella, A., Natta, A.: The shape of the inner rim in proto-planetary disks. Astron. Astrophys. **438**, 899–907 (2005). doi:10.1051/0004-6361:20052773. arXiv:astro-ph/0503635

22. Kraus, S., Hofmann, K.H., Benisty, M., Berger, J.P., Chesneau, O., Isella, A., Malbet, F., Meilland, A., Nardetto, N., Natta, A., Preibisch, T., Schertl, D., Smith, M., Stee, P., Tatulli, E., Testi, L., Weigelt, G.: The origin of hydrogen line emission for five Herbig Ae/Be stars spatially resolved by VLTI/AMBER spectro-interferometry. Astron. Astrophys. **489**, 1157–1173 (2008) doi:10.1051/0004-6361:200809946. arXiv:0807.1119.

23. Kraus, S., Preibisch, T., Ohnaka, K.: Detection of an inner gaseous component in a Herbig Be star accretion disk: near- and mid-infrared spectrointerferometry and radiative transfer modeling of MWC 147. Astrophys. J. **676**, 490–508 (2008) doi:10.1086/527427. arXiv:0711.4988.

24. Lachaume, R., Malbet, F., Monin, J.L.: The vertical structure of T Tauri accretion discs, III: consistent interpretation of spectra and visibilities with a two-layer model. Astron. Astrophys. **400**, 185–202 (2003). doi:10.1051/0004-6361:20021037. arXiv:astro-ph/0206307

25. Leinert, C., van Boekel, R., Waters, L.B.F.M., Chesneau, O., Malbet, F., Köhler, R., Jaffe, W., Ratzka, T., Dutrey, A., Preibisch, T., Graser, U., Bakker, E., Chagnon, G., Cotton, W.D., Dominik, C., Dullemond, C.P., Glazenborg-Kluttig, A.W., Glindemann, A., Henning, T., Hofmann, K.H., de Jong, J., Lenzen, R., Ligori, S., Lopez, B., Meisner, J., Morel, S., Paresce, F., Pel, J.W., Percheron, I., Perrin, G., Przygodda, F., Richichi, A., Schöller, M., Schuller, P., Stecklum, B., van den Ancker, M.E., von der Lühe, O., Weigelt, G.: Mid-infrared sizes of circumstellar disks around Herbig Ae/Be stars measured with MIDI on the VLTI. Astron. Astrophys. **423**, 537–548 (2004). doi:10.1051/0004-6361:20047178

26. Li Causi, G., Antoniucci, S., Tatulli, E.: De-biasing interferometric visibilities in VLTI-AMBER data of low SNR observations. Astron. Astrophys. **479**, 589–595 (2008). doi:10.1051/0004-6361:20077629. arXiv:0711.4938

27. Malbet, F.: Inner disk regions revealed by infrared interferometry. In: Bouvier, J., Appenzeller, I. (eds.) IAU Symposium, vol. 243, pp. 123–134 (2007). doi:10.1017/S1743921307009489

28. Malbet, F., Bertout, C.: Detecting T Tauri disks with optical long-baseline interferometry. Astron. Astrophys. Suppl. Ser. **113**, 369 (1995). arXiv:astro-ph/9505061

29. Malbet, F., Berger, J.P., Colavita, M.M., Koresko, C.D., Beichman, C., Boden, A.F., Kulkarni, S.R., Lane, B.F., Mobley, D.W., Pan, X.P., Shao, M., van Belle, G.T., Wallace, J.K.: FU Orionis resolved by infrared long-baseline interferometry at a 2 AU scale. Astrophys. J. Lett. **507**, L149–L152 (1998). doi:10.1086/311688. arXiv:astro-ph/9808326

30. Malbet, F., Lachaume, R., Berger, J.P., Colavita, M.M., di Folco, E., Eisner, J.A., Lane, B.F., Millan-Gabet, R., Ségransan, D., Traub, W.A.: New insights on the AU-scale circumstellar structure of FU Orionis. Astron. Astrophys. **437**, 627–636 (2005). doi:10.1051/0004-6361:20042556. arXiv:astro-ph/0503619

31. Malbet, F., Benisty, M., de Wit, W.J., Kraus, S., Meilland, A., Millour, F., Tatulli, E., Berger, J.P., Chesneau, O., Hofmann, K.H., Isella, A., Natta, A., Petrov, R.G., Preibisch, T., Stee, P., Testi, L., Weigelt, G., Antonelli, P., Beckmann, U., Bresson, Y., Chelli, A., Dugué, M., Duvert, G., Gennari, S., Glück, L., Kern, P., Lagarde, S., Le Coarer, E., Lisi, F., Perraut, K., Puget, P., Rantakyrö, F., Robbe-Dubois, S., Roussel, A., Zins, G., Accardo, M., Acke, B., Agabi, K., Altariba, E., Arezki, B., Aristidi, E., Baffa, C., Behrend, J., Blöcker, T., Bonhomme, S., Busoni, S., Cassaing, F., Clausse, J.M., Colin, J., Connot, C., Delboulbé, A., Domiciano de Souza, A., Driebe, T., Feautrier, P., Ferruzzi, D., Forveille, T., Fossat, E., Foy, R., Fraix-Burnet, D., Gallardo, A., Giani, E., Gil, C., Glentzlin, A., Heiden, M., Heininger, M., Hernandez Utrera, O., Kamm, D., Kiekebusch, M., Le Contel, D., Le Contel, J.M., Lesourd, T., Lopez, B., Lopez, M., Magnard, Y., Marconi, A., Mars, G., Martinot-Lagarde, G., Mathias, P., Mège, P., Monin, J.L., Mouillet, D., Mourard, D., Nussbaum, E., Ohnaka, K., Pacheco, J., Perrier, C., Rabbia, Y., Rebattu, S., Reynaud, F., Richichi, A., Robini, A., Sacchettini, M., Schertl, D., Schöller, M., Solscheid, W., Spang, A., Stefanini, P., Tallon, M., Tallon-Bosc, I., Tasso, D., Vakili, F., von der Lühe, O., Valtier, J.C., Vannier, M., Ventura, N.: Disk and wind interaction in the young stellar object MWC 297 spatially resolved with AMBER/VLTI. Astron. Astrophys. **464**, 43–53 (2007). doi:10.1051/0004-6361:20053924. arXiv:astro-ph/0510350

32. Millan-Gabet, R., Schloerb, F.P., Traub, W.A.: Spatially resolved circumstellar structure of Herbig Ae/Be stars in the near-infrared. Astrophys. J. **546**, 358–381 (2001). doi:10.1086/318239. arXiv:astro-ph/0008072

33. Millan-Gabet, R., Monnier, J.D., Akeson, R.L., Hartmann, L., Berger, J.P., Tannirkulam, A., Melnikov, S., Billmeier, R., Calvet, N., D'Alessio, P., Hillenbrand, L.A., Kuchner, M., Traub, W.A., Tuthill, P.G., Beichman, C., Boden, A., Booth, A., Colavita, M., Creech-Eakman, M., Gathright, J., Hrynevych, M., Koresko, C., Le Mignant, D., Ligon, R., Mennesson, B., Neyman, C., Sargent, A., Shao, M., Swain, M., Thompson, R., Unwin, S., van Belle, G., Vasisht, G., Wizinowich, P.: Keck interferometer observations of FU Orionis objects. Astrophys. J. **641**, 547–555 (2006). doi:10.1086/500313. arXiv:astro-ph/0512230

34. Millan-Gabet, R., Monnier, J.D., Berger, J.P., Traub, W.A., Schloerb, F.P., Pedretti, E., Benisty, M., Carleton, N.P., Haguenauer, P., Kern, P., Labeye, P., Lacasse, M.G., Malbet, F., Perraut, K., Pearlman, M., Thureau, N.: Bright localized near-infrared emission at 1–4 AU in the AB

Aurigae disk revealed by IOTA closure phases. Astrophys. J. Lett. **645**, L77–L80 (2006). doi:10.1086/506153. arXiv:astro-ph/0606059

35. Millan-Gabet, R., Malbet, F., Akeson, R., Leinert, C., Monnier, J., Waters, R.: The circumstellar environments of young stars at AU scales. In: Reipurth, B., Jewitt, D., Keil, K. (eds.) Protostars and Planets V, pp. 539–554 (2007)

36. Monnier, J.D., Millan-Gabet, R.: On the interferometric sizes of young stellar objects. Astrophys. J. **579**, 694–698 (2002). doi:10.1086/342917. arXiv:astro-ph/0207292

37. Muzerolle, J., D'Alessio, P., Calvet, N., Hartmann, L.: Magnetospheres and disk accretion in Herbig Ae/Be stars. Astrophys. J. **617**, 406–417 (2004). doi:10.1086/425260. arXiv:astro-ph/0409008

38. Natta, A., Prusti, T., Neri, R., Wooden, D., Grinin, V.P., Mannings, V.: A reconsideration of disk properties in Herbig Ae stars. Astron. Astrophys. **371**, 186–197 (2001). doi:10.1051/0004-6361:20010334

39. Quanz, S.P., Henning, T., Bouwman, J., Ratzka, T., Leinert, C.: FU Orionis: the MIDI VLTI perspective. Astrophys. J. **648**, 472–483 (2006). doi:10.1086/505857. arXiv:astro-ph/0605382

40. Ratzka, T., Leinert, C., Henning, T., Bouwman, J., Dullemond, C.P., Jaffe, W.: High spatial resolution mid-infrared observations of the low-mass young star TW Hydrae. Astron. Astrophys. **471**, 173–185 (2007). doi:10.1051/0004-6361:20077357. arXiv:0707.0193

41. Renard, S., Malbet, F., Benisty, M., Thiébaut, E., Berger, J.P.: Milli-arcsecond images of the Herbig Ae star HD 163296. Astron. Astrophys. **519**, A26 (2010). doi:10.1051/0004-6361/201014910. arXiv:1007.2930

42. Schegerer, A.A., Wolf, S., Ratzka, T., Leinert, C.: The T Tauri star RY Tauri as a case study of the inner regions of circumstellar dust disks. Astron. Astrophys. **478**, 779–793 (2008). doi:10.1051/0004-6361:20077049. arXiv:0712.0696

43. Shu, F., Najita, J., Ostriker, E., Wilkin, F., Ruden, S., Lizano, S.: Magnetocentrifugally driven flows from young stars and disks, I: a generalized model. Astrophys. J. **429**, 781–796 (1994). doi:10.1086/174363

44. Shu, F.H., Galli, D., Lizano, S., Cai, M.J.: Magnetization, accretion, and outflows in young stellar objects. In: Bouvier, J., Appenzeller, I. (eds.) IAU Symposium, vol. 243, pp. 249–264 (2007). doi:10.1017/S1743921307009611

45. Tannirkulam, A., Harries, T.J., Monnier, J.D.: The inner rim of YSO disks: effects of dust grain evolution. Astrophys. J. **661**, 374–384 (2007). doi:10.1086/513265. arXiv:astro-ph/0702044

46. Tatulli, E., Isella, A., Natta, A., Testi, L., Marconi, A., Malbet, F., Stee, P., Petrov, R.G., Millour, F., Chelli, A., Duvert, G., Antonelli, P., Beckmann, U., Bresson, Y., Dugué, M., Gennari, S., Glück, L., Kern, P., Lagarde, S., Le Coarer, E., Lisi, F., Perraut, K., Puget, P., Rantakyrö, F., Robbe-Dubois, S., Roussel, A., Weigelt, G., Zins, G., Accardo, M., Acke, B., Agabi, K., Altariba, E., Arezki, B., Aristidi, E., Baffa, C., Behrend, J., Blöcker, T., Bonhomme, S., Busoni, S., Cassaing, F., Clausse, J.M., Colin, J., Connot, C., Delboulbé, A., Domiciano de Souza, A., Driebe, T., Feautrier, P., Ferruzzi, D., Forveille, T., Fossat, E., Foy, R., Fraix-Burnet, D., Gallardo, A., Giani, E., Gil, C., Glentzlin, A., Heiden, M., Heininger, M., Hernandez Utrera, O., Hofmann, K.H., Kamm, D., Kiekebusch, M., Kraus, S., Le Contel, D., Le Contel, J.M., Lesourd, T., Lopez, B., Lopez, M., Magnard, Y., Mars, G., Martinot-Lagarde, G., Mathias, P., Mège, P., Monin, J.L., Mouillet, D., Mourard, D., Nussbaum, E., Ohnaka, K., Pacheco, J., Perrier, C., Rabbia, Y., Rebattu, S., Reynaud, F., Richichi, A., Robini, A., Sacchettini, M., Schertl, D., Schöller, M., Solscheid, W., Spang, A., Stefanini, P., Tallon, M., Tallon-Bosc, I., Tasso, D., Vakili, F., von der Lühe, O., Valtier, J.C., Vannier, M., Ventura, N.: Constraining the wind launching region in Herbig Ae stars: AMBER/VLTI spectroscopy of HD 104237. Astron. Astrophys. **464**, 55–58 (2007). doi:10.1051/0004-6361:20065719. arXiv:astro-ph/0606684

47. Tatulli, E., Malbet, F., Ménard, F., Gil, C., Testi, L., Natta, A., Kraus, S., Stee, P., Robbe-Dubois, S.: Spatially resolving the hot CO around the young Be star 51 Ophiuchi. Astron. Astrophys. **489**, 1151–1155 (2008). doi:10.1051/0004-6361:200809627. arXiv:0806.4937

48. Thi, W.F., van Dalen, B., Bik, A., Waters, L.B.F.M.: Evidence for a hot dust-free inner disk around 51 Oph. Astron. Astrophys. **430**, L61–L64 (2005). doi:10.1051/0004-6361:

200400132. arXiv:astro-ph/0412514
49. van Boekel, R., Min, M., Leinert, C., Waters, L.B.F.M., Richichi, A., Chesneau, O., Dominik, C., Jaffe, W., Dutrey, A., Graser, U., Henning, T., de Jong, J., Köhler, R., de Koter, A., Lopez, B., Malbet, F., Morel, S., Paresce, F., Perrin, G., Preibisch, T., Przygodda, F., Schöller, M., Wittkowski, M.: The building blocks of planets within the 'terrestrial' region of protoplanetary disks. Nature **432**, 479–482 (2004). doi:10.1038/nature03088
50. Vinković, D., Jurkić, T.: Relation between the luminosity of young stellar objects and their circumstellar environment. Astrophys. J. **658**, 462–479 (2007). doi:10.1086/511327. arXiv:astro-ph/0612039
51. Vinković, D., Ivezić, Ž., Miroshnichenko, A.S., Elitzur, M.: Discs and haloes in pre-main-sequence stars. Mon. Not. R. Astron. Soc. **346**, 1151–1161 (2003). doi:10.1111/j.1365-2966.2003.07159.x. arXiv:astro-ph/0309037
52. Weigelt, G., Grinin, V.P., Groh, J.H., Hofmann, K.H., Kraus, S., Miroshnichenko, A.S., Schertl, D., Tambovtseva, L.V., Benisty, M., Driebe, T., Lagarde, S., Malbet, F., Meilland, A., Petrov, R., Tatulli, E.: VLTI/AMBER spectro-interferometry of the Herbig Be star MWC 297 with spectral resolution 12 000. Astron. Astrophys. **527**, A103 (2011). doi:10.1051/0004-6361/201015676. arXiv:1101.3695
53. Zhu, Z., Hartmann, L., Calvet, N., Hernandez, J., Tannirkulam, A.K., D'Alessio, P.: Long-wavelength excesses of FU Orionis objects: flared outer disks or infalling envelopes? Astrophys. J. **684**, 1281–1290 (2008). doi:10.1086/590241. arXiv:0806.3715

Chapter 7
An Introduction to Accretion Disks

E. Alecian

Abstract In this chapter I first present the pre-main sequence stars around which accretion disks are observed and summarise the main observed characteristics of the classical T Tauri and Herbig Ae/Be stars. I then review the theoretical and empirical reasons that lead many scientists to conclude that the objects of this class are surrounded by accretion disk. Finally I review the basic characteristics of the disks of these stars, and expose the observations that lead us to derive them.

7.1 Introduction

7.1.1 The Star Formation

It is now well accepted among the scientific community that stars form inside initially stable molecular clouds. An external trigger breaks the stability of the cloud that collapses under gravity forces. The contraction of the matter gives birth to a class 0 object: a proto-star strongly accreting matter from a massive envelope (with a mass larger than the mass of the proto-star). The matter from the envelope is accreted onto the growing proto-star and also onto a disk to subsequently form a class I object constituted of a proto-star accreting matter mainly from a disk, the whole surrounded by an envelope much less massive than the stellar mass. Little by little the envelope is dissipating through jets, winds or accretion, and a star surrounded by an accretion disk forms a class II object. The accretion disk will also dissipate by accretion onto the star, coagulation onto proto-planets, photoevaporation, or other clearing mechanisms to finally give a class III object constituted by a star with a debris disk. At this stage the accretion stops.

During the proto-stellar (class 0 and I) phase, the central objects are highly obscured, and the systems radiate essentially from the submillimetric to the mid-infrared. This phase lasts about 10^5–10^6 years. At these stages, we believe that there

E. Alecian (✉)
LESIA-Observatoire de Paris, CNRS, UPMC Univ., Paris-Diderot Univ., 5 place Jules Janssen, 92195 Meudon Principal Cedex, France
e-mail: evelyne.alecian@obspm.fr

J.-P. Rozelot, C. Neiner (eds.), *The Environments of the Sun and the Stars*, Lecture Notes in Physics 857, DOI 10.1007/978-3-642-30648-8_7, © Springer-Verlag Berlin Heidelberg 2013

Fig. 7.1 PMS theoretical
evolutionary tracks computed
by Behrend and Maeder [13]
and plotted in an
Hertzsprung–Russell (HR)
diagram. The tracks start on
the birthline and end on the
Zero-Age Main-Sequence
(ZAMS). The transport of
energy inside the star
(radiative or convective or
both) is indicated with
different broken lines. The
zones surrounded with *blue*
(*brown*) *lines* represent the
region where the Herbig
Ae/Be (T Tauri) stars are
situated (adapted from [13])

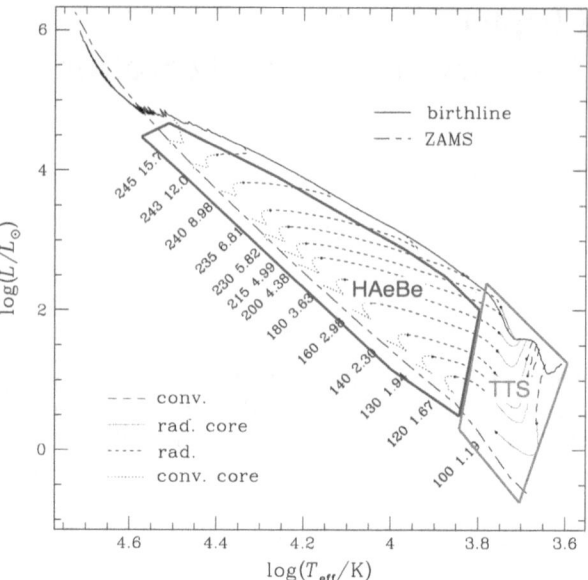

is no strong difference between low- and high-mass stars, as even at high-mass it is
very likely that stars are forming from a low-mass proto-star that will accrete much
more matter than future low-mass stars (e.g., [13]). Once the major accreting phase
is done and the central star radiates mainly in the visible bands, the star starts its
pre-main sequence phase (class II and III).

7.1.2 The Pre-Main Sequence Phase of Stellar Evolution

At this stage, it is very convenient to follow the evolution of the star inside the
Hertzsprung–Russell (HR) diagram, i.e. a plot of the stellar luminosity in function
of its effective temperature (Fig. 7.1). A star that comes just out from the proto-
stellar phase is situated on the birthline, then follows a pre-main sequence (pre-
MS) evolutionary tracks corresponding to its mass, and finally reaches the zero-age
main-sequence (ZAMS), starting the longest phase of the evolution of a star, the
main-sequence (MS). The time that a star spends on the pre-MS is highly dependent
of its mass and varies from ∼100 Myr (at 1 M⊙) to ∼0.15 Myr (at 15 M⊙).

During the pre-MS, a star radiates mainly energy from its gravitational contraction. The nuclear reaction that will transform hydrogen into helium in the core of
the star will only start at the end of the pre-MS phase. Above 20 M⊙, a star on
the birthline has already started to burn hydrogen in its core and therefore does not
undergo a pre-MS phase.

The transfer of energy inside a pre-MS star changes with time and is highly dependent of the mass of the star. At low-mass, up to ∼1.4 M⊙, the star first descends

the Hayashi track in the HR diagram with a fully convective interior. Then a radiative zone develops in the core and grows progressively until the star reaches the ZAMS. Above 1.2 $M_\odot$, the radiative core grows even bigger, to form a totally radiative star. Later, a convective core forms once the nuclear reaction are triggered. Above 1.4 $M_\odot$, the star does not go through the Hayashi phase, while above 4 $M_\odot$, the star starts its pre-MS already totally radiative.

The presence of a convective zone under the surface of the star plays an important role on its activity along the pre-MS (and also later on the MS) and hence on its interaction with its environment. A convective envelope can engender magnetic fields through dynamo processes. These magnetic fields can only exist at low mass and allow the exchange of matter of the star with its environment through magnetospheric accretion and winds. At higher mass, exchange occurs mainly through winds and only occasionally through accretion. This exchange of matter occurs in regions very close to the star (within few astronomical units (AU)).

Among the class II and III objects, we distinguish the T Tauri stars from the Herbig Ae/Be stars. The T Tauri stars have cool photospheres ($\sim$3000–6000 K) that present specific observed properties revealing their young age such as IR-excess, spectral lines in emission, irregular light-variation, and an association with a bright or dark nebulosity [52]. The Herbig Ae/Be stars have similar properties and are considered the analogues of the T Tauri stars at higher temperature (above 6000 K, see [46, 77]). The approximate situation of both classes of objects in the HR diagram is schematically represented in Fig. 7.1. One of the major differences between Herbig Ae/Be stars and T Tauri stars is the magnetic activity. The T Tauri stars have masses below 1.5 Msun, and almost all of them have a convective envelope and therefore can create a magnetic field dynamo. All Herbig Ae/Be stars are mainly radiative and cannot create such magnetic fields. This results in that the activity close to the stellar surfaces is different in both kind of stars. Note however that fossil magnetic fields have been discovered in a small portion of Herbig Ae/Be stars (e.g. [4, 5]). The role of these fields in the stellar activity is not clear at the moment.

In the following we will concentrate on the environment situated at about 1 AU and above, where the circumstellar matter is concentrated into disks that surround the class II objects and the transition objects from class II to class III.

7.1.3 The Class II Objects

In this section I will describe the main properties of the T Tauri and Herbig Ae/Be stars in order to give the reader a global view of the behaviour of these objects that helped the scientists to propose models for the environment of these young stars.

Among the T Tauri stars, we distinguish two sub-classes: the classical T Tauri (CTTS|see classical T Tauri stars) stars, showing strong Hα emission, and the weak-line T Tauri stars (WTTS) with only faint spectral emission. Typical optical spectra of CTTS and WTTS are shown in Fig. 7.2. The classical T Tauri stars show very strong emission in the Balmer lines Hα and Hβ and also in may metallic lines, contrary to the WTTS. The Balmer jump on the blue part of the spectrum is in emission

Fig. 7.2 Optical spectrum of CTTS and WTTS. Note the differences between CTTS and WTTS spectra in the Balmer lines Hα and Hβ, in the Balmer jump and in the continuum that is veiled for the CTTS (adapted from [14])

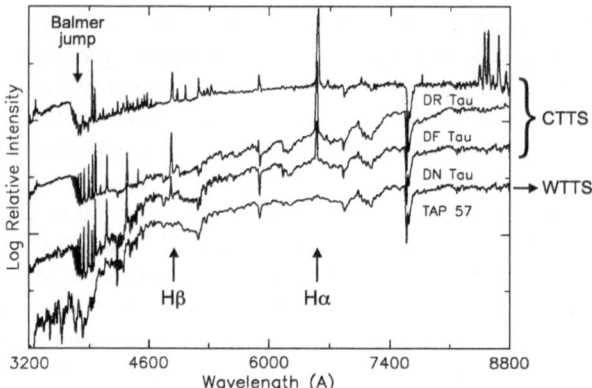

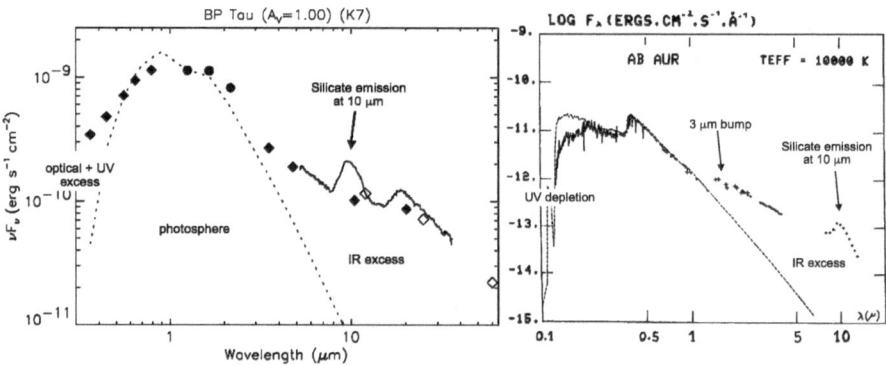

Fig. 7.3 Typical SEDs of CTT (*left*) and HAeBe (*right*) stars. The *crosses*, *diamonds* and *full lines* represent the observations. The theoretical spectrum of the photosphere of the star computed with the stellar temperature are overplotted with *dashed lines*. The observations illustrate the typical IR excess and deviations from a purely photospheric spectrum in the UV. Typical features are also indicated with arrows (adapted from [38], and [21])

in CTTS and in absorption in WTTS (more similar to normal MS stars). The continuum level of some of the CTTS is higher than in normal stars. This means that a continuum component is superimposed to the stellar continuum. This additional component is called veiling and is not observed among WTTS. Finally, the Ca-II IR triplet on the red part of the spectrum is sometimes seen in emission in CTTS (e.g. [14]).

The spectrum of the Herbig Ae/Be stars is very similar to the CTTS ones with strong emission in the Balmer and metallic lines and occasional emission in the Ca II IR triplet. However, no veiling has been detected so far in HAeBe spectra [37, 40], and the Balmer jump is not seen in emission, but generally a smaller discontinuity is present than in normal stars of same spectral types (e.g. [39]).

In Fig. 7.3 the typical spectral energy distributions (SEDs) of CTT and HAeBe stars are shown. This plot represents the light flux in function of wavelength. The squares, diamonds and crosses represent the observations obtained with photometric or low-spectral-resolution instrumentations. The dashed lines represent the predictions from models of stellar atmospheres calculated with suitable temperature in each example. Compared to a normal MS star, both CTT and HAeBe stars have strong IR-excess, and many times a silicate emission at 10 μm is observed. A "3 μm IR bump" is observed in the near-IR spectrum of many HAeBe stars, but never in TT stars (e.g. [32, 45]). On the blue side of the SED, an optical and UV excess is observed in CTT stars, while a UV depletion is sometimes observed in HAeBe stars.

The classical T Tauri stars can also be strongly photometric variable with amplitude of variability very different from one star to another, ranging from 0.1 to 2 mag. The variability is found irregular in most of the CTTS, but quasi-periodicity has been observed in the AATAU-type stars, called after their prototype AA Tau. In this sub-class of objects, the maximum flux is roughly at the same level, while the minimum flux is strongly varying from one cycle to another. The periods vary from 1 to 20 days (e.g. [16]).

The Herbig Ae/Be stars are usually not photometrically variable [47]. However a small portion is found to show strong drop of luminosity up to 3 mag. These UXOR stars have been named after their prototype UX Ori and concern only HAe stars. This drop in luminosity is accompanied with a blueing effect of the star that is not yet understood (e.g. [79, 82]).

The emission spectra of CTT and HAeBe stars are strongly variable. Hα can show sometimes double-picked emission, often associated with circumstellar disks, and sometimes P Cygni profiles, revealing winds in the close stellar environment (e.g. [6, 23]). Among the Herbig Ae objects, the spectral lines of the UXOR-type stars are superimposed with transient absorption components (TACs) highly variable that could be formed by circumstellar clouds (e.g. [44]). These variabilities are not found periodic, except in few Herbig Ae stars that show periodic features inside the Hα line profile (e.g. [22]).

Finally X-ray emission are very often detected from CTTS. These emission are highly variable with flares, and reveal hot plasma in the close environment (e.g. [50]). In the low-mass classical T Tauri stars the X-ray emission are believed to come from the magnetic activity present at the surface of the star. X-ray emission have only been firmly established coming from HAeBe stars in only a small number of them (e.g. [76]). While no correlation has been observed between the presence of X-rays and magnetic fields, their origin is still highly debated.

The characteristics of the CTTS and HAeBe stars summarised in this section reveal a complex structure and geometry surrounding the young stars that affect the full electromagnetic spectrum, from X-rays to IR, and even up to radio as we will see below. The following of this chapter will only concentrate on material from the disk which is cooler than material closer to the star and will therefore focus mainly on the stellar properties observed in the IR and at larger wavelength.

The aim of the following sections is to introduce the reader to what are called *accretion* disks, and what are our basic knowledge on the structure, composition and evolution of the disks. The paper does not aim to produce an exhaustive review on accretion disk, and I would redirect the reader to the following reviews for more

Fig. 7.4 Schematic
representation of angular
momentum and mass
transport inside an accretion
disk as described by
Lynden-Bell and Pringle [59]

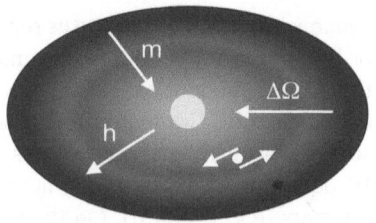

details on accretion/protoplanetary disks and on the recent work that has been done
is this field: Williams and Cieza [84] for a global review on protoplanetary disks
and their evolution, Dullemond and Monnier [33] for the inner region of the disks,
Armitage [9] for the dynamics of disks, Zinnecker and Yorke [86] for massive star
disks, van Dishoeck [80] for IR spectroscopy of gas and dust in the disks, Balbus
[12] for the angular momentum transport in accretion disks, and Wyatt [85] for the
final stages of protoplanetary disks.

7.2 Accretion Disk

7.2.1 Definition and Evidence

When the molecular cloud contracts to form a star because of the presence of angular
momentum, the contraction will not be entirely spherical. While the central protostar
grows roughly spherically by accreting matter from the envelope, the outer part of
the cloud tends to be mainly distributed into a disk due to the action of the gravity
and centrifugal forces. In presence of rotation the orbit of least energy is circular.
The matter located in the disk is rotating with a Keplerian motion around the central
star. The Keplerian angular velocity of a particle orbiting in the disk is given by

$$\Omega_K = \left(\frac{GM_\star}{r^3}\right)^{1/2},\tag{7.1}$$

where G is the universal gravitational constant, $M_\star$ is the mass of the star, and r
is the distance between the centre of the star and the particle (see a schematic representation of the problem in Fig. 7.4). As a result, the angular velocity decreases
outwards, while the specific angular momentum that scales as $h = r^2\Omega$ increases
outwards (the specific angular momentum is the angular momentum per unit of
mass). If the disk is viscous, the inner parts that are rotating faster than the outer
parts shear past the outer parts. A transport of angular momentum is established
from the inner part to the outer part. According to Lynden-Bell and Pringle [59],
a simple energy balance shows that a viscous disk can lower its energy by transporting mass towards lower radius and angular momentum towards larger radius. The
Keplerian disks surrounding young stars are therefore very likely accreting mass
from the outer parts to the inner parts of the disk, while angular momentum is evac-
uated towards the outer part of the disk.

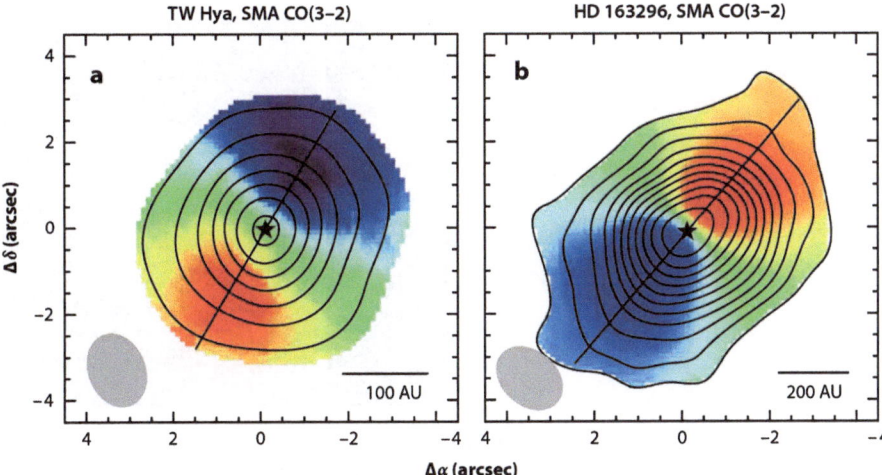

Fig. 7.5 CO(3–2) emission from the disk of the T Tauri star TW Hya (**a**) and the Herbig Ae star HD 163296 (**b**). The colors represent the velocity field of the gas with respect to us and show clearly matter coming towards us (*in blue*) and moving away from us (*red*) consistent with a Keplerian rotation of material in a disk that would be inclined with respect to the line of sight (from [84])

Evidence of material in Keplerian rotation within elongated structure has been obtained by many authors by measuring the radio emission from the molecular gas in the environments of young stars (Fig. 7.5, see also [61]). Evidence of hot disks has also been obtained using various observing techniques such as coronography (e.g. [10, 41]), adaptive optic (e.g. [69]), interferometry [35], or IR imagery (e.g. [43, 68]). From these observations it seems now evident that the material situated from small AU to large AU is distributed into a disk.

7.2.2 The Disk Structure and Composition

7.2.2.1 Passive Versus Accretion Disks

One of the first models of accretion disk has been proposed by Lynden-Bell and Pringle [59] to understand the evolution of viscous disks surrounding stars. They propose that in such Keplarian viscous disks the dissipation lead to shining. The light emitted from the disk can be divided in two components: from the disc itself radiating energy at a temperature that varies with the distance from the central star and a boundary layer that is heated by the interaction of the inner part of the disk and the surface of the star. This "boundary-layer model" was predicting the presence of IR excess and UV excess, as observed in many T Tauri stars.

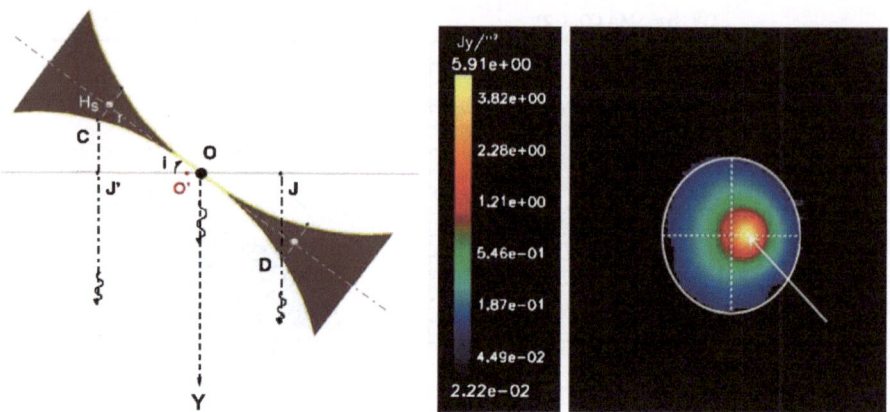

Fig. 7.6 *Left*: Schematic of the flared disk of HD 97048. The disk is seen from below with an inclination *i* between the line of sight and the disk axis. C and D are two points situated at the surface of the disk and at the same distance from the star (in *O*). Due to geometrical effect, we expect the centre of emission from C and D to be in *O'* that is offset from the geometrical center *O*. *Right*: IR observation of the disk around HD 97048. The photo-centre of the IR emission (indicated by the *arrow*) does not coincide with the geometrical center of the image (surrounded by a *white ellipse*) as was predicting the sketch of the left panel (from [57])

Many authors used this initial idea to reproduce the SED of T Tauri stars and in particular the IR slope of the SED. Adams et al. [2] defined the spectral indices as

$$n = -\frac{d \log \lambda F_\lambda}{d \log \lambda}, \tag{7.2}$$

where λ is the wavelength, and F_λ is the flux from the object. n is evaluated in the near- to mid-IR portion of the spectrum (i.e. $\lambda \sim 1$–10 μm). The classical T Tauri stars present typical spectral indices from 4/3 to nearly 0 (flat IR spectrum). A classical Keplerian accretion disk as proposed by Lynden-Bell and Pringle [59] predicts a spectral index of 4/3. On the other hand, Adams et al. [2] showed that a "passive" disk (i.e. without accretion and therefore viscous dissipation) with the reprocessing of about 25 % of the stellar light being the only source of energy, can also produce a flux with a spectral index of 4/3.

7.2.2.2 The Flaring Disk

The spectral indices of T Tauri stars are mostly found to be higher than 4/3, implying that an additional source to the viscous accretion or to the light reprocessing has to be present. Kenyon and Hartmann [53] show that a flat IR spectrum can be reproduced with a passive reprocessing disk if it flares slightly as radial distance increases (Fig. 7.6, left). They argue that if the disk is optically thick, reprocessing must exist and therefore a significant fraction of the IR emission has to come from reprocessing. A disk in hydrostatic equilibrium tends to flare: in the direction perpendicular to the mid-plane of the disk, the component of the gravitational force of the star is balanced by the gas pressure gradient. So in the disk, at a radius r from the centre of the star and at a distance z above the mid-plane, the pressure of the gas (p) is given by

$$\frac{dp}{p} = \frac{\Omega_K^2}{v_s^2} z\, dz, \tag{7.3}$$

where v_s is the speed of sound, and Ω_K is the Keplerian angular velocity as defined before. From this equation we can derive the scale height of the disk (e.g. Kenyon and Hartman 1987):

$$\frac{h}{r} = \left(\frac{v_s^2 r}{G M_\star}\right)^{1/2}. \tag{7.4}$$

If the internal temperature of the disk $T_{int}(r) \propto v_s^2$ falls off more slowly than r^{-1}, then the relative thickness of the disk will increase outwards: the surface of the disk is therefore concave. The predicted internal disk temperature of passive reprocessing or active accretion flat disks is $T_{int}(r) \propto r^{-3/4}$ [2, 59]. If the disk is vertically isothermal, the surface temperature (T_s) should have the same temperature as the interior, and the scale height would be proportional to $r^{1/8}$, which makes the flared disk a plausible hypothesis.

A protostellar disk is composed of gas and dust with a ratio of about 100:1 in mass (assumed similar to the interstellar medium). While the flaring structure is supported by the gas, the dust is absorbing and reemitting the stellar light. In the case of a vertically isothermal disk with dust well mixed with the gas, Kenyon and Hartmann find that the effective height $H_s(r)$ of the disk is about $3h$. More recent models give $H_s(r) \sim 4h$ (e.g. [27]). $H_s(r)$ can therefore be written as the power law $H_s(r) = H_0(r/R_\star)^z$, where $z = (3 - \gamma)/2$, and where γ is defined as $T_s(r) \propto r^{-\gamma}$. If optically thick, a flared disk will intercept a larger fraction of the stellar radiation than a completely flat disk. Kenyon and Hartmann find that for a temperature distribution $T_s(r) \propto r^{-1/2}$, about 50 % of the stellar light is reprocessed and can reproduce flat IR spectrum as observed in the most extreme T Tauri stars. However, Adams et al. [3] argue that because of vertical dust settling within the disk, Kenyon and Hartmann overestimate the reprocessing efficiency, and in order to reproduce their large-IR excess and reach a temperature distribution of $T_s(r) \propto r^{-1/2}$, an additional non-viscous activity within the disk must be present.

7.2.2.3 The Two-Layer Model

Calvet et al. [19, 20] performed more detailed calculation of a steady, viscous, geometrically thin, optically thick flared accretion disk. In their models they assume that the internal energy of the disk has both sources, the irradiation from the star and accretion. The outer layers of the disk form an atmosphere, and the disk structure is therefore the combination of an optically thick viscous disk with an optically thin non-viscous atmosphere in radiative equilibrium. The atmosphere is irradiated by the star, which absorbs a fraction of the stellar flux and scatter the other fraction.

Its temperature is determined by heating from the viscous interior and the stellar radiation absorption as it travels down the atmosphere. They find that the irradiated disk dominates the flux at wavelength larger than 1 μm whatever the stellar effective temperature or the mass accretion rate inside the disk.

They performed radiative transfer inside the atmosphere by including specific electronic, vibrational and rotational transitions of molecular and atomic elements (H, H_2, C, Si, Mg, CO, TiO, OH, H_2O and silicate) that are assumed to be the main source of opacity in the near- and mid-IR. They find that the spectral signatures of CO and TiO, and the silicate feature at 10 μm is strongly dependent on the mass accretion rate and the stellar effective temperature, as the former affects the viscous energy injected into the atmosphere and the latter affects the amount of radiation heating the atmosphere. These features are formed at low optical depth in the upper layer of the atmosphere mostly affected by the stellar radiation: the lower the accretion rate and the higher the stellar temperature, the more these features appear in emission. For high accretion rate and low stellar temperature, they would appear in absorption. Therefore Calvet et al. propose that the observation of these features would give an indication on the accretion rate inside the disk.

7.2.2.4 The Disk Inner Gap

Chiang and Goldreich [25, 26] investigate the case of a passive disk in hydrostatic and radiative equilibrium. Their calculations are similar to those performed by Calvet et al., except that the optically thick disk interior is heated by the radiation of the stellar light reemitted by the dust grains in the surface layer. The dust absorbing the incident stellar ration radiates equal amounts of IR radiation into the inward (the disk interior) and outward hemispheres. Their models aim to reproduce the nearly flat mid- to far-IR spectrum of T Tauri stars. Their models allowed the line-of-sight inclination angle to vary, and they argue that due to the flaring if the inclination is high (the disk is seen near equator-on), the outer part of the disk could occult the inner regions and reduce significantly the flux shortward 30 μm, and could explain the lack of IR emission around 12 μm observed in many T Tauri stars. From the fitting of the SED of the stars their model derives the inclination angle, the surface density of the disk, the emissivity of the grains, their typical size, and the inner and outer radii of the disk. Figure 7.7 shows the result of the fit of the SED of the T Tauri star GM Aur. They find that even with a inclined disk, the disk must be truncated between the star and the inner rim of the disk up to 4.8 AU, which is about 60 times larger than the dust sublimation radius (radius below which the temperature is too high for the dust to persist). They therefore cast some doubt about the reality of such a large gap in the centre of the disk. Inner holes are also required in disks around Herbig Ae/Be stars but with a much smaller size. Hillenbrand et al. [51] have fitted the SED of group I Herbig Ae/Be stars (i.e. stars with spectral indices of 4/3) with an optically thick, geometrically flat accretion disk. They can well reproduce the spectral shape of these stars under the condition that the inner part of the radius (shortward 2.2 μm, i.e. few stellar radii) is optically thin. Lada and

Fig. 7.7 Observed SED
(*filled circles*) of the T Tauri
star GM Aur superimposed
with the inclined passive disk
model (*full line*). The best
disk model has been found
with an inner edge at 4.8 AU,
an outer edge at 311 AU, and
an inclination angle of 59°
(from [26]). Note that the
x-axis is in frequency (and
not in wavelength as
generally found in the
literature)

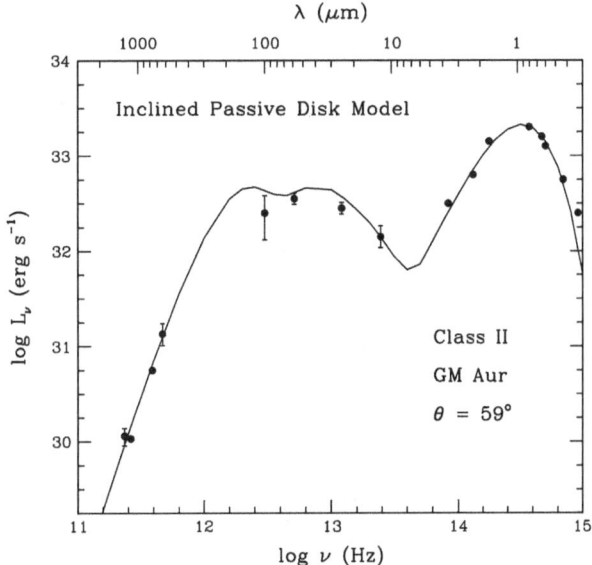

Adams [56] reach the same conclusion by studying the distribution of HAeBe stars in the *JHK* infrared color–color diagram. In particular, they find that to reproduce the observations, the inner edges of the disk should have temperature between 2000 and 3000 K, close to the dust sublimation temperature.

7.2.2.5 Evidence of Grain Growth

D'Alessio et al. [29, 30] have developed models of irradiated accretion disks. The disk is assumed to be geometrically thin and in steady state with a constant mass accretion rate and in vertical hydrostatic equilibrium. The disk is heated by viscous dissipation, radioactive decay, cosmic rays and stellar radiation. They find that the irradiation from the central star is the main source of heating of the disk except in the innermost regions, within $\sim$2 AU from the central star, where the stellar light has a grazing incidence, and hence the viscous dissipation is the dominant source of heating. They used their model to fit the SEDs of a large sample of T Tauri stars. Their model can well reproduce the observed near-IR fluxes, but they predict fluxes that are too high at far-IR and too low at millimeter wavelength (Fig. 7.8, upper left), indicating that their model is too geometrically thick at large radii. They propose that dust settling and coagulation could have already occurred, which would reduce the geometrical thickness of the disk. D'Alessio et al. [28] allowed in their models dust grain growth with power-law size distributions, such as the size distribution of dust grains is given by $n(a) = n_0 a^{-p}$, where a is the size of the grains, n_0 is the normalisation constant, and p is a free parameter. They find that compared to models with a dust disk composition similar to the interstellar medium (assumed similar to the dust composition of molecular clouds), their models reduce the vertical dust

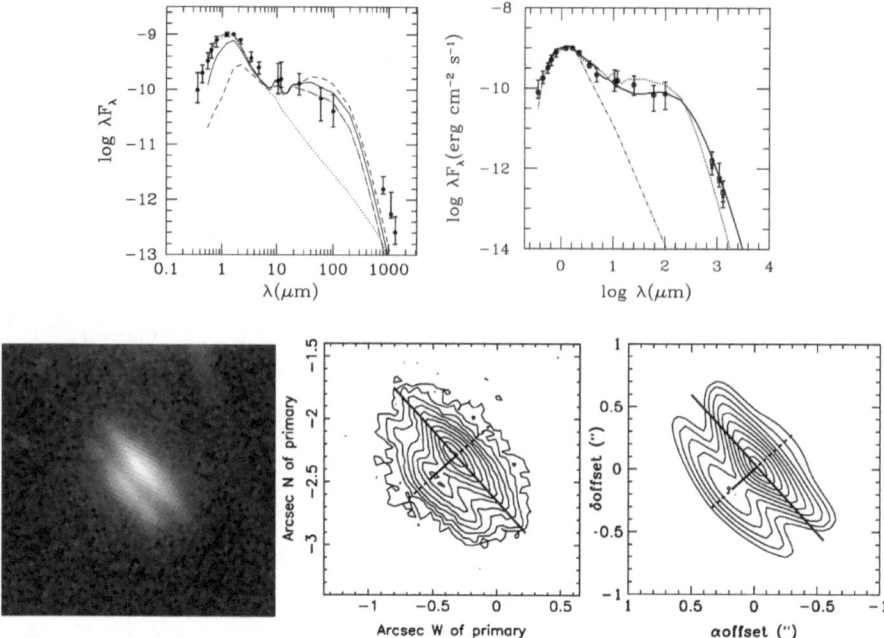

Fig. 7.8 This group of figures illustrates the necessity to take into account a non-uniform distribution of grain sizes in models of disk in order to better reproduce the observations. *Upper left*: The median observed SED (*filled circle*) of the PMS stars of the Taurus-Aurigae molecular cloud superimposed with the models of D'Alessio et al. [29] of three different disk radii assuming grains distribution similar to the interstellar medium (ISM). Whatever the disk size, they cannot reproduce at the same time the far-IR and mm observations (from [29]). *Upper right*: The *circles* are the same data point as in the left panel. The *dot-dashed line* is the photospheric spectrum. The *dotted-line* is the ISM-dust model, while the *solid line* is a dust model that allowed the dust distribution to vary as a power law with $p = 3.5$ (see text) and a maximum grain size of 1 mm. This model is capable of a better agreement between the far-IR and mm observations of T Tauri stars (from [28]). *Bottom left and middle*: *Hubble Space Telescope* (*HST*) visible image of the companion to HK Tau, designed as HK Tau/c, and its surface brightness map (from [73]). *Bottom right*: Surface brightness map of a disk model [28]. Note the strong similarities between the observations (*bottom middle*) and the model (*bottom right*)

height, and hence the irradiation heating is reduced, while the emission of the disk at mm and submillimeter wavelengths is enhanced. This model agrees better with the SEDs and IR imaging of T Tauris stars (Fig. 7.8, upper right and bottom).

7.2.2.6 Dust Composition

Chiang et al. [27] improved their model of passive disk by accounting a range of particle sizes as a function of grain nature and therefore as a function of local temperature. Figure 7.9 (upper right) is a schematic representation of the disk showing the zones of grain composition. In their model, Chiang et al. have limited the com-

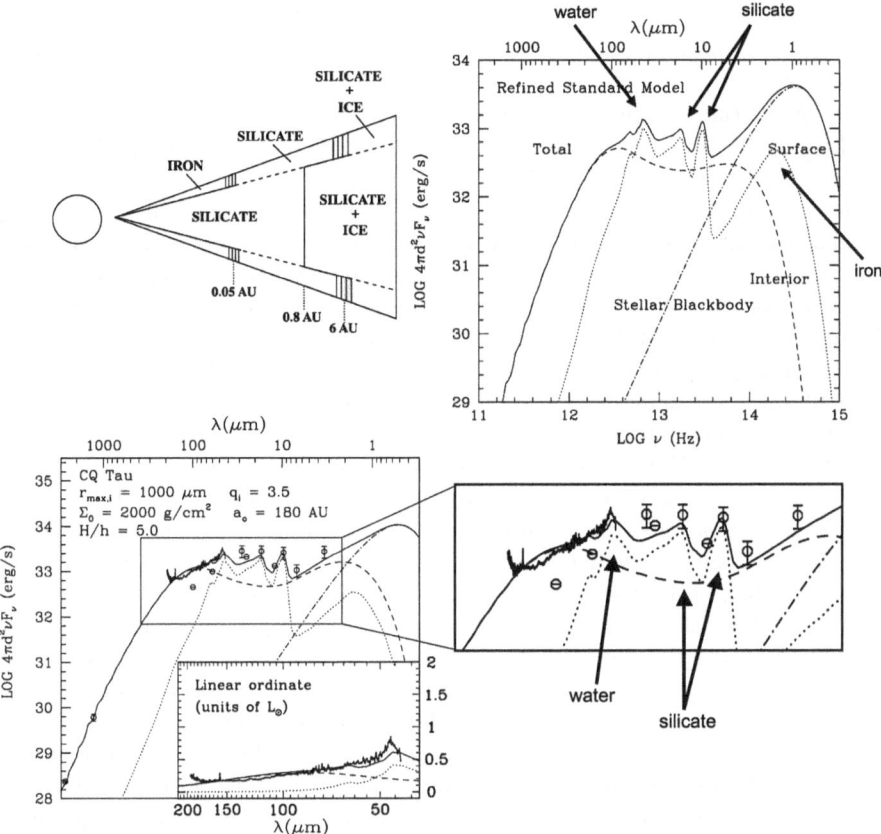

Fig. 7.9 *Upper left*: Schematic of zones of grain composition in the interior and in the atmosphere of the disk as computed in the models of Chiang et al. [27]. *Upper right*: Modelled SED of a passive disk taking into account a range of particle sizes (*solid line*). The water and silicate features are easily identifiable. However, the iron feature is lost in the stellar photospheric spectrum. The different contribution to the total SED from the stellar photosphere, the disk interior and the disk atmosphere are plotted with broken lines. (Note that the flux is plotted as a function of frequency and not wavelength.) *Bottom*: In the *left panel* there is plotted the SED of the Herbig Ae star CQ Tau superimposed with the best-fit model of Chiang et al. [27]. A close-up view is shown in the *right panel*. The water and silicate features are detected in this star (adapted from [27])

putation to only three types of grains: metallic iron, amorphous olivine (silicate) and water ice. They have tested that these grains are dominating the disk physical processes at temperature going from about 100 to 2000 K. In their models, where the local dust temperatures (that is assumed similar to the gas temperature) are lower than $\sim$150 K (that is the water ice sublimation temperature, and corresponds to a distance from a typical T Tauri star of about 6 AU in the atmosphere of the disk and 0.8 AU in the interior of the disk), the grains are mainly silicate covered with water ice. At temperature between $\sim$150 K and $\sim$1500 K (the silicate sublimation temperature, corresponding to $\sim$0.05 AU in the atmosphere and $\sim$2 R_* in the interior), they assume that the grains are mainly silicate only. Finally, between 1500 K and 2000 K (the iron sublimation temperature, corresponding to $\sim 2\ R_*$), the grains are mainly iron.

In Fig. 7.9 (upper right) we see the detailed relative contribution from the stellar blackbody, the disk interior and the disk surface emissions to the total SED of T Tauri stars as predicted by the models of Chiang et al. They find that two emission features from the silicate could be detected at 10 and 18 μm, and water ice features could also be observed at 45 and 62 μm. A dearth of emission between 2 and 8 μm from the silicate is predicted and explained by the fact that the silicate grains have a typical size lower than 1 μm and are therefore transparent at these wavelengths. Chiang et al. [27] used their models to fit the IR spectra of two T Tauri and three Herbig Ae stars from the Long-Wave-Spectrometer (LWS, 43–195 μm) onboard the Infrared Space Observatory (ESO). They were able to identify the water feature at 45 μm and both silicate features as predicted by their models, in both T Tauri stars and in the coldest Herbig Ae star considered in their study. No evidence of water could be found in the ISO spectra of the two hottest Herbig Ae stars. Later, using data from the Spitzer Infrared Spectrometer onboard the space telescope Spitzer, Pontoppidan et al. [71] were not able to find evidence of water in the spectra of the 25 A or B stars of their sample, while more than half of the late-type (M-K) stars have detectable water emission. This suggests that the temperature of the central star has a strong impact on the excitation and/or chemistry at the surface of the disks. On the other side, evidence of polycyclic aromatic hydrocarbons (PAHs) has been observed in the IR spectrum of many Herbig Ae/Be stars, but not in T Tauri stars. The PAHs bands are excited by the UV radiation from the star and explain why they are found stronger in the most massive stars and are absent in T Tauri stars. Besides amorphous silicate, crystalline silicates are also found in Herbig Ae/Be stars. However, no correlation with age has been found, and it is not clear what is the origin of these crystalline silicates and what are their evolution with the age.

7.2.2.7 The Puffed-Up Inner Rim

While the models detailed in the previous sections have been mainly developed for low-mass T Tauri stars, they have been relatively successfully applied to Herbig Ae/Be stars, and it is now evident that the IR excess observed in these objects is emitted by their dusty disks. However, one of the characteristics of the SED, specific to Herbig Ae/Be stars, is still difficult to interpret with the simple models described above: the near-IR or "3 μm bump", characterised by an inflection of the SED at wavelength shortwards ~ 2.2 μm.

Hillenbrand et al. [51] were able to fit the SED of HAeBe stars using a non-irradiated geometrically-flat, optically thick disks with a very large accretion rate, and an optically thin inner hole. However Hartmann et al. [45] argue that an optically thin inner disk is inconsistent with a large accretion rate. Instead, they propose that the NIR excess emission originates from a dusty circumstellar envelope containing very small grains transiently heated by the UV flux of the star. However, this

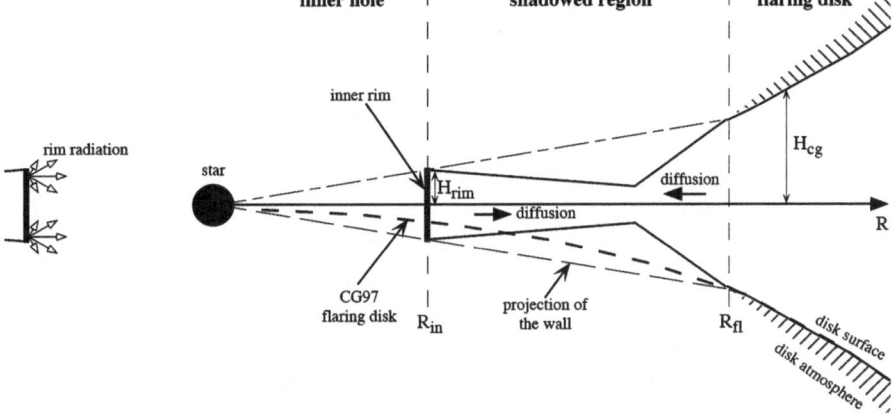

Fig. 7.10 Schematic representing the formation of the puffed-up inner rim, the shadowed region right behind the rim and the remaining flaring disk not shadowed by the rim (from [32])

hypothesis has been ruled out for theoretical [66] and observational reasons [62]. Natta et al. [67] proposed instead a model of passive flared disk with an optically thin inner hole. As the dust sublimation temperature is around 2000 K, corresponding to few stellar radii, the inner part of the disk is naturally depleted of dust. The inner rim of the disk is therefore directly irradiated by the star and is therefore hotter than what would be a full flared disk at the same radii (close to the star, below ∼1 AU, the stellar light irradiates the disk with a grazing angle). As the rim is hotter, its vertical scale increases, and it puffs up. This hot inner wall could explain the near-IR bump observed in many Herbig Ae/Be stars. Dullemond et al. [32] go further and try to understand the impact of the presence of this inner wall to the rest of the disk. In particular, they investigate the impact of the shadow of the rim to the rest of the flaring disk. The shadowed region can cover a large part of the disk, and in some cases the whole disk could reside inside the shadow. Therefore, in some cases the 10 μm silicate feature could disappear, which would explain its absence in some of the HAeBe SEDs [62]. In the other cases, the part of the disk that is directly illuminated by the star continues to flare and to be mainly heated by the stellar radiation. In Fig. 7.10 the models proposed by Dullemond et al. [32] are schematically represented. In a self-shadowed disk, the disk is not irradiated by the central star anymore, and therefore the evidence of PAHs in the stellar spectra should be strongly reduced. The ISO spectra analysis of 46 HAeBe stars performed by Acke and van den Ancker [1] confirm these expectations.

Recent observational evidence of inner hole have been obtained recently [42, 63] in Herbig Ae/Be stars. Furthermore, the recent determination of mass accretion rate in Herbig Ae/Be stars revealed values that are sufficiently low to ensure an optically thin inner gaseous disk at radii shorter than the dust sublimation radius. These results are therefore in agreement with the Dullemond et al. model.

7.2.2.8 The Inner Disk Warp of Classical T Tauri Stars

As presented in Sect. 7.1.3, a sub-class of classical T Tauri stars present cyclic photometric variability, the AATAU-type stars. Bouvier et al. have deeply studied the spectroscopic, photometric and polarimetric characteristics of their prototype and proposed a model that can explain all of them. Bouvier et al. [17] have found that the photometric light-curve of AA Tau displays a constant brightness interrupted by quasi-cyclical drops. The drops can change in shape and duration from one cycle to another but regularly appear every 8.2 days. They also observe that the veiling increases and the polarisation is larger when the star faints. The low-mass CTT stars can develop magnetic fields at their surface and experience magnetospheric accretion. The large-scale magnetic field of AA Tau is similar to an inclined dipole that disrupts the inner part of the disk to form a magnetosphere. The matter from the inner part of the disk is channeled along the field lines to reach the surface of the star close to the magnetic pole, where accretion shocks are formed, which are causing the veiling largely observed among T Tauri stars. Bouvier et al. propose that the interaction of the magnetic fields with the inner part of the disk causes the part of the disk that is closer to the magnetic pole (i.e. where the magnetic field is more intense) to wrap. A schematic representation of their model is presented in Fig. 7.11. As the star is seen nearly edge-on, the wrap crosses the line of sight and fades the star regularly as the star rotates. The polarisation is therefore increasing as a larger quantity of material is scattering the stellar light during fading events, and the veiling also increases as the stellar spots face us during the fading events.

The system AA Tau experiences some quiescent episode during which the fading disappears. Bouvier et al. [18] argue that it is the result of the dynamic of the magnetic field of AA Tau that can be temporarily disrupted. The magnetosphere of the star extends from the stellar surface to the inner edge of the disk that is rotating with a Keplerian velocity different from the stellar rotation velocity. The differential rotation from the foot-points of the magnetic field lines that are anchored in the stellar surface on one side and in the inner edge of the disk on the other side could shear the field lines until it reaches a critical point where disruption occurs. After few cycles, the large-scale magnetic field is reconstructed with or without the same configuration as before the disruption. The difference observed from one drop to another is the result of a combination of the dynamic interaction of the magnetic field with the disk, of the dynamical movement inside the disk, and of the dynamical distortion that the magnetic fields experience inside the magnetosphere.

AA Tau seems to not be a peculiar object. Recent observations of the satellite CoRoT that aims at observing the same field of the sky for long time, and compiling light-curve of many objects obtained on many successive months without interruption, have revealed many AATAU-type light-curves in the very young cluster NGC 2264 [7]. This is a very nice confirmation that high dynamical processes are acting in very young star-disk systems and play an important role on the shaping of the magnetosphere and the inner part of the disk.

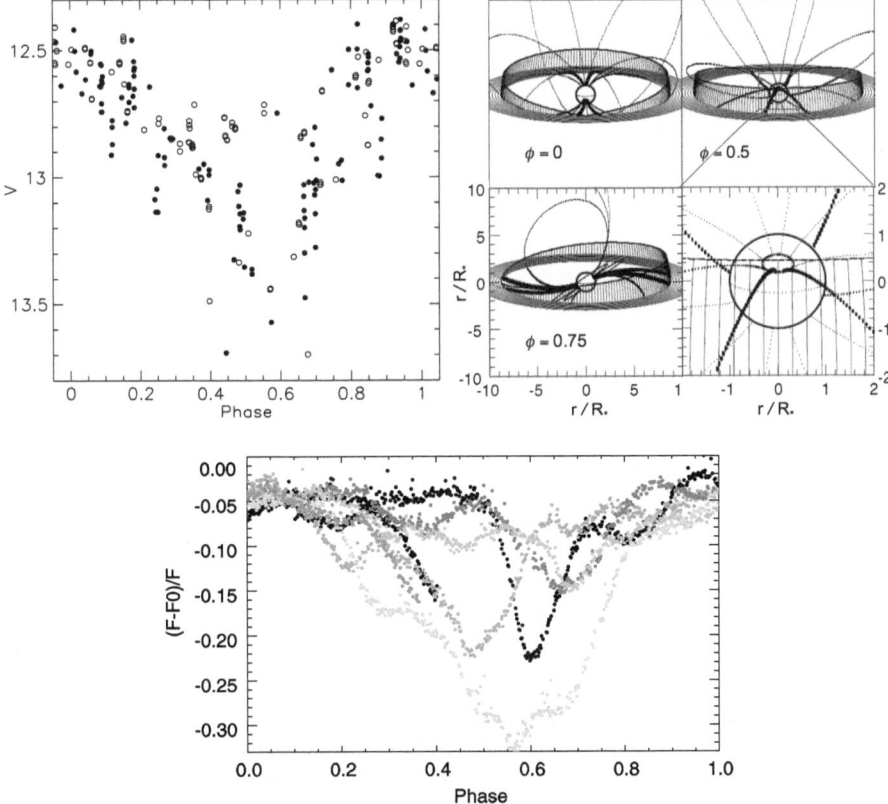

Fig. 7.11 *Upper left*: AA Tau light-curve folded in phase. The different symbols represent different cycles (from [17]). *Upper right*: Illustration of the model of the inner disk wrap proposed to explain the light variations of AATAU-type stars. *Three panels* represent the star-disk system at three different rotation phases (0., 0.5, 0.75). The *last panel* is a close-up view of the panel at phase = 0.5 (from [17]). *Bottom*: CoRoT light-curve folded in phase for one of the AATAU-type NGC 2264 star. The different colors represent different cycles (from [7])

7.2.2.9 Clumpiness of the Near-Stellar HAe Dust

Among the Herbig Ae stars, a sub-class shows very strong photometric variability. These stars are called the UXOR objects. These objects show in their spectra transient absorption components (TACs) that consist in absorption lines superimposed with the photospheric lines that appear and disappear within few days (see Fig. 7.12, left, [64, 65]). These TACs are observed in Balmer lines and metallic lines as well. These TACs reveal the presence of clumps of dust and gas that cross the line of sight. It is believed that these clumps are formed from hydrodynamical fluctuations inside the disk. If the disk is viewed close to edge-one, then the manifestation is therefore observable through photometry and spectroscopy. At lower inclination, no evidence of these clumps can be detected, which explains why the UXOR phenom-

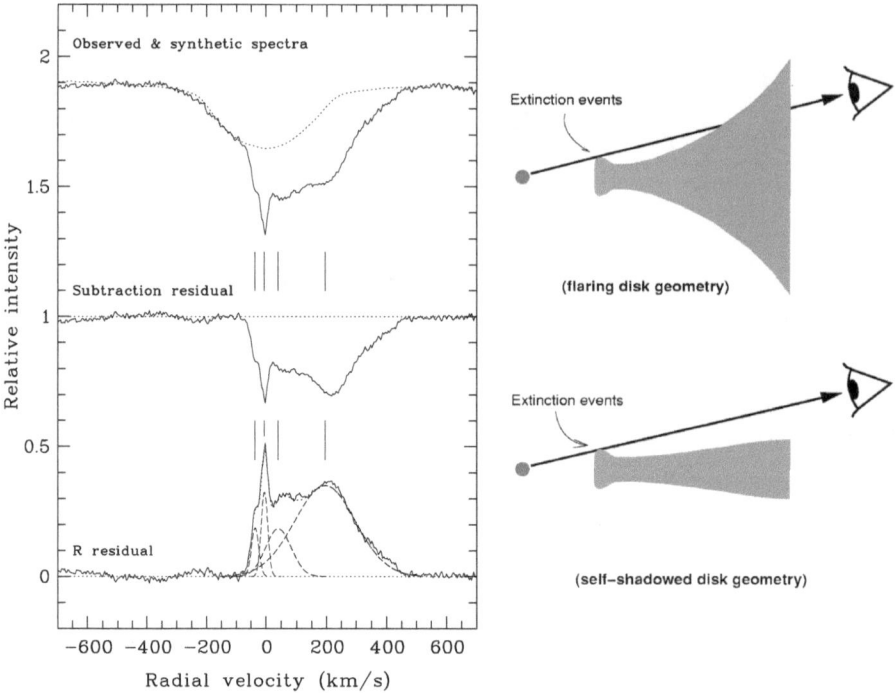

Fig. 7.12 *Left*: Illustration of the transient absorption components (TACs) observed in UXOR type star. *On the top* there is represented the Ca II K line as observed in the spectrum of UX Ori. The predicted photospheric line profile is overplotted with a *dotted line*. *In the middle* the predicted photospheric profile has been subtracted to the whole profile. *At the bottom* the normalised residual ($R = 1$—observed/photospheric) is plotted. The four individual TACs have been identified with *dashed lines* (from [65]). *Right*: Illustration of the nearly-edge one Dullemond et al. model. *On the top* is represented a flaring disk that is obscuring the star and the inner edge of disk, forbidding the observer to detect the fading events of UXOR-type stars. *In the bottom*, a self-shadowed disk seen nearly edge-on could exhibit these events to the observer (from [34])

ena are observed only in a small fraction of the Herbig Ae stars. Dullemond et al. [34] argue that their model of puffed-up inner rim that shadows the disk can explain how these blobs are observable. They explain that if the disk is flaring and the disk is seen near edge-on, with an inclination that allows the clumps to cross the line of sight, the outer part of the disk would occult the star. However, in the case of self-shadowed disks, with an edge-on inclination the inner part of the disk where the blobs are formed can be aligned with the line of sight (Fig. 7.12, right).

Other hypotheses have been proposed to explain the UXOR phenomena, such as an occulting screen that would occult the star and the close circumstellar gas, or magnetospheric accretion, or gaseous evaporation bodies [15, 64, 72]. A combination of some of these theories could also be possible. Many more observations and models need to be performed in order to fully understand the origin of the UXOR phenomenon.

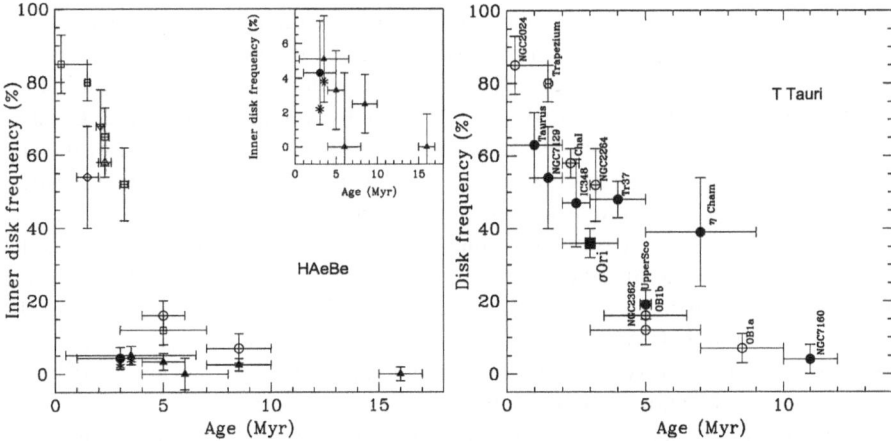

Fig. 7.13 *Left*: Inner disk frequency of HAeBe stars in nearby OB association as a function of age (from [49]). *Right*: Disk frequency of T Tauri stars of young clusters and associations (from [48])

7.2.3 The Disk Evolution

The observation of very young OB associations, where sequential formation occurred, revealed that in low-mass T Tauri star the disk frequency is dependent on the age suggesting a disk lifetime around 10 Myr. In the more massive Herbig Ae/Be star the lifetime is even reduced to about 5 Myr ([31, 48–50], Fig. 7.13). Hernandez et al. [49] find that the inner disk is disappearing faster than the outer part with a lifetime for the inner part that is around 3 Myr for Herbig Ae/Be stars.

Central holes, nearly devoid of dust, have been inferred from IR observations in the disks of T Tauri and Herbig Ae/Be stars with sizes of the order of 10 AU, i.e. much larger than the dust sublimation radius (e.g. [36, 54, 74, 78]). These disks are found around stars that are slightly older than stars surrounded with full dusty disks and that are younger than the young main-sequence stars surrounded with debris disk (Vega-like stars [11]). These disks are therefore assumed to be a transitional phase between accretion disk and debris disk and are called *transitional* disks.

These results suggest that the disk clearing process is from inside to out. There exist many theories of clearing processes that could explain that the inner part of the dusty disk disappear before the outer part. Weidenschilling [83] has shown that the growth of grains by coagulation (due to Brownian motion, turbulence and settling in the high-density internal disk) is faster in the inner disk than in the outer disk. If grains grow in the inner part, the opacity at short wavelengths decreases, which would explain the lack of near-IR excess in transitional disks. Another explanation is the formation of a planet. A planet of high-enough mass can clear its surrounding via torque and create a gap that will progressively extend up to the surface of the star to produce an inner hole as observed. The gap and the hole would be the result of an enhanced angular moment transport and accretion induced by spiral waves (e.g. [81]). Magneto-rotational instability (MRI) has also been proposed to explain the disk clearing. The stellar X-rays would ionise the disk. A surface layer would be formed and well coupled with the inertial magnetic field of the disk. MRI would therefore be activated, and either a disk wind or a enhanced accretion would allow the gas and dust to leave the disk [24, 75]. Finally, photoevaporation could also explain the disk clearing from inside to outside. The UV radiation field of the star ionise the outer part flaring disk (where the incident angle is higher). Above a critical radius, the matter leaves the disk through an evaporating wind. The mass accretion rate inside the disk decreases with time. At one point it reaches a value smaller than the mass loss rate of the evaporating wind. The outer part is not able anymore to replenish the inner part of the disk that continuously accretes matter onto the star until it is emptied. When the inner disk disappears within about 10^5 years, the ionising stellar radiation can penetrate the outer part of the disk, and about 10^5 years later the gas in the whole disk is lost by photoevaporation [8].

There is not yet any favoured scenario for the clearing of circumstellar disk. A combination of many of those presented here is certainly possible. Evidence of dust coagulation has been obtained in many disks (e.g. [58]), and (proto)planets have been detected in the transitional disk of young stars (e.g. [55]). It is generally believed that the natural evolution of the dust inside an accretion disk is to grow to form large grains that will continue to grow to form planetesimals and then planets. The question that still needs to be answered is whether these processes play an important role in the disk clearing. MRI could also play an important role but only in low-mass T Tauri stars as X-rays are required. In low-mass stars, X-rays are produced by the dynamo activity at the surface of the star, which is absent in higher-mass stars. On the contrary, the stellar UV flux increases with the stellar mass, which would favour the photoevaporation scenario for the Herbig Ae/Be star.

7.3 Conclusion

Today, most of the scientific community is convinced that the circumstellar gas and dust of the Herbig Ae/Be and T Tauri stars are mainly distributed into a disk. We did not discuss winds in this paper, but evidence of winds is also largely found in HAeBe and TT stars. The environments of class II stars is therefore mainly constituted of winds and disks. From theoretical and observational evidences it is very likely that the disks are flaring, at least in the low-mass stars were the UV radiation field is too low to photoevaporate the disk, and at least at one point during the history of the accretion disk. It appears from many theories that the irradiation from the star is the main source of energy. In T Tauri stars there is observational evidence that matter is accreted from the inner part of the disk to the stellar surface. The inner part of the disk must therefore be replenished, which implies accretion within the disk. Accretion is therefore likely occurring in the inner part of the disk; however, it is very difficult to access to observational indices from that part of the disk. In the case of Herbig Ae/Be stars, accretion is not so evident. Magnetospheric accretion is absent, and we can therefore wonder if accretion from the inner part of the disk to the stellar surface is really happening. More work needs to be done to answer this question.

Most of our knowledge on the properties of accretion disks reside on the analyses of the SED of the star, which is a great tool but not sufficient to fully understand the physics of the disk. Recent works using multi-wavelength and multi-techniques of observations allowed a big step in our knowledge of the disks surrounding young stars (e.g. [70]). The new generation of instruments on the VLTI and the recent launch of the spatial telescope Herschel will bring new tremendous observational constraints that will challenge the disk theories and motivate new ones.

Acknowledgements   I acknowledge financial support from "Programme National de Physique Stellaire" (PNPS) of CNRS/INSU, France.

References

1. Acke, B., van den Ancker, M.E.: Astron. Astrophys. **426**, 151 (2004)
2. Adams, F.C., Lada, C.J., Shu, F.H.: Astrophys. J. **312**, 788 (1987)
3. Adams, F.C., Lada, C.J., Shu, F.H.: Astrophys. J. **326**, 865 (1988)
4. Alecian, E., Wade, G.A., Catala, C., et al.: Mon. Not. R. Astron. Soc. (2012, submitted)
5. Alecian, E., Wade, G.A., Catala, C., et al.: Mon. Not. R. Astron. Soc. (2012, submitted)
6. Alencar, S.H.P., Johns-Krull, C.M., Basri, G.: Astron. J. **122**, 3335 (2001)
7. Alencar, S.H.P., Teixeira, P.S., Guimarães, M.M., et al.: Astron. Astrophys. **519**, A88 (2010)
8. Alexander, R.D., Clarke, C.J., Pringle, J.E.: Mon. Not. R. Astron. Soc. **369**, 229 (2006)
9. Armitage, P.J.: Annu. Rev. Astron. Astrophys. **49**, 195 (2011)
10. Augereau, J.C., Lagrange, A.M., Mouillet, D., Ménard, F.: Astron. Astrophys. **365**, 78 (2001)
11. Backman, D.E., Paresce, F.: In: Levy, E.H., Lunine, J.I. (eds.) Protostars and Planets III, pp. 1253–1304 (1993)
12. Balbus, S.A.: Annu. Rev. Astron. Astrophys. **41**, 555 (2003)
13. Behrend, R., Maeder, A.: Astron. Astrophys. **373**, 190 (2001)
14. Bertout, C.: Annu. Rev. Astron. Astrophys. **27**, 351 (1989)
15. Beust, H., Karmann, C., Lagrange, A.-M.: Astron. Astrophys. **366**, 945 (2001)
16. Bouvier, J., Alencar, S.H.P., Boutelier, T., et al.: Astron. Astrophys. **463**, 1017 (2007)
17. Bouvier, J., Chelli, A., Allain, S., et al.: Astron. Astrophys. **349**, 619 (1999)
18. Bouvier, J., Grankin, K.N., Alencar, S.H.P., et al.: Astron. Astrophys. **409**, 169 (2003)
19. Calvet, N., Magris, G.C., Patino, A., D'Alessio, P.: Rev. Mex. Astron. Astrofís. **24**, 27 (1992)
20. Calvet, N., Patino, A., Magris, G.C., D'Alessio, P.: Astrophys. J. **380**, 617 (1991)
21. Catala, C.: In: Reipurth, B. (ed.) European Southern Observatory Conference and Workshop Proceedings, vol. 33, pp. 471–489 (1989)
22. Catala, C., Donati, J.F., Böhm, T., Landstreet, J., Henrichs, H.F.: Astron. Astrophys. **345**, 884 (1999)
23. Catala, C., Felenbok, P., Czarny, J., Talavera, A., Boesgaard, A.M.: Astrophys. J. **308**, 791 (1986)
24. Chiang, E., Murray-Clay, R.: Nat. Phys. **3**, 604 (2007)
25. Chiang, E.I., Goldreich, P.: Astrophys. J. **490**, 368 (1997)
26. Chiang, E.I., Goldreich, P.: Astrophys. J. **519**, 279 (1999)
27. Chiang, E.I., Joung, M.K., Creech-Eakman, M.J., et al.: Astrophys. J. **547**, 1077 (2001)
28. D'Alessio, P., Calvet, N., Hartmann, L.: Astrophys. J. **553**, 321 (2001)
29. D'Alessio, P., Calvet, N., Hartmann, L., Lizano, S., Cantó, J.: Astrophys. J. **527**, 893 (1999)
30. D'Alessio, P., Canto, J., Calvet, N., Lizano, S.: Astrophys. J. **500**, 411 (1998)
31. Dent, W.R.F., Greaves, J.S., Coulson, I.M.: Mon. Not. R. Astron. Soc. **359**, 663 (2005)

32. Dullemond, C.P., Dominik, C., Natta, A.: Astrophys. J. **560**, 957 (2001)
33. Dullemond, C.P., Monnier, J.D.: Annu. Rev. Astron. Astrophys. **48**, 205 (2010)
34. Dullemond, C.P., van den Ancker, M.E., Acke, B., van Boekel, R.: Astrophys. J. Lett. **594**, L47 (2003)
35. Eisner, J.A., Lane, B.F., Hillenbrand, L.A., Akeson, R.L., Sargent, A.I.: Astrophys. J. **613**, 1049 (2004)
36. Espaillat, C., Muzerolle, J., Hernández, J., et al.: Astrophys. J. Lett. **689**, L145 (2008)
37. Folsom, C.P., Bagnulo, S., Wade, G.A., et al.: Mon. Not. R. Astron. Soc. **422**, 2072 (2012)
38. Furlan, E., Hartmann, L., Calvet, N., et al.: Astrophys. J. Suppl. Ser. **165**, 568 (2006)
39. Garrison, L.M. Jr.: Astrophys. J. **224**, 535 (1978)
40. Ghandour, L., Strom, S., Edwards, S., Hillenbrand, L.: In: The, P.S., Perez, M.R., van den Heuvel, E.P.J. (eds.) The Nature and Evolutionary Status of Herbig Ae/Be Stars. Astronomical Society of the Pacific Conference Series, vol. 62, p. 223 (1994)
41. Grady, C.A., Polomski, E.F., Henning, T., et al.: Astron. J. **122**, 3396 (2001)
42. Grady, C.A., Woodgate, B., Heap, S.R., et al.: Astrophys. J. **620**, 470 (2005)
43. Grasdalen, G.L., Strom, S.E., Strom, K.M., et al.: Astrophys. J. Lett. **283**, L57 (1984)
44. Grinin, V.P., Kozlova, O.V., Natta, A., et al.: Astron. Astrophys. **379**, 482 (2001)
45. Hartmann, L., Kenyon, S.J., Calvet, N.: Astrophys. J. **407**, 219 (1993)
46. Herbig, G.H.: Astrophys. J. Suppl. Ser. **4**, 337 (1960)
47. Herbst, W., Shevchenko, V.S.: Astron. J. **118**, 1043 (1999)
48. Hernández, J., Calvet, N., Briceño, C., et al.: Astrophys. J. **671**, 1784 (2007)
49. Hernández, J., Calvet, N., Hartmann, L., et al.: Astron. J. **129**, 856 (2005)
50. Hernández, J., Hartmann, L., Megeath, T., et al.: Astrophys. J. **662**, 1067 (2007)
51. Hillenbrand, L.A., Strom, S.E., Vrba, F.J., Keene, J.: Astrophys. J. **397**, 613 (1992)
52. Joy, A.H.: Astrophys. J. **102**, 168 (1945)
53. Kenyon, S.J., Hartmann, L.: Astrophys. J. **323**, 714 (1987)
54. Kim, K.H., Watson, D.M., Manoj, P., et al.: Astrophys. J. **700**, 1017 (2009)
55. Kraus, A.L., Ireland, M.J.: Astrophys. J. **745**, 5 (2012)
56. Lada, C.J., Adams, F.C.: Astrophys. J. **393**, 278 (1992)
57. Lagage, P.-O., Doucet, C., Pantin, E., et al.: Science **314**, 621 (2006)
58. Lommen, D.J.P., van Dishoeck, E.F., Wright, C.M., et al.: Astron. Astrophys. **515**, A77 (2010)
59. Lynden-Bell, D., Pringle, J.E.: Mon. Not. R. Astron. Soc. **168**, 603 (1974)
60. Mannings, V., Sargent, A.I.: Astrophys. J. **490**, 792 (1997)
61. Mannings, V., Sargent, A.I.: Astrophys. J. **529**, 391 (2000)
62. Meeus, G., Waters, L.B.F.M., Bouwman, J., et al.: Astron. Astrophys. **365**, 476 (2001)
63. Monnier, J.D., Millan-Gabet, R.: Astrophys. J. **579**, 694 (2002)
64. Mora, A., Eiroa, C., Natta, A., et al.: Astron. Astrophys. **419**, 225 (2004)
65. Mora, A., Natta, A., Eiroa, C., et al.: Astron. Astrophys. **393**, 259 (2002)
66. Natta, A., Kruegel, E.: Astron. Astrophys. **302**, 849 (1995)
67. Natta, A., Prusti, T., Neri, R., et al.: Astron. Astrophys. **371**, 186 (2001)
68. Okamoto, Y.K., Kataza, H., Honda, M., et al.: Astrophys. J. **706**, 665 (2009)
69. Pantin, E., Waelkens, C., Lagage, P.O.: Astron. Astrophys. **361**, L9 (2000)
70. Pinte, C., Padgett, D.L., Ménard, F., et al.: Astron. Astrophys. **489**, 633 (2008)
71. Pontoppidan, K.M., Salyk, C., Blake, G.A., et al.: Astrophys. J. **720**, 887 (2010)
72. Rodgers, B., Wooden, D.H., Grinin, V., Shakhovsky, D., Natta, A.: Astrophys. J. **564**, 405 (2002)
73. Stapelfeldt, K.R., Krist, J.E., Menard, F., et al.: Astrophys. J. Lett. **502**, L65 (1998)
74. Strom, K.M., Strom, S.E., Edwards, S., Cabrit, S., Skrutskie, M.F.: Astron. J. **97**, 1451 (1989)
75. Suzuki, T.K., Inutsuka, S.-I.: Astrophys. J. Lett. **691**, L49 (2009)
76. Testa, P., Huenemoerder, D.P., Schulz, N.S., Ishibashi, K.: Astrophys. J. **687**, 579 (2008)
77. The, P.S., de Winter, D., Perez, M.R.: Astron. Astrophys. Suppl. Ser. **104**, 315 (1994)
78. Uchida, K.I., Calvet, N., Hartmann, L., et al.: Astrophys. J. Suppl. Ser. **154**, 439 (2004)
79. van den Ancker, M.E., The, P.S., de Winter, D.: Astron. Astrophys. **309**, 809 (1996)
80. van Dishoeck, E.F.: Annu. Rev. Astron. Astrophys. **42**, 119 (2004)

81. Varnière, P., Blackman, E.G., Frank, A., Quillen, A.C.: Astrophys. J. **640**, 1110 (2006)
82. Waters, L.B.F.M., Waelkens, C.: Annu. Rev. Astron. Astrophys. **36**, 233 (1998)
83. Weidenschilling, S.J.: Icarus **127**, 290 (1997)
84. Williams, J.P., Cieza, L.A.: Annu. Rev. Astron. Astrophys. **49**, 67 (2011)
85. Wyatt, M.C.: Annu. Rev. Astron. Astrophys. **46**, 339 (2008)
86. Zinnecker, H., Yorke, H.W.: Annu. Rev. Astron. Astrophys. **45**, 481 (2007)

Chapter 8
Stellar Winds, Magnetic Fields and Disks

Asif ud-Doula

Abstract All main sequence stars lose mass via stellar winds. The winds of cool stars like the sun are driven by gas pressure gradient. However, the winds of hot massive stars which tend to be luminous are driven by emitted by the star radiation pressure. Mass loss from such winds are significantly higher. In this article, I describe the nature of such radiatively driven winds and show how they interact with rotation and magnetic fields leading to stellar spindown and large-scale disk-like structures. In particular, I show that the overall degree to which the wind is influenced by the field depends largely on a single, dimensionless, "wind magnetic confinement parameter", η_* ($=B_{eq}^2 R_*^2 / \dot{M} v_\infty$), which characterizes the ratio between magnetic field energy density and kinetic energy density of the wind.

8.1 The Theory of Stellar Winds

There is a big variation in the nature of stars we see in the sky. What distinguishes them the most is their effective surface temperatures that divide them into two distinct categories, hot (also known as massive) and cool stars. What unites all these stars is that they all lose mass via a continuous outflow of material from stellar surface called stellar wind. Sun-like cool stars which have convective envelopes lose mass via gas pressure. Massive, hot stars which lack convective envelopes lose mass via radiation pressure due to their enormous intrinsic brightness. I describe physics of these two types of stellar winds next.

8.1.1 Gas Pressure Driven Isothermal Winds: The Solar Wind

8.1.1.1 The Expansion of the Solar Corona

The first model of a solar corona that extends beyond the earth's orbit was proposed by Chapman and Zirin [2], who assumed a static atmosphere where the energy is

A. ud-Doula (✉)
Penn State Worthington Scranton, 120 Ridge View Drive, Dunmore, PA 18512, USA
e-mail: asif@psu.edu

J.-P. Rozelot, C. Neiner (eds.), *The Environments of the Sun and the Stars*,
Lecture Notes in Physics 857,
DOI 10.1007/978-3-642-30648-8_8, © Springer-Verlag Berlin Heidelberg 2013

transferred by thermal conduction alone. For a steady state, the heat flux across any Gaussian surface is constant, and if we assume spherical symmetry, then

$$4\pi r^2 \kappa \frac{dT}{dr} = C,$$ (8.1)

where C is a constant, r is the radius, and $\kappa \, dT/dr$ is the heat flux density with the thermal conduction $\kappa = \kappa_0 T^{5/2}$ [19]. The above equation can be readily integrated to yield:

$$T = T_0 \left(\frac{R_0}{r} \right)^{2/7}$$ (8.2)

under the assumption that $T = T_0$ at $r = R_0$ and temperature T goes to zero at large radii. Moreover, this corona is subject to the hydrostatic equilibrium condition,

$$\frac{1}{\rho} \frac{dP}{dr} = \frac{GM_\odot}{r^2}.$$ (8.3)

Using the ideal gas law, the pressure can be written as $P = \rho a^2$ where $a^2 = a_0^2 (R_0/r)^{2/7}$ with the isothermal sound speed a_0 at the base of the corona. The hydrostatic equation can be now integrated:

$$P = P_0 \exp \left[\frac{7R_0}{5H_0} \left[\left(\frac{R_0}{r} \right)^{5/7} - 1 \right] \right],$$ (8.4)

where P_0 is the pressure at the coronal base $r = R_0$, $H_0 = a_0^2 R_0^2 / GM_\odot$ is the scale height. Clearly, as $r \to \infty$, the pressure approaches a finite value,

$$P(\infty) = P_0 \exp \left[-\frac{7R_0}{5H_0} \right].$$ (8.5)

Here,

$$a \frac{R_0}{H_0} = R_0 \frac{GM_\odot}{a_0^2 R_0^2}$$

$$\approx \frac{14}{(T_0/10^6 \text{ K})}$$ (8.6)

for R_0 close to the solar surface. This leads to

$$\log \left(\frac{P_0}{P_\infty} \right) \approx \frac{8.5}{(T_0/10^6)}.$$ (8.7)

In practice, P_∞ is the pressure of the interstellar medium (ISM) that surrounds the heliosphere, and we observe this ratio $\log (P_0/P_{ISM}) \approx 12$. So, if we wish to have a hydrostatic corona that is contained by the gas pressure of the ISM, the temperature of the corona must be

$$T_0 < 0.7 \times 10^6 \text{ K.} \qquad (8.8)$$

But the observations show that the temperature of the solar corona is a few million degrees K. Clearly, P_∞ obtained from this model cannot be matched by the interstellar medium pressure. In addition to this shortcoming, in the limit as $r \to \infty$, the temperature approaches zero, and since the pressure is finite at large radii, the density becomes arbitrarily large. Because of these inconsistencies, the solar corona cannot remain static.

8.1.2 Parker's Solution to the Solar Wind

Parker [16] resolved the above problem by suggesting that the solar corona is continuously expanding outwards simply because there is no "lid" that can contain the corona. He called this outflow as the "solar wind."

Parker considered a spherically symmetric solar wind that is in a steady, time-independent motion where all the properties vary with distance r only. Then the equation of conservation of momentum for a fluid parcel becomes

$$v\frac{dv}{dr} = -\frac{1}{\rho}\frac{dP}{dr} - \frac{GM_\odot}{r^2}, \qquad (8.9)$$

where $M_\odot$ is the mass of the sun. The equation of mass conservation is

$$\frac{1}{r^2}\frac{d}{dr}\left(r^2\rho v\right) = 0. \qquad (8.10)$$

If we assume that the mass loss rate $\dot{M}$ is a constant, then

$$\dot{M} = 4\pi r^2 \rho(r) v(r). \qquad (8.11)$$

In principle, as the gas expands, it will cool, and unlike the hot-star winds where intense radiation from the star keeps the wind nearly isothermal, the solar wind may not remain at constant temperature. But the hot solar coronal material has a very high conductivity due to the mobile electrons, implying thus a small temperature gradient. As such, for simplicity, it is assumed that the solar wind behaves like a perfect gas maintained at a constant temperature T, which is of the order of a few MK as mentioned above. Using the perfect gas law and Eq. (8.11), we can now replace the pressure gradient term in the continuity equation (8.9) to obtain

$$\frac{1}{\rho}\frac{dP}{dr} = a^2\frac{1}{\rho}\frac{d\rho}{dr}$$

$$= -\frac{a^2}{v}\frac{dv}{dr} - \frac{2a^2}{r}. \qquad (8.12)$$

Fig. 8.1 The solution topology for the isothermal solar wind equations. The Y-axis u/a represents velocity in the unit of sound speed, and the X-axis is the radius in the units of critical radius, r_c. Types 3 and 4 solutions are supersonic at the base of the corona and thus are physically inadmissible. Type 1 represents the "breeze" solutions that never reach supersonic values, contrary to observations. The solution type 2 (*bold line*) is in agreement with observations

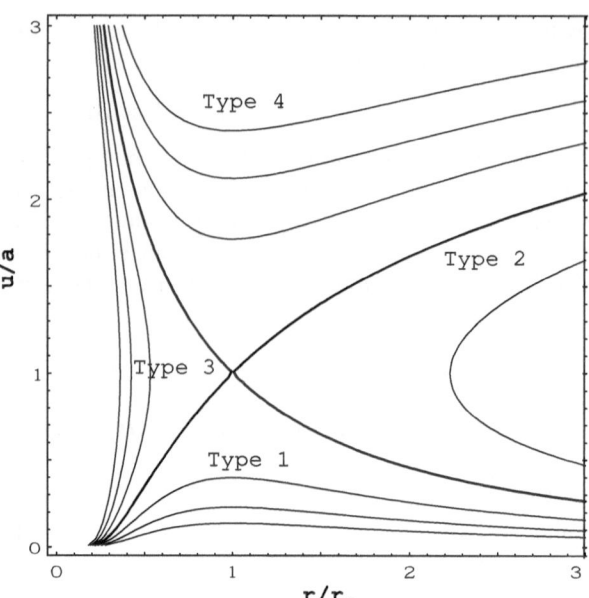

Substituting this into Eq. (8.9) yields

$$\left[v^2 - a^2\right]\frac{1}{v}\frac{dv}{dr} = \frac{2a^2}{r} - \frac{GM_\odot}{r^2}. \tag{8.13}$$

We obtain the critical solution at $r = r_c$, with

$$r_c = \frac{GM_\odot}{2a^2}. \tag{8.14}$$

This defines the "critical radius" and can be obtained either when

$$v(r_c) = a \tag{8.15}$$

or

$$\frac{1}{v}\frac{dv}{dr}(r = r_c) = 0. \tag{8.16}$$

We are interested in continuous and single-valued solutions of $v(r)$ and dv/dr. If condition (8.15) is satisfied, then dv/dr has the same sign for all r or else (by the mean value theorem) there would have been a radius r_c where dv/dr would attain zero. This implies that $v(r)$ is either monotonically increasing or monotonically decreasing function of radius. On the other hand, if condition (8.16) is met, $(v^2 - a^2)$ has the same sign for all r, implying that at radius $r = r_c$ the speed $v(r_c)$ attains a maximum or a minimum.

The solution topology to this isothermal solar wind problem is sketched in Fig. 8.1. Each of the solution types fits a different set of boundary conditions. Mathematically, all these solutions are acceptable, but physically not all of them are plausible. For instance, the solution types 3 and 4 can be ruled out as acceptable on the basis of the fact that they all start off at the base of the corona at supersonic speeds. We have no physical or observational evidence to believe that the solar corona is expanding supersonically at the base. The remaining two types of solutions have small speeds at the base, but they differ at large radii. The wind in solution type 2 is supersonic at the outer radii, while in the solution type 1, often called "breeze" solution, the wind speed approaches zero asymptotically implying a finite pressure at $r = \infty$. But this finite pressure cannot be matched by the interstellar pressure. Thus, the solutions of type 2 remain as the most plausible solutions.

8.1.3 Line-Force in Stellar Winds

We just showed that the solar wind arises from pressure expansion of the hot corona which is superheated to temperatures of a few million degrees by the mechanical energy dumped by the convection near the surface of the relatively cool sun. By contrast, the hot stars with surface temperatures 10,000–100,000 K are considered not to have convection zones, a necessary mechanism to have hot coronae. Their winds remain nearly at surface temperatures of the star. Thus, they lack the very high pressure needed to drive the wind against the gravity in a similar fashion to the solar wind. But the hot stars are very luminous, and it is the pressure of radiation that drives their winds.

Intuitively light is not a good source for momentum transfer. This is mainly because the momentum of a photon is determined by division of its energy by the maximum possible speed, c, the speed of light. But the hot stars are not ordinary objects. They are very massive and very luminous. In their case, the photons become the dominating factor in controlling the physics of the continuous outflow of material from the stellar surface. The typical flow speeds of hot-star winds can reach as high as 3000 km/s, much faster than the 400–700 km/s speed of the solar wind. At the same time, typical mass loss rates for the hot stars can reach up to $\sim 10^{-6}\ M_\odot/\mathrm{yr}$, which are up to a factor of a billion higher than that of the sun.

In this section we discuss the interesting question of how photons can drive hot-star winds. First, we present the theory of the line-driving and then illustrate solutions with some numerical simulations.

8.1.3.1 Radiation Force for a Point-Source Star

The line force per unit mass (acceleration) due to radiation at a radius r is given generally by

$$\mathbf{g}_{\mathrm{rad}} = \frac{1}{c} \oint \int_{\nu=0}^{\infty} \kappa_\nu I_\nu(r, \hat{n}) \hat{\mathbf{n}}\, d\Omega\, d\nu, \tag{8.17}$$

where the line strength has been characterized by isotropic κ_ν, the total mass extinction coefficient that includes both absorption and scattering. The monochromatic radiative intensity I_ν along the direction $\hat{n}$ is integrated over the solid angle Ω.

The total mass extinction coefficient can be separated into two parts:

$$\kappa_\nu = \kappa_e + \kappa_L, \tag{8.18}$$

where κ_e is the mass absorption coefficient (sometimes called "opacity") due to scattering of continuum photons by free electrons, given by $\kappa_e = \sigma_e/\mu_e$. Here σ_e $(=0.66 \times 10^{-24} \text{ m}^2)$ is the classical Thompson cross-section, and μ_e is the mean atomic mass per free electron defined as $\mu_e = 2m_H/(1 + X)$, with m_H and X the mass and mass-fraction of hydrogen, and κ_L the absorption coefficient due to the bound electrons of a line. Thus, the radiation force can be separated in two parts as well:

$$g_{\text{rad}} = g_e + g_L, \tag{8.19}$$

where g_e is the radiation force due scattering of continuum photons by free electrons, and g_L is the line force.

8.1.3.2 Force Due to Electron Scattering

The continuum processes in hot stars are presumed to be dominated by electron scattering. Now, we can write

$$\mathbf{g}_e = \frac{\kappa_e}{c} \oint \int_{\nu=0}^{\infty} I_\nu(r, \hat{n}) \hat{\mathbf{n}} \, d\Omega \, d\nu. \tag{8.20}$$

In the case of spherical symmetry and a point source star, the integrand gives just the total radiation flux. Thus,

$$g_e = \frac{\kappa_e}{c} \frac{L_*}{4\pi r^2}, \tag{8.21}$$

where L_* is the total bolometric luminosity of the star. Here we assumed that the wind is optically thin to the continuum radiation.

Note that g_e is inversely proportional to r^2, just like the gravity. Therefore, it is useful to compare g_e with gravity g by defining

$$\Gamma \equiv \frac{g_e}{g} = \frac{\kappa_e L_*}{4\pi G M_* c}, \tag{8.22}$$

where Γ is often referred as the Eddington parameter, G is the gravitational constant, and M_* is the mass of the star. Clearly, if the value of Γ exceeds unity, the star cannot remain in hydrostatic equilibrium. For the massive OB-type stars, $\Gamma \approx 0.5$. Thus, g_e essentially reduces the effective gravity by $1 - \Gamma$, i.e., $g_{\text{eff}} = g(1 - \Gamma)$.

8.1.3.3 Radiation Force Due to a Single Line

In formulating the radiation force due to a single line, we will follow mostly Cranmer [4], which is based on the original formulation line force by Castrol et al. [1, CAK hereafter]. In general,

$$\mathbf{g}_L = \frac{\kappa_L}{c} \oint \int_{v=0}^{\infty} \tilde{\phi}(v - v') I_v(r, \hat{n}) \hat{\mathbf{n}} \, d\Omega \, dv, \qquad (8.23)$$

where κ_L is the mass absorption coefficient for a single line, and $\tilde{\phi}(v)$ is a normalized line profile function. Here v' is the frequency of the line in the comoving frame of the gas and can be related to the emitted frequency v_0 by

$$v' = v_0 \left(1 + \frac{\hat{\mathbf{n}} \cdot \mathbf{v}(\mathbf{r})}{c} \right), \qquad (8.24)$$

with $\mathbf{v}(\mathbf{r})$ the flow velocity, non-relativistic in the domain of our interest.

Now, for convenience, let us make a change of variables and define the frequency in the unit of Doppler widths:

$$x = \frac{v - v_0}{\Delta v_D}, \qquad (8.25)$$

where the Doppler width $\Delta v_D = v_0 v_{\text{th}}/c$ with the ion thermal speed in the gas v_{th}. Thus, the single line force can be rewritten as

$$\mathbf{g}_L = \frac{\kappa_L \Delta v_D}{c} \oint \int_{x=-\infty}^{\infty} \phi\left(x - \frac{\hat{\mathbf{n}} \cdot \mathbf{v}(\mathbf{r})}{v_{\text{th}}} \right) I_v(r, \hat{n}) \, \hat{\mathbf{n}} \, d\Omega \, dx. \qquad (8.26)$$

Note that the lower limit for $x = -c/v_{\text{th}}$ has been extended to $x = -\infty$. We can do this because the line opacity is mostly concentrated around the line center and $-c/v_{\text{th}}$ is many Doppler widths away from it. The error introduced by this approximation is negligible, but it makes the further analysis much simpler.

The intensity $I_v(r, \hat{n})$ will, in general, have a contribution from the direct intensity from the core and a contribution from the diffuse component arising from the radiation scattered (or created) within the wind,

$$I_v = I_v^{\text{dir}} + I_v^{\text{diff}}. \qquad (8.27)$$

In this work, the treatment of the line force is based on the *pure absorption* model. In this model the force is computed by integration of the frequency-dependent attenuation of the direct radiation. The contribution to the line force from the diffuse radiation is ignored since in the smooth and supersonic wind we consider here, the line force arising from it is nearly fore-aft symmetric yielding net zero force. Now, the direct component of the intensity can be written as

$$I_v^{\text{dir}}(r) = I_* e^{-\tau_v(r)}, \qquad (8.28)$$

where I_* is the core intensity, assumed to be constant throughout the stellar surface, and τ_ν is the frequency-dependent optical depth along a path of length s:

$$\tau_\nu(r) = \int_{R_*}^{r} \kappa \rho(r') \phi\left(x - \frac{\hat{\mathbf{n}} \cdot \mathbf{v}(r')}{v_{\text{th}}}\right) dr'. \tag{8.29}$$

In general, Eq. (8.28) involves *nonlocal* integration over space, but one can localize the radiation transport by invoking the *Sobolev approximation* [18]. The key idea of this approximation is to turn this spatial integral into an integral over comoving frame frequency. In order to do this, Sobolev assumed that the line profile function is very narrow, δ-function like, and over a length scale L_{Sob} which is small compared to hydrodynamic scale lengths, the density remains fairly constant. As such, the optical depth integral becomes a function of *local* variables only,

$$\tau_\nu(r) = \kappa \rho(r) \int_{R_*}^{r} \phi\left(x - \frac{\hat{\mathbf{n}} \cdot \mathbf{v}(r')}{v_{\text{th}}}\right) dr'. \tag{8.30}$$

Here, let us transform the frequency variable into the comoving frame frequency:

$$x' = x - \frac{\hat{\mathbf{n}} \cdot \mathbf{v}(r')}{v_{\text{th}}}, \tag{8.31}$$

$$dx' = -\frac{1}{v_{\text{th}}} (\hat{\mathbf{n}} \cdot \nabla)(\hat{\mathbf{n}} \cdot \mathbf{v}(r)) \, dr'. \tag{8.32}$$

With these changes, the integral for the optical depth becomes

$$\tau_\nu(r) = \frac{\kappa_L v_{\text{th}} \rho(r)}{(\hat{\mathbf{n}} \cdot \nabla)(\hat{\mathbf{n}} \cdot \mathbf{v}(r))} \int_{x - \frac{\hat{\mathbf{n}} \cdot \mathbf{v}(r)}{v_{\text{th}}}}^{\infty} \phi(x') \, dx'. \tag{8.33}$$

Like before, the limit of integration is extended to infinity. We can define the constant part of the integral, the Sobolev optical depth,

$$\tau_S \equiv \frac{\kappa_L v_{\text{th}} \rho(r)}{(\hat{\mathbf{n}} \cdot \nabla)(\hat{\mathbf{n}} \cdot \mathbf{v}(r))}, \tag{8.34}$$

and

$$\Phi \equiv \int_{x - \frac{\hat{\mathbf{n}} \cdot \mathbf{v}(r)}{v_{\text{th}}}}^{\infty} \phi(x') \, dx'. \tag{8.35}$$

Now the equation for the single line force can be rewritten in a convenient form:

$$\begin{aligned}
\mathbf{g}_L &= \frac{\kappa_L \Delta v_D}{c} \oint \int_{x=-\infty}^{\infty} \phi(x') I_* e^{-\tau_S \Phi(x,r)} \hat{\mathbf{n}} \, d\Omega \, dx' \\
&= \frac{\kappa_L \Delta v_D}{c} \oint \int_{x=-\infty}^{\infty} I_* e^{-\tau_S \Phi(x,r)} \hat{\mathbf{n}} \, d\Omega \, d\Phi(x,r). \tag{8.36}
\end{aligned}$$

This can be easily integrated to give

$$\mathbf{g}_L = \frac{\kappa_L \Delta \nu_D}{c} \left(\oint I_* \hat{\mathbf{n}} d\Omega \left[\frac{1 - e^{-\tau_S}}{\tau_S} \right] \right). \tag{8.37}$$

The term in the open bracket is the total flux of radiation at frequency ν and distance r, and for point-like star model, we have:

$$g_L = \frac{\kappa_L \nu_{th}}{c^2} \frac{\nu_0 L_\nu}{4\pi r^2} \left[\frac{1 - e^{-\tau_S}}{\tau_S} \right]. \tag{8.38}$$

In the approximation of a point-like star, the Sobolev optical depth collapses into a simpler form,

$$\tau_S = \frac{\kappa_L \nu_{th} \rho(r)}{\partial \nu_r / \partial r}. \tag{8.39}$$

For most hot stars, the lines in consideration are near the peak of the continuum spectrum; as such, we can assume $\nu_0 L_\nu \approx L_*$. Thus,

$$g_L = \frac{\kappa_L \nu_{th}}{c^2} \frac{L_*}{4\pi r^2} \left[\frac{1 - e^{-\tau_S}}{\tau_S} \right]. \tag{8.40}$$

8.1.3.4 Force Due to an Ensemble of Lines

In the previous section we formulated the radiation force due to a single line in the stellar wind. In reality, the total radiative acceleration is produced by a large number of lines. In order to account for the contributions from all the lines, one has to sum over all of them:

$$g_{\text{lines}} = \sum_{\text{lines}} \frac{\kappa_L \nu_{th}}{c^2} \frac{L_*}{4\pi r^2} \left[\frac{1 - e^{-\tau_S}}{\tau_S} \right], \tag{8.41}$$

where g_{lines} is the cumulative force due to the ensemble of lines.

In practice, the number of lines is huge and can be expressed as a statistical distribution. In its original formulation, CAK considered an extensive list of subordinate lines of C^{+3} and parameterized the line-force via a power law fit in terms of *force multiplier*, $M(t)$, as a function of the electron-scattering optical depth, t:

$$g_{\text{lines}} \propto M(t) \propto kt^{-\alpha}, \tag{8.42}$$

where $t = \kappa_e \rho \nu_{th} / (\partial \nu_r / \partial r)$, k defines the overall strength of the lines, and α determines the fraction of optically thick lines. Owocki et al. [15, OCR hereafter] generalized this and formulated a number distribution of lines as an exponentially-truncated power law,

$$\frac{dN(\kappa)}{d\kappa} = \frac{1}{\kappa_0} \left(\frac{\kappa_L}{\kappa_0} \right)^{\alpha - 2} e^{-\kappa / \kappa_{max}}, \tag{8.43}$$

where κ_0 is a normalization constant that is related to the CAK k parameter by $\kappa_0 = \Gamma(\alpha)(v_{th}/c)(\kappa_0/\kappa_e)^{1-\alpha}/(1-\alpha)$, the cutoff κ_{max} limits the maximum line strength (OCR), κ_e is the electron scattering coefficient, and $\Gamma(\alpha)$ is the complete Gamma function. With this distribution of lines, the summation in the above equation can be replaced by an integral over κ:

$$g_{lines} = \sum_{lines} g_L$$

$$= \int_0^\infty g_L \frac{dN}{d\kappa_L} d\kappa_L$$

$$= \int_0^\infty \frac{\kappa_L v_{th}}{c^2} \frac{L_*}{4\pi r^2} \left[\frac{1-e^{-\tau_S}}{\tau_S} \right] \frac{1}{\kappa_0} \left(\frac{\kappa_L}{\kappa_0} \right)^{\alpha-2} e^{-\kappa/\kappa_{max}} d\kappa_L. \quad (8.44)$$

This integral can be readily integrated using the definition of τ_S, Eq. (8.39), yielding

$$g_{lines} = \frac{\Gamma(\alpha)}{1-\alpha} \frac{\kappa_0 v_{th}}{c} \frac{L_*}{4\pi r^2 c} \left[\frac{\kappa_0 v_{th} \rho(r)}{\partial v_r/\partial r} \right]^{-\alpha} \left(\frac{(\tau_{max}+1)^{1-\alpha}-1}{\tau_{max}^{1-\alpha}} \right) \quad (8.45)$$

with $\tau_{max} \equiv \kappa_{max} v_{th} \rho(r)/(\partial v_r/\partial r)$. For $\tau_{max} \gg 1$, the term in the round bracket equals unity, and we retrieve the original CAK force. In this work, for the sake of simplicity, we shall ignore this correction term in what follows since its contribution is negligible in any case. Note however that the expression above contains the thermal velocity of the ion, v_{th}. The quantity κ_0, which is related to the line strength, is dependent on v_{th}. As such, the appearance of the thermal speed is redundant.

Gayley [8] recast the CAK line-force in terms of $\bar{Q}$ that is closely related to the classical oscillator strength and eliminates v_{th} from the line-force expression. He assumed the following identity:

$$\frac{\kappa_0 v_{th}}{c} \equiv \bar{Q} \kappa_e \Gamma(\alpha)^{-\frac{1}{1-\alpha}}. \quad (8.46)$$

Here $\bar{Q}$ remains fairly constant at ~ 1000 for most O and B stars. With a little algebra, we can rewrite the line-force as

$$g_{lines} = \frac{1}{(1-\alpha)} \frac{\bar{Q}\kappa_e L_*}{4\pi r^2 c} \left(\frac{\partial v/\partial r}{\rho c \bar{Q}\kappa_e} \right)^\alpha. \quad (8.47)$$

Throughout this article, we will follow Gayley's formalism of the CAK force.

8.1.4 Solution to the 1-D CAK Wind

In this section we outline the possible solutions to a wind that is subject to the CAK force as derived in Eq. (8.47). Since the line force is very dominant in hot-star winds, the gas pressure plays virtually no role, and one can ignore it altogether to a good approximation. Thus, the momentum equation for 1D time-steady radial flow of line-driven wind becomes

$$
v\frac{dv}{dr} = -(1 - \Gamma)\frac{GM}{r^2} + \frac{\bar{Q}\kappa_e L}{4\pi r^2 c}\frac{1}{1-\alpha}\left[\frac{dv/dr}{\bar{Q}\kappa_e \rho c}\right]^\alpha
$$

$$
= -(1 - \Gamma)\frac{GM}{r^2} + \frac{\bar{Q}\kappa_e L}{4\pi r^2 c}\frac{1}{1-\alpha}\left[\frac{4\pi r^2 v\, dv/dr}{\bar{Q}\kappa_e \dot{M} c}\right]^\alpha. \tag{8.48}
$$

In the second line in the above equation we used the mass continuity equation $\dot{M} = 4\pi r^2 \rho v$. To simplify this equation, let us divide both sides by the effective gravity $(1 - \Gamma)GM/r^2$ and define inertial acceleration in this unit:

$$
w \equiv \frac{r^2 v\, dv/dr}{(1-\Gamma)GM}. \tag{8.49}
$$

Thus,

$$
w = -1 + \frac{L}{c^{1+\alpha}(1-\alpha)\dot{M}^\alpha}\left[\frac{4\pi GM(1-\Gamma)}{\bar{Q}\kappa_e}\right]^{\alpha-1} w^\alpha
$$

$$
\equiv -1 + Cw^\alpha. \tag{8.50}
$$

Here, the constant $C \sim 1/\dot{M}^\alpha$ determines the mass loss rate, and the solution to the above dimensionless equation will depend on this constant. To show this, let us first rearrange the above equation:

$$
Cw^\alpha = w + 1. \tag{8.51}
$$

Figure 8.2 illustrates the graphical solution to the above equation for various values of C. The left-hand side (LHS) of the equation represents the line force, and the right-hand side (RHS) tells us how much of it goes into inertia (w) vs. gravity. For high $\dot{M}$ or small C, there are no solutions, while for small $\dot{M}$ or high C, there are two solutions. The intermediate of these two is the critical solution that corresponds to maximal CAK mass loss rate solution. It requires that Cw^α line intersect the $1 + w$ line tangentially, i.e.,

$$
C_c\alpha w^{\alpha-1} = 1. \tag{8.52}
$$

If we solve for w, we obtain

$$
w = \frac{\alpha}{1-\alpha} = \text{constant}. \tag{8.53}
$$

Now the constant C_c becomes

$$
C_c = \frac{(1-\alpha)^{\alpha-1}}{\alpha^\alpha}. \tag{8.54}
$$

Fig. 8.2 Graphical solutions
of the dimensionless equation
of motion () representing 1D
CAK wind solutions. If $\dot{M}$ is
big, there are no solution; if
$\dot{M}$ is small, there are two
solutions; only the critical
value of $\dot{M} = \dot{M}_{\text{CAK}}$ gives a
single solution

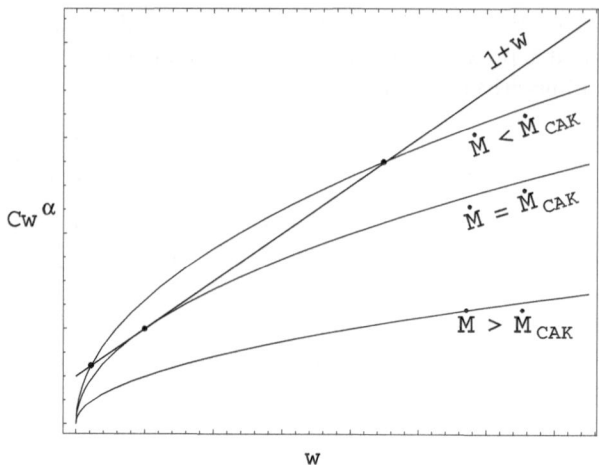

Using the definition for C and after some simple algebraic manipulations, one finds
that

$$\dot{M}_{\text{CAK}} = \frac{L}{c^2} \frac{\alpha}{1-\alpha} \left[\frac{\bar{Q}\Gamma}{1-\Gamma} \right]^{(1-\alpha)/\alpha} . \tag{8.55}$$

Notice that the complete Gamma function, $\Gamma(\alpha)$, disappeared. This is due to the
definition of $\bar{Q}$ in Eq. (8.46). If we assume that $\alpha = 1/2$, we can integrate Eq. (8.53)
to obtain the following velocity law for line-driven hot-star winds:

$$v(r) = \sqrt{\frac{\alpha}{1-\alpha}} v_{\text{esc}} \left(1 - \frac{R_*}{r} \right)^{1/2} , \tag{8.56}$$

where $v_{\text{esc}} = \sqrt{2GM(1-\Gamma)/R_*}$ is the escape speed from the stellar surface. Note
that the terminal speed here is nearly equal to escape speed, $v_\infty = \sqrt{\alpha/(1-\alpha)} v_{\text{esc}}$.
In fact, for $\alpha = 1/2$, they are exactly equal. This is the classical CAK solution of
line-driven winds. For the general case with $0 < \alpha < 1$, one can fit the velocity
solution with

$$v(r) = v_\infty \left(1 - \frac{R_*}{r} \right)^\beta , \tag{8.57}$$

where the exponent β determines how steeply the velocity reaches the terminal
speed. The solution we derived above (Eq. (8.56)) is widely known in the literature as the $\beta = 1/2$ velocity law. It underestimates observed terminal velocities for
most hot stars. In the next section we discuss an extension of this model that takes
into account the finite size of the star.

8.1.5 Finite Disk Correction Factor

In the above analysis, we assumed that the star is a point like source. That enabled us to derive a relatively simple form of the line force. In reality, stars have finite size, and especially near the surface it has significant dynamical effects on the wind outflow. This arises due to the presence of nonradial rays from the stellar disk. In this section we will derive this effect of the finite size of the star on the line force. We assume spherical symmetry with $\hat{\mathbf{n}}$ along the z-axis and purely radial wind outflow,

$$\hat{\mathbf{n}} = \mu'\hat{\mathbf{r}} + \sqrt{1-\mu'^2}\hat{\theta}', \tag{8.58}$$

$$\mathbf{v} = v_r\hat{\mathbf{r}}, \tag{8.59}$$

where $\mu' = \cos\theta'$. The general velocity gradient can be derived as

$$(\hat{\mathbf{n}}\cdot\nabla)(\hat{\mathbf{n}}\cdot\mathbf{v}(\mathbf{r})) = \mu'^2\frac{\partial v_r}{\partial r} + \frac{v_r}{r}(1-\mu'^2) \tag{8.60}$$

$$= \frac{\partial v_r}{\partial r}\frac{\sigma\mu'^2+1}{\sigma+1}, \tag{8.61}$$

where we have defined

$$\sigma \equiv \frac{\partial\ln v_r}{\partial\ln r} - 1. \tag{8.62}$$

The bolometric intensity can be expressed as

$$I_*(r,\theta',\phi') = \frac{L_*}{4\pi R_*^2}D(\mu',r) \tag{8.63}$$

with the limb darkening function $D(\mu',r)$. For simplicity, we will assume a uniformly bright disk,

$$D(\mu',r) = \begin{cases} 0 & \text{if } -1 \le \mu' < \mu_*, \\ 1/\pi & \text{if } \mu_* < \mu' \le +1, \end{cases}$$

where

$$\mu_* \equiv \sqrt{1 - \frac{R_*^2}{r^2}} \tag{8.64}$$

defines the cosine of the angle subtended by the stellar disk. The radial line force we have derived in Eq. (8.47) now can be rewritten as

$$g_{\text{lines}} = \frac{1}{(1-\alpha)}\frac{\bar{Q}\kappa_e L_*}{4\pi R_*^2 c}\left(\frac{1}{\rho c\bar{Q}\kappa_e}\right)^\alpha 2\pi\int_{-1}^{+1}D(\mu',r)\left[\frac{\partial v_r}{\partial r}\frac{\sigma\mu'^2+1}{\sigma+1}\right]^\alpha d\mu'. \tag{8.65}$$

The integration is straightforward and yields

$$g_{\text{lines}} = \frac{1}{(1-\alpha)} \frac{\bar{Q}\kappa_e L_*}{4\pi R_*^2 c} \left(\frac{\partial v_r/\partial r}{\rho c \bar{Q}\kappa_e}\right)^\alpha \frac{(\sigma+1)^{\alpha+1} - (\mu_*^2\sigma+1)^{\alpha+1}}{\sigma(\alpha+1)(\sigma+1)^\alpha}. \qquad (8.66)$$

We can recast the above expression by rewriting R_*^2 as $r^2(1-\mu_*^2)$:

$$g_{\text{lines}} = \frac{1}{(1-\alpha)} \frac{\bar{Q}\kappa_e L_*}{4\pi r^2 c} \left(\frac{\partial v_r/\partial r}{\rho c \bar{Q}\kappa_e}\right)^\alpha \left[\frac{(\sigma+1)^{\alpha+1} - (\mu_*^2\sigma+1)^{\alpha+1}}{\sigma(1-\mu_*^2)(\alpha+1)(\sigma+1)^\alpha}\right]. \qquad (8.67)$$

The term in the square bracket in the above equation is the only extra term compared to the expression of the line-force for the case of a point source (Eq. (8.47)). In the literature this is called as *finite disk correction* factor, f_d, and appears as Eq. (50) in [1],

$$f_d = \frac{(\sigma+1)^{\alpha+1} - (\mu_*^2\sigma+1)^{\alpha+1}}{\sigma(1-\mu_*^2)(\alpha+1)(\sigma+1)^\alpha}. \qquad (8.68)$$

This plays an important role in the dynamics of hot-star winds wherein it reduces the total mass loss by factor of about two and increases the terminal velocity by factor 2–3. Thus, the neglegence of this factor may be rather misleading.

8.2 The Effects of Magnetic Fields on Massive Star Winds

We saw earlier that the solutions to CAK wind imply steady, smooth, and spherically symmetric outflows. However, there is evidence that massive star winds have extensive structures and variability on a range of spatial and temporal scales.

Relatively small-scale, stochastic structure—e.g., as evidenced by often quite constant soft X-ray emission [11], or by UV lines with extended black troughs understood to be a signature of a nonmonotonic velocity field [12]—seems most likely a natural result of the strong, intrinsic instability of the line-driving mechanism itself [5, 14]. But larger-scale structure—e.g., as evidence by explicit UV line profile variability in even low signal-to-noise IUE spectra [9, 10]—seems instead likely to be the consequence of wind perturbation by processes occurring in the underlying star. For example, the photospheric spectra of many hot stars show evidence of radial and/or nonradial pulsation, and in a few cases there is evidence linking this with observed variability in UV wind lines [13, 20].

An alternate scenario—as explored by ud-Doula and Owocki [25] and briefly summarized below—is that, in at least some hot stars, surface magnetic fields could perturb, and perhaps even channel, the wind outflow, leading to rotational modulation of wind structure that is diagnosed in UV line profiles, and perhaps even to magnetically confined wind-shocks with velocities sufficient to produce the relatively hard X-ray emission seen in some hot stars.

Next, we summarize this basic paradigm and show that a dimensionless number called "wind magnetic confinement" parameter can well characterize such a magnetized wind.

8.2.1 Magneto-Hydrodynamic Equations

Massive star winds are mainly composed of fully ionized hydrogen and helium, and
trace amounts of partially ionized metals ($\sim 10^{-4}$). These winds are dense enough
that the mean free paths between collisions are relatively small, and the plasma
can be treated as a single fluid. The time-dependent MHD (magnetohydrodynamic)
equations governing the system include the conservation of mass,

$$\frac{D\rho}{Dt} + \rho \nabla \cdot \mathbf{v} = 0, \tag{8.69}$$

and the equation of motion

$$\rho \frac{D\mathbf{v}}{Dt} = -\nabla p + \frac{1}{4\pi}(\nabla \times \mathbf{B}) \times \mathbf{B} - \frac{GM\hat{\mathbf{r}}}{r^2} + \mathbf{g}_{\text{external}}, \tag{8.70}$$

where ρ, p, and $\mathbf{v}$ are the mass density, gas pressure, and velocity of the fluid flow,
and $D/Dt = \partial/\partial t + \mathbf{v} \cdot \nabla$ is the advective time derivative. The gravitational constant G and stellar mass M set the radially directed ($\hat{\mathbf{r}}$) gravitational acceleration.
The term $\mathbf{g}_{\text{external}}$ represents the total external force that may include the centrifugal
force. In our case it is the line-force and the force due to scattering of the stellar
luminosity L by the free electron opacity κ_e. The magnetic field $\mathbf{B}$ is constrained to
be divergence free:

$$\nabla \cdot \mathbf{B} = 0, \tag{8.71}$$

and, under our assumption of an idealized MHD flow with infinite conductivity (e.g.,
[17]), its inductive generation is described by

$$\frac{\partial \mathbf{B}}{\partial t} = \nabla \times (\mathbf{v} \times \mathbf{B}). \tag{8.72}$$

The one last equation that we should, in principle, consider solving is an explicit
equation for conservation of energy, e.g., a purely adiabatic energy equation,

$$\rho \frac{D}{Dt}\left(\frac{e}{\rho}\right) = -p\nabla \cdot v. \tag{8.73}$$

But the massive star winds are relatively dense, and the energy balance is primarily
dominated by radiative processes that are rapid enough to keep the wind at nearly
constant temperature. Thus, one can assume an explicitly isothermal flow with $T =
T_{\text{eff}}$, where T_{eff} is the effective stellar temperature. In such a case, the sound speed
$a = \sqrt{kT/\mu}$ with k Boltzmann's constant and μ the mean atomic weight of the gas
is a constant. The perfect gas law can be used to compute the pressure:

$$p = \rho a^2. \tag{8.74}$$

8.2.2 The Wind Magnetic Confinement Parameter

In an interplay between magnetic field and stellar wind, the dominance of the field
is determined how strong the field is relative to the wind. In order to understand
the competition between these two, let us first define a characteristic parameter for
the relative effectiveness of the magnetic fields in confining and/or channeling the
wind outflow. Specifically, consider the ratio between the energy densities of field
vs. flow,

$$
\begin{aligned}
\eta(r, \theta) &\equiv \frac{B^2/8\pi}{\rho v^2/2} \\
&\approx \frac{B^2 r^2}{\dot{M} v} \\
&= \left[\frac{B_*^2(\theta) R_*^2}{\dot{M} v_\infty} \right] \left[\frac{(r/R_*)^{-2n}}{1 - R_*/r} \right],
\end{aligned}
\tag{8.75}
$$

where the latitudinal variation of the surface field has the dipole form given
by $B_*^2(\theta) = B_o^2(\cos^2\theta + \sin^2\theta/4)$. In general, a magnetically channeled outflow
will have a complex flow geometry, but for convenience, the second equality in
Eq. (8.75) simply characterizes the wind strength in terms of a spherically symmetric mass loss rate $\dot{M} = 4\pi r^2 \rho v$. The third equality likewise characterizes the
radial variation of outflow velocity in terms of the phenomenological velocity
law $v(r) = v_\infty(1 - R_*/r)$, with v_∞ the wind terminal speed; this equation furthermore models the magnetic field strength decline as a power-law in radius,
$B(r) = B_*(R_*/r)^{(n+1)}$, where, e.g., for a simple dipole, $n = 2$.

With the spatial variations of this energy ratio thus isolated within the right square
bracket, we see that the left square bracket represents a dimensionless constant that
characterizes the overall relative strength of field vs. wind. Evaluating this in the
region of the magnetic equator ($\theta = 90°$), where the tendency toward a radial wind
outflow is in most direct competition with the tendency for a horizontal orientation
of the field, one can thus define a equatorial wind magnetic confinement parameter,

$$
\begin{aligned}
\eta_* &\equiv \frac{B_*^2(90°) R_*^2}{\dot{M} v_\infty} \\
&= 0.4 \frac{B_{100}^2 R_{12}^2}{\dot{M}_{-6} v_8},
\end{aligned}
\tag{8.76}
$$

where $\dot{M}_{-6} \equiv \dot{M}/(10^{-6} M_\odot/\text{yr})$, $B_{100} \equiv B_0/(100 \text{ G})$, $R_{12} \equiv R_*/(10^{12} \text{ cm})$, and
$v_8 \equiv v_\infty/(10^8 \text{ cm/s})$. As these stellar and wind parameters are scaled to typical
values for an OB supergiant, e.g., ζ Pup, the last equality in Eq. (8.76) immediately
suggests that for such winds, significant magnetic confinement or channeling should
require fields of order $\sim$100 G. By contrast, in the case of the sun, the much weaker
mass loss ($\dot{M}_\odot \sim 10^{-14} M_\odot/\text{yr}$) means that even a much weaker global field ($B_0 \sim$

1 G) is sufficient to yield $\eta_* \simeq 40$, implying a substantial magnetic confinement of the solar coronal expansion. This is consistent with the observed large extent of magnetic loops in optical, UV, and X-ray images of the solar corona.

8.2.3 Alfvén Radius

What determines the extent of the effectiveness of magnetic, is the Alfvén radius, R_A, where flow and Alfvén velocities are equal. This can be derived from Eq. (8.75), where the second square bracket factor shows the overall radial variation, n is the power-law exponent for radial decline of the assumed stellar field, e.g., $n = 2$ for a pure dipole, and β is the velocity-law index, with typically $\beta \approx 1$. For a star with a nonzero field, we have $\eta_* > 0$, and so given the vanishing of the flow speed at the atmospheric wind base, this energy ratio always starts as a large number near the stellar surface, $\eta(r \to R_*) \to \infty$. But from there outward it declines quite steeply, asymptotically as r^{-4} for a dipole, crossing unity at the Alfvén radius defined implicitly by $\eta(R_A) \equiv 1$.

For a canonical $\beta = 1$ wind velocity law, explicit solution for R_A along the magnetic equator requires finding the appropriate root of

$$\left(\frac{R_A}{R_*}\right)^{2n} - \left(\frac{R_A}{R_*}\right)^{2n-1} = \eta_*, \tag{8.77}$$

which for integer $2n$ is just a simple polynomial, specifically a quadratic, cubic, or quartic for $n = 1$, 1.5, or 2. Even for noninteger values of $2n$, the relevant solutions can be approximated (via numerical fitting) to within a few percent by the simple general expression

$$\frac{R_A}{R_*} \approx 1 + (\eta_* + 1/4)^{1/(2n)} - (1/4)^{1/(2n)}. \tag{8.78}$$

For weak confinement, $\eta_* \ll 1$, we find $R_A \to R_*$, while for strong confinement, $\eta_* \gg 1$, we obtain $R_A \to \eta_*^{1/(2n)} R_*$. In particular, for the standard dipole case with $n = 2$, we expect the strong-confinement scaling $R_A/R_* \approx 0.3 + \eta_*^{1/4}$.

Clearly R_A represents the radius at which the wind speed v exceeds the local Alfvén speed V_A. It also characterizes the maximum radius where the magnetic field still dominates over the wind.

8.2.4 2D MHD Simulations of Massive Star Winds

Initial MHD simulations carried out by ud-Doula and Owocki [25] focused on a parameter study of the wind magnetic confinement as defined above. For simplicity, these assumed isothermal wind with no stellar rotation.

A key result is that the overall degree to which the wind is influenced by the field depends largely on a single, dimensionless, wind magnetic confinement parameter, η_* ($=B_{eq}^2 R_*^2/\dot{M}v_\infty$), which characterizes the ratio between magnetic field energy density and kinetic energy density of the wind (see Fig. 8.3). For weak confinement $\eta_* \leq 1$, the field is fully opened by the wind outflow, but nonetheless for confinements as small as $\eta_* = 1/10$ can have a significant back-influence in enhancing the density and reducing the flow speed near the magnetic equator. For stronger confinement $\eta_* > 1$, the magnetic field remains closed over a limited range of latitude and height about the equatorial surface but eventually is opened into a nearly radial configuration at large radii. Within closed loops, the flow is channeled toward loop tops into shock collisions that are strong enough to produce hard X-rays, with the stagnated material then pulled by gravity back onto the star in quite complex and variable inflow patterns. Within open field flow, the equatorial channeling leads to oblique shocks that are again strong enough to produce X-rays and also lead to a thin, dense, slowly outflowing "disk" at the magnetic equator. The polar flow is characterized by a faster-than-radial expansion that is more gradual than anticipated in previous 1D flow-tube analyses, and leads to a much more modest increase in terminal speed (<30 %), consistent with observational constraints.

Subsequent simulations [7] use full energy equation to simulate the wind of the magnetic star θ^1 Ori C. The magnetically channeled wind shock model provides excellent agreement with the diagnostics from the phase-resolved Chandra spectroscopy of θ^1 Ori C. The modest line widths are consistent with the predictions of the MHD simulations. The X-ray light curve and He-like f/i values indicate that the bulk of the X-ray-emitting plasma is located at approximately $1.5R_*$, very close to the photosphere, which is also consistent with the MHD simulations of the MCWS mechanism. The simulations also correctly predict the temperature and total luminosity of the X-ray-emitting plasma. Quite remarkably, the only inputs to the MHD simulations were the magnetic field strength, effective temperature, radius, and mass-loss rate of θ^1 Ori C, all of which are fairly well constrained by observation.

8.3 The Effects of Field-Aligned Rotation on Magnetized Massive Star Winds

More generally massive stars tend to have quite rapid rotation, as evidenced both by the substantial broadening in photospheric spectral lines [3, 6], which indicate projected rotation speeds of hundreds of km/s, and by the relatively short-period of observed modulations for some stars, e.g., the magnetic Bp star σ Ori E, for which the inferred rotation period is about 1.2 d [26]. Both lines of evidence suggest that hot-star rotation rates are commonly a substantial fraction of the "critical" rate at which the equatorial surface would be in Keplerian orbit. Since this implies centrifugal forces that are comparable to the inward pull of gravity, it is clear that such levels of rotation could significantly influence the magnetic channeling of a stellar wind.

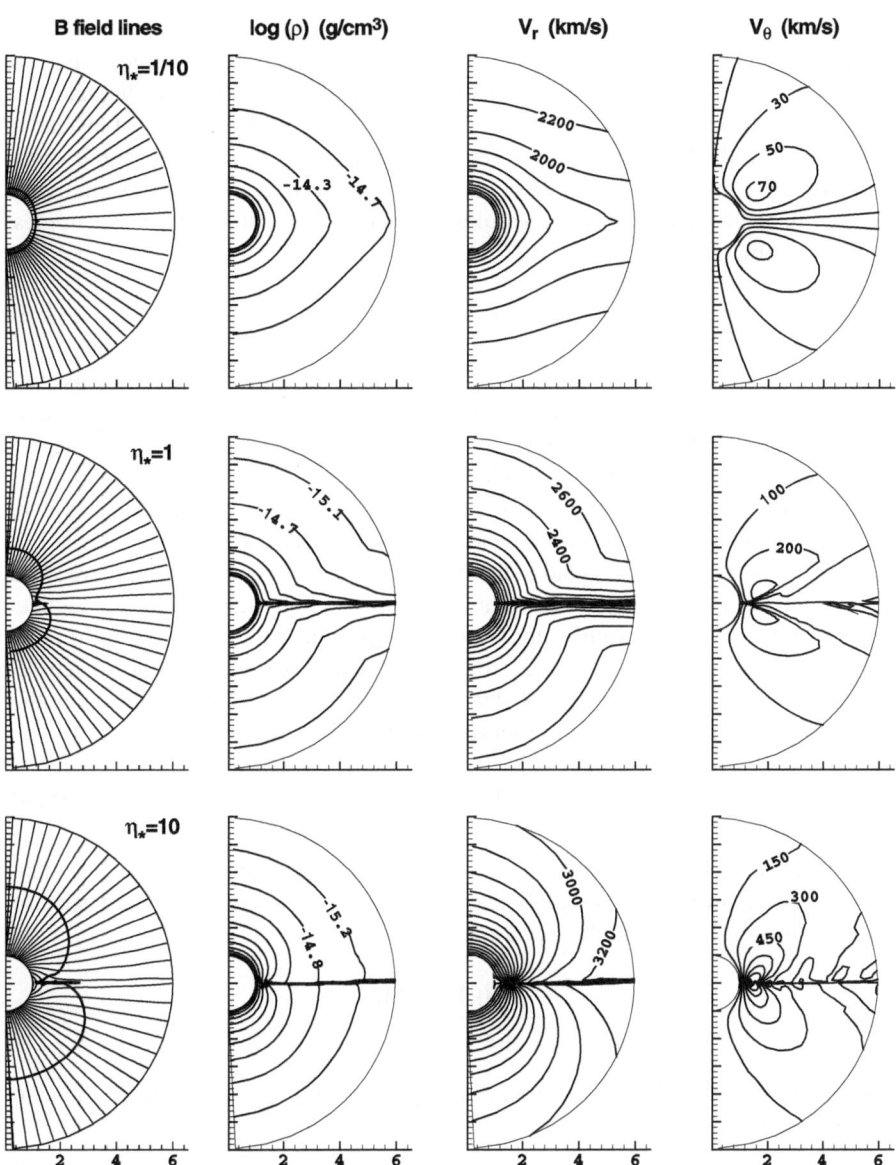

Fig. 8.3 Comparison of overall properties at the final simulation time ($t = 450$ s) for 3 MHD models, chosen to span a range of magnetic confinement from small (*top row*; $\eta_* = 1/10$), to medium (*middle row*; $\eta_* = 1$), and to large (*bottom row*; $\eta_* = 10$). The *leftmost panels* show magnetic field lines, together with the location (*bold contour*) of the Alfvén radius, where the radial flow speed equals the Alfvén speed. From left to write, the remaining columns show contours of log(density), radial velocity, and latitudinal velocity

8.3.1 Rotational Parameter

To facilitate the discussion about how rotation affects the winds of magnetic massive stars, one can introduce a new "rotational parameter", W. This can be characterized in terms of a speed, namely the equatorial surface rotation speed $V_{\rm rot}$. But instead of relating that to the flow speed or Alfvén speed in the stellar *wind*, the stellar origin of rotation suggests that it may be better to compare it to a speed representative of the gravity at the stellar *surface*. Specifically, let us thus define our dimensionless rotation parameter as

$$W \equiv \frac{V_{\rm rot}}{V_{\rm orb}}, \tag{8.79}$$

where $V_{\rm orb} \equiv \sqrt{GM/R_*}$ is the *orbital* speed near the equatorial surface.[1] This characterizes the azimuthal speed needed for the outward centrifugal forces to balance the stellar surface gravity. It is only a factor $1/\sqrt{2}$ less than the speed $V_{\rm esc}$ needed to fully *escape* the star's surface gravitational potential.

For a nonmagnetic rotating star, conservation of angular momentum in a wind outflow causes the azimuthal speed near the equator to decline outward as $v_\phi \sim 1/r$, meaning that rotation effects tend to be of diminishing importance in the outer wind.

By contrast, in a rotating star with a sufficiently strong magnetic field, magnetic torques on the wind can spin it up; for some region near the star, i.e., up to about the maximum loop closure radius R_c (closely related to Alfvén radius R_A discussed earlier), they can even maintain a nearly rigid-body rotation, for which the azimuthal speed now *increases* outward in proportion to the radius:

$$v_\phi(r) = V_{\rm rot}\frac{r}{R_*}; \quad r \lesssim R_c. \tag{8.80}$$

As such, even for a star with surface rotation below the orbital speed, $W < 1$, maintaining rigid rotation will eventually lead to a balance between the outward centrifugal force from rotation and the inward force of gravity:

$$\frac{v_\phi^2(R_K)}{R_K} = \frac{GM}{R_K^2}. \tag{8.81}$$

Combining this with Eqs. (8.79) and (8.80) gives a simple expression for the associated "Kepler radius",

$$R_K = W^{-2/3}R_*. \tag{8.82}$$

---

[1] This is closely related to the commonly used rotation parameter $\omega \equiv \Omega/\Omega_{\rm crit}$, defined by the star's angular rotating frequency Ω relative to the value this would have as the star approaches "critical" rotation, Ω_c. The choice here more directly relates to the additional local speed needed to propel material into Keplerian orbit, and avoids some subtle assumptions (e.g. rigid-body rotation using a Roche potential for gravity) about how the global stellar envelope structure adjusts to approaching the critical rotation limit.

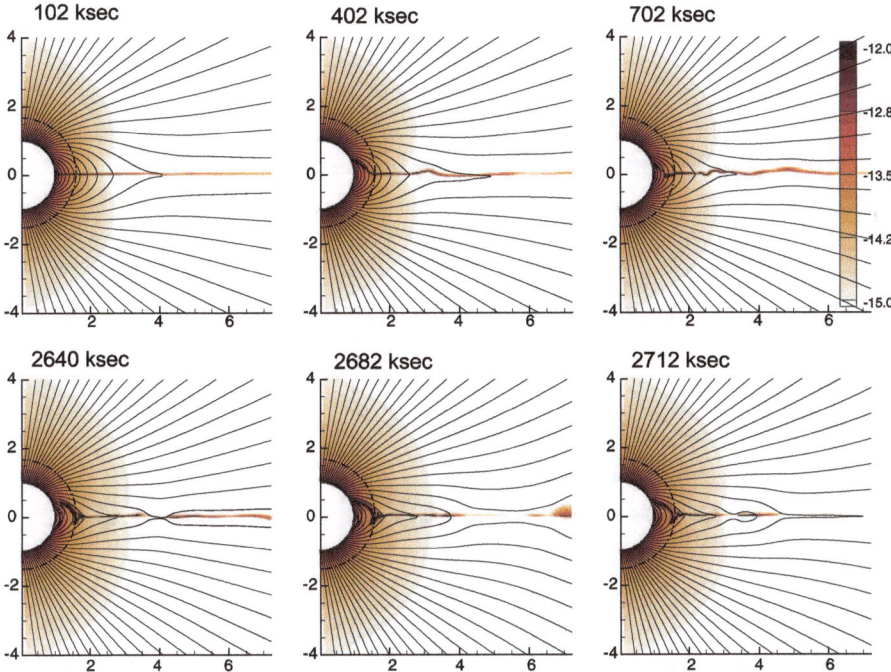

Fig. 8.4 Snapshots of density (in cgs units on a logarithmic color-scale) and field lines (*solid lines*) at the labeled time intervals for the model with $\eta_* = 100$ and $W = 1/2$. The *top panels* show the model during a relatively quiescent period, when a dense rigidly rotating disk is being gradually built up. Note, however that material below the Kepler co-rotation radius (R_K, shown as a *dashed circle*) falls back onto the stellar surface, due to the lack of sufficient centrifugal support. The extent of the disk in this phase is determined by the magnetic field strength and extends up to the Alfvén radius $R_A \approx 3.4 R_*$, which is somewhat above the maximum outer radius of closed magnetic loops, R_c. The *bottom panels* show the model later in the evolution, during one of the episodic centrifugal breakout events

Unsupported material at radii $r < R_K$ will tend to fall back toward the star, but any material maintained in rigid-rotation to radii $r > R_K$ will have a centrifugal force that *exceeds* gravity and so will tend to be propelled further outward. Indeed, any corotating material above an "escape radius", which is only slightly beyond the Kepler radius,

$$R_E = 2^{1/3} R_K, \qquad (8.83)$$

will have sufficient rotational energy to escape altogether the local gravitational potential, unless, of course, temporarily held down by the magnetic field.

8.3.2 Disk Formation

This paradigm is well demonstrated by ud-Doula et al. [23] using 2D MHD simulations assuming field-aligned rotation. Figures 8.4 show a series of time snapshots

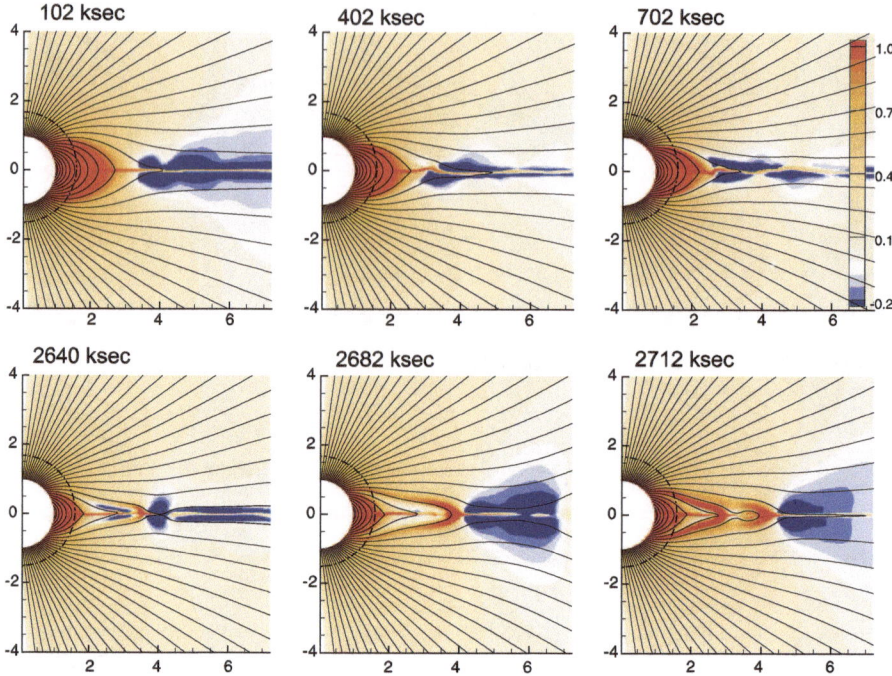

Fig. 8.5 For the same model and same time snapshots as in Fig. 8.4, the azimuthal speed scaled by the local co-rotation speed, $\chi \equiv v_\phi / \Omega r \sin \theta$. During the quiescent period, the magnetosphere extends nearly to the closure radius $R_c \approx 2.7 R_*$, and corotates almost rigidly with the star (*top panels*). However, during episodic centrifugal breakouts (*bottom panels*), this rigidly rotating magnetosphere shrinks nearly to $R_K \approx 1.6 R_*$, represented by the dashed circle. In addition to the mass lost to the breakout, this also helps some of the disk mass to leak inward as infall back onto the star

of the 2D spatial configuration of the magnetic field (solid lines), with the color scale representing logarithm of density; Fig. 8.5 give a similar time sequence for the azimuthal flow speed, scaled relative to the value that would occur in rigid-rotation, i.e., $\chi \equiv v_\phi(r) / \Omega r \sin \theta$, where $\Omega \equiv V_{rot}/R_*$ is the star's angular rotation frequency. The time snapshots were chosen to illustrate both relatively quiescent intervals (top panels) and phases with dynamic centrifugal breakout (bottom panels). The dashed circle represents the Kepler corotation radius at the equator ($R_K \approx 1.6 R_*$).

In the evolution immediately following the initial condition, the magnetic field channels wind material toward the tops of closed loops near the equator, where the collision with the opposite stream leads to a dense disk-like structure (see top panels). But the gas is also generally torqued by the field, with, as can be seen in the upper panels of Fig. 8.5, material in the closed magnetosphere up to $R_c \approx 2.7 R_*$ kept nearly in rigid-body corotation with the star. Note that these closed, rigidly rotating loops thus extend through and beyond the Kepler radius. For any material trapped on loops below R_K, the outward centrifugal support is less than the inward pull of gravity; since much of this material is compressed into clumps that are too dense to be significantly line-driven, it thus eventually falls back to the star following complex patterns along the closed-field loops.

By contrast, the dense material *above* the dashed line at R_K has a net radially *outward* force from the centrifugal acceleration vs. gravity. Still, during the initial build-up of this material at the tops of loops above R_K, the magnetic field provides tension that is strong enough to hold it down, forming then a segment of the *rigidly rotating disk* predicted in the analytic Rigidly Rotating Magnetosphere (RRM) analysis by [21]. However, eventually material in the outer region of this RRM accumulates to sufficient density to force open the magnetic field, leading to the kind of centrifugally driven breakout events simulated in [22]. This is illustrated here in the bottom panels of Fig. 8.4.

8.3.3 Spindown

Another natural consequence of interaction of magnetic field with wind and rotation is the increase in angular momentum loss from the stellar surface. It turns out that the angular momentum is carried away by not only the gas, but also by the field itself in the form of Maxwellian stress tensor. In a classical analysis, Weber and Davis [27, hereafter WD] showed that the total angular momentum loss $\dot{J} = (2/3)\dot{M}\Omega R_A^2$, where $\dot{M}$ is the mass loss rate, Ω is the stellar angular velocity, and R_A is a characteristic Alfvén radius.

However, there were two key shortcomings in that analysis. Firstly, it assumed a monopole field. Secondly, it never provided a way to compute Alfvén radius. In the context of massive star winds, ud-Doula et al. [24] analyzed the angular momentum loss in massive stars. A key distinction was that for a dipole field they used, the Alfvén radius has a strong-field scaling $R_A/R_* \approx \eta_*^{1/4}$, instead of the scaling $R_A/R_* \sim \sqrt{\eta_*}$ for a monopole field used by WD. This leads to a slower stellar spindown time that in the dipole case scales as $\tau_{\rm spin} = \tau_{\rm mass} 1.5k/\sqrt{\eta_*}$, where $\tau_{\rm mass} \equiv M/\dot{M}$ is the characteristic mass loss time, and k is the dimensionless factor for stellar moment of inertia, typically $k \sim 0.1$ for most massive stars. The full numerical scaling relation gives typical spindown times of order 1 Myr for several known magnetic massive stars.

8.4 Summary

In this chapter, I discussed briefly the nature of gas pressure driven winds and compared this to line-driven massive star winds. Despite being few in numbers, massive stars play an important role in enriching interstellar medium. So, it is important to study how these stars lose their mass. In particular, I outlined how magnetic field and rotation affect such mass loss.

I demonstrate that the overall degree to which the wind is influenced by the field depends largely on a single, dimensionless, wind magnetic confinement parameter, η_*. Stars with moderate magnetic confinement $\eta_* > 1$ will have substantial modulation in the wind due to the field that can lead to density enhanced equatorial regions and observable latitudinal variation of wind speed. Such magnetically confined winds can also be responsible for X-ray production. One example is slowly rotating θ^1 Ori C for which MHD simulations reproduce well-observed X-ray properties.

Strong magnetic confinement, $\eta_* \gg 1$, in combination with rotation leads to rigidly rotating disks that MHD simulations cannot easily model but can be well explained by semi-analytic RRM formulation, as is the case for the Bp star σ Ori E. One of the main shortcomings of all the MHD simulations discussed in this article is that they were all 2D. For proper modeling of stellar wind, 3D is necessary. Such 3D MHD models are currently under way.

Acknowledgements   The author thanks CNRS for the invitation to the school devoted on the topic "From solar environment to stellar environment" held in Roscoff (F).

References

1. Castor, J.I., Abbott, D.C., Klein, R.I.: Astrophys. J. **195**, 157 (1975)
2. Chapman, S., Zirin, H.: Smithson. Contrib. Astrophys. **2**, 1 (1957)
3. Conti, P.S., Ebbets, D.: Astrophys. J. **213**, 438 (1977)
4. Cranmer, S.R.: Ph.D. thesis (1996)
5. Feldmeier, A.: Astron. Astrophys. **299**, 523 (1995)
6. Fukuda, I.: Publ. Astron. Soc. Pac. **94**, 271 (1982)
7. Gagné, M., Oksala, M.E., Cohen, D.H., et al.: Astrophys. J. **628**, 986 (2005)
8. Gayley, K.G.: Astrophys. J. **454**, 410 (1995)
9. Howarth, I.D., Smith, K.C.: Astrophys. J. **439**, 431 (1995)
10. Kaper, L., Henrichs, H.F., Nichols, J.S., Snoek, L.C., Volten, H., Zwarthoed, G.A.A.: Astron. Astrophys. Suppl. Ser. **116**, 257 (1996)
11. Long, K.S., White, R.L.: Astrophys. J. Lett. **239**, L65 (1980)
12. Lucy, L.B.: Astrophys. J. **255**, 286 (1982)
13. Mathias, P., Aerts, C., Briquet, M., De Cat, P., Cuypers, J., Van Winckel, H., Flanders, Le Contel, J.M.: Astron. Astrophys. **379**, 905 (2001)
14. Owocki, S.P.: Astrophys. Space Sci. **221**, 3 (1994)
15. Owocki, S.P., Castor, J.I., Rybicki, G.B.: Astrophys. J. **335**, 914 (1988)
16. Parker, E.N.: Astrophys. J. **128**, 664 (1958)
17. Priest, E.R., Hood, A.W.: Advances in Solar System Magnetohydrodynamics. Cambridge University Press, Cambridge (1991)
18. Sobolev, V.V.: Sov. Astron. **1**, 332 (1957)
19. Spitzer, L. Jr.: Astrophys. J. **124**, 20 (1956)
20. Telting, J.H., Aerts, C., Mathias, P.: Astron. Astrophys. **322**, 493 (1997)
21. Townsend, R.H.D., Owocki, S.P.: Mon. Not. R. Astron. Soc. **357**, 251 (2005)
22. ud-Doula, A., Townsend, R.H.D., Owocki, S.P.: Astrophys. J. Lett. **640**, L191 (2006)
23. ud-Doula, A., Owocki, S.P., Townsend, R.H.D.: Mon. Not. R. Astron. Soc. **385**, 97 (2008)
24. ud-Doula, A., Owocki, S.P., Townsend, R.H.D.: Mon. Not. R. Astron. Soc. **392**, 1022 (2009)
25. ud-Doula, A., Owocki, S.P.: Astrophys. J. **576**, 413 (2002)
26. Walborn, N.R.: Astrophys. J. Lett. **243**, L37 (1981)
27. Weber, E.J., Davis, L. Jr.: Astrophys. J. **148**, 217 (1967)

Chapter 9
Magnetic Field and Convection in the Cool Supergiant Betelgeuse

P. Petit, M. Aurière, R. Konstantinova-Antova, A. Morgenthaler, G. Perrin, T. Roudier, and J.-F. Donati

Abstract We present the outcome of a highly-sensitive search for magnetic fields on the cool supergiant Betelgeuse. A time-series of six circularly polarized spectra was obtained using the NARVAL spectropolarimeter at Télescope Bernard Lyot (Pic du Midi Observatory (F)), between March and April 2010. Zeeman signatures were repeatedly detected in cross-correlation profiles, corresponding to a longitudinal component of about 1 G. The time-series unveils a smooth increase of the longitudinal field from 0.5 to 1.5 G, correlated with radial velocity fluctuations. We observe a strong asymmetry of Stokes V signatures, also varying in correlation with the radial velocity. The Stokes V line profiles are red-shifted by about 9 km s^{-1} with respect to the Stokes I profiles, suggesting that the observed magnetic elements may be concentrated in the sinking components of the convective flows.

9.1 Introduction

The widespread signatures of magnetic activity in cool stars are believed to originate from the coexistence of convection and rotation in their outer layers. Since the first dynamo models of Parker [20], rotation is generally accepted as a central ingredient of these stellar dynamos in a two-level action. The stellar spin is first involved in the generation of radial and latitudinal shears that are able to wind up the field lines of a seed magnetic field around the rotation axis, resulting in the creation of a strong toroidal field component. Stellar rotation is also acting on the vertical plasma flows through the Coriolis force that succeeds at generating helical motions able to twist again the field lines of the toroidal field and generate a poloidal field component.

P. Petit (✉) · M. Aurière · A. Morgenthaler · T. Roudier · J.-F. Donati
IRAP, CNRS & Université de Toulouse, 14 avenue Edouard Belin, 31400 Toulouse, France
e-mail: petit@ast.obs-mip.fr

R. Konstantinova-Antova
Institute of Astronomy, Bulgarian Academy of Science, 72 Tsarigradsko shose, 1784 Sofia, Bulgaria

G. Perrin
Observatoire de Paris, LESIA, 5 Place Jules Janssen, 92190 Meudon, France

J.-P. Rozelot, C. Neiner (eds.), *The Environments of the Sun and the Stars*,
Lecture Notes in Physics 857,
DOI 10.1007/978-3-642-30648-8_9, © Springer-Verlag Berlin Heidelberg 2013

If many details of the physical processes involved in such large-scale dynamos are still a matter of debate (see e.g. [6] for a review), this theoretical framework is now widely accepted to interpret the cyclical activity behaviour of the Sun and solar-type stars. However, other models suggest that convection alone is able to sustain a dynamo, without any rotational effects involved (e.g. [5, 32]). Turbulent dynamo action may be responsible for the smallest-scale magnetic elements observed on the solar surface [16], although intranetwork magnetic elements may also result from the decay of larger magnetic regions (created by the global dynamo) through the continuous convective mixing of the solar upper layers.

Since both global and small-scale dynamos may be simultaneously active in the Sun, it is not easy to disentangle the respective magnetic outcome of these two different processes. A promising way to reach this goal consists in observing a star with no rotation at all, or at least a star rotating so slowly that the onset of a global dynamo in its internal layers is unlikely. If stellar spectropolarimetry is our best asset to detect a magnetic field in a non-rotating star, the polarimetric detection of Zeeman signatures is mostly insensitive to small-scale magnetic elements as those expected to be generated by a local dynamo. This issue is unescapable for solar-type dwarfs, on which millions of photospheric convective cells are visible at any time, resulting in a highly tangled intranetwork field pattern. Cool supergiant stars may offer a rare opportunity to circumvent this problem, since their convective cells are expected to be much larger than on the Sun, with only a few of them covering the stellar surface [7, 27], so that the spatial scale of convection may be sufficiently large to limit the mutual cancellation of Zeeman signatures of close-by magnetic elements with opposite polarities.

To test the feasibility of magnetic field detection in cool supergiant stars, we have concentrated on Betelgeuse (α Orionis), one of the brightest members of this class. We briefly summarize a few well-known properties of this object, present the spectropolarimetric observations gathered with NARVAL and discuss the first outcome of this project.

9.2 The Cool Supergiant α Orionis

With an M2Iab spectral type classification, Betelgeuse can be taken as the proto-type of cool supergiant stars. Thanks to its proximity to the Earth, it was the first star to have its radius determined using interferometry [18], and a recent value of $645 \pm 129 R_\odot$ was derived in near-infrared interferometry [21]. A significant scatter is observed in radius estimates depending on the adopted wavelength domain, because of the extension and thermal structure of the stellar atmosphere (see e.g. Uitenbroek et al. [30] and references therein). The efficient mass-loss of Betelgeuse is at the origin of an inhomogeneous distribution of gas and dust, detected up to a few tens of stellar radii [12].

Using spatially-resolved, high-resolution UV spectroscopy with the HST, Uiten-broek et al. [30] were able to propose a rotation period of about 17 yr and a low inclination of the rotation axis, of about 20°. The basic ingredients of a global dynamo, primarily based on rotation, are therefore likely missing in α Orionis.

9.3 Surface Magnetic Field

Spectropolarimetric observations of Betelgeuse were obtained using the NARVAL spectropolarimeter [1], installed at Télescope Bernard Lyot,[1] Pic du Midi Observatory (France), in a highly sensitive search for a weak surface magnetic field [4]. The data were collected during 6 nights in 2010, from 14 March to 17 April. They consist of high-resolution ($R = 65,000$) echelle spectra in light intensity (Stokes I) and circular polarization (Stokes V), offering an almost contiguous coverage of the wavelength interval between 370 nm and 1,000 nm. Each Stokes V spectrum is built from a sequence of four exposures taken with different azimuths of the polarimetric optics. Using a different combination of the sub-exposures, a "null" control spectrum is also computed, which should not contain any detectable signature. To reach a high signal-to-noise ratio and avoid any saturation of the detector, a sequence of 16 to 20 spectra was acquired for every individual night and later averaged.

To lower further the noise level and improve the detectability of Zeeman signatures, all spectra were processed using the Least-Squares-Deconvolution (LSD) cross-correlation technique [8, 13]. By doing so, mean line profiles were computed, using a list of 15,000 photospheric atomic lines corresponding to the atmospheric parameters of Betelgeuse. The resulting time-series of Stokes V profiles is plotted in Fig. 9.1. Circularly polarized signatures, most likely generated through the Zeeman effect, are detected above noise level at the radial velocity of the star.

Using the centre-of-gravity technique [26], the Stokes V signatures can be translated into estimates of the line-of-sight component (B_l) of the magnetic field [4]. The series of B_l measurements are plotted in Fig. 9.2. They display an average value of the order of 1 G. If the variability of Stokes V profiles is barely visible to the naked eye, B_l estimates enable one to unveil a first type of variability of the Zeeman signatures, with a regular increase of the field strength over our observing window, from 0.5 to 1.5 G.

9.4 Surface Convection and Magnetic Activity

The wide spectral coverage and high spectral resolution of NARVAL spectra give access to a number of classical activity indicators that provide us with a useful set of additional measurements, carrying information that complement the longitudinal field estimates.

---

[1]The Bernard Lyot Telescope is operated by the Institut National des Sciences de l'Univers of the French Centre National de la Recherche Scientifique.

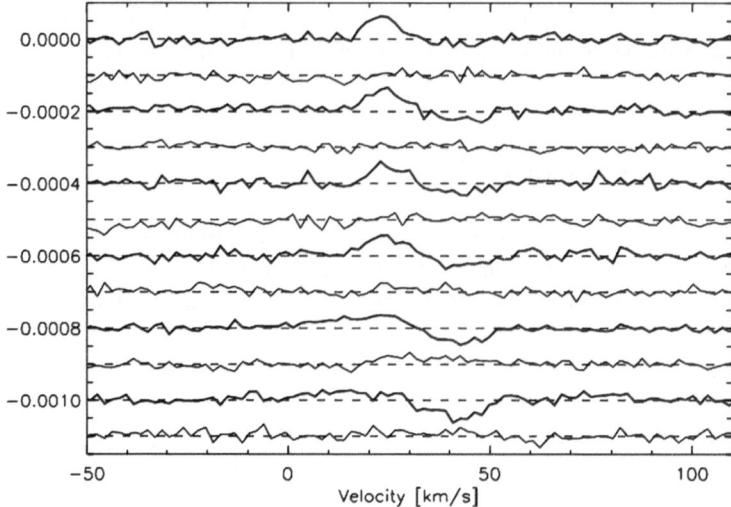

Fig. 9.1 Time-series of Stokes V (*bold lines*) and their corresponding "null" line profiles (*thin line below*). The successive profiles are shifted vertically for display clarity. After [4]

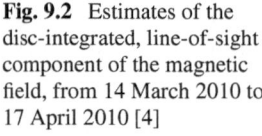

Fig. 9.2 Estimates of the disc-integrated, line-of-sight component of the magnetic field, from 14 March 2010 to 17 April 2010 [4]

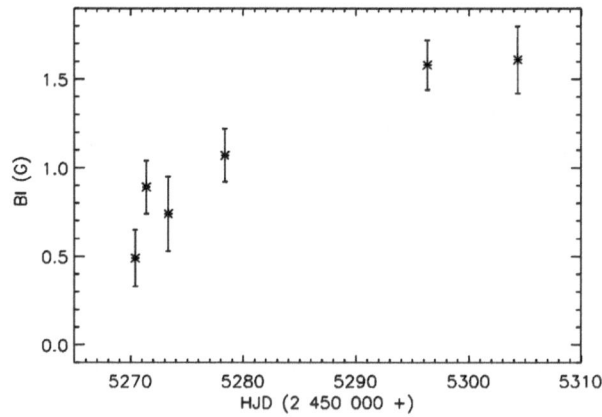

9.4.1 Chromospheric Emission

The stellar chromospheric flux can be inferred from the various chromospheric lines showing up in NARVAL intensity spectra. We choose here to concentrate on the Ca II infrared triplet, located in a spectral domain where the signal-to-noise ratio is generally high in Betelgeuse spectra.

An emission index is constructed by integrating the fluxes Ca_1, Ca_2 and Ca_3 in three rectangular bandpasses, centred around the three components of the Ca II triplet (at 849.8, 854.2 and 866.2 nm), each one with a width of 0.2 nm. We also integrate the flux in two rectangular bands C_1 and C_2 with a width of 0.5 nm in the

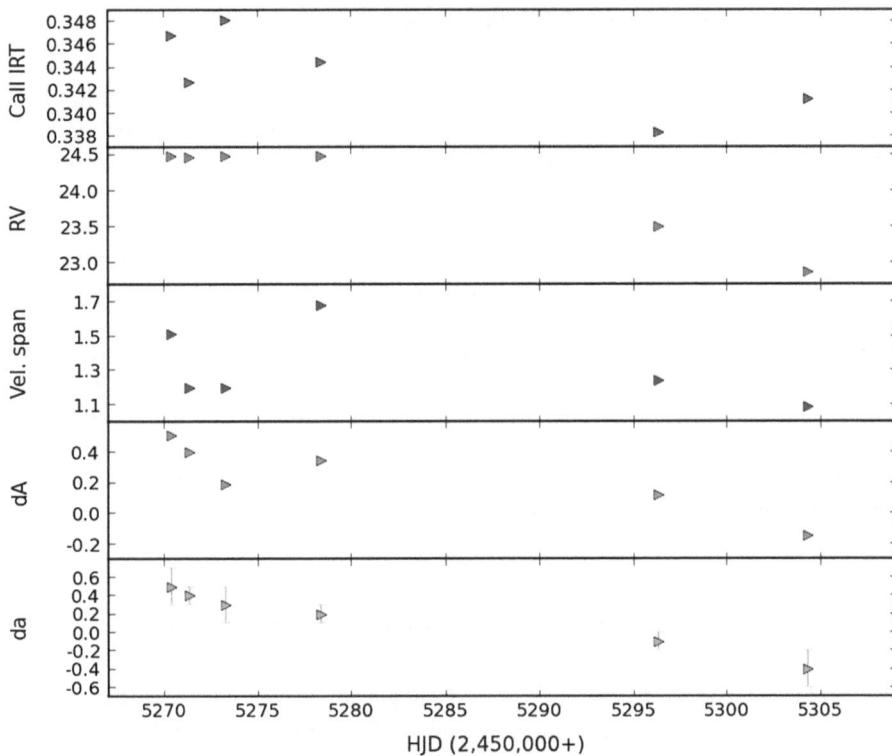

Fig. 9.3 Additional activity indicators derived from NARVAL spectra. From top to bottom, we plot the Ca II IRT index, the mean radial velocity of Stokes I LSD profiles (RV) in $km\,s^{-1}$, the velocity span of bisectors calculated from Stokes I LSD profiles ($km\,s^{-1}$), the relative *surface* asymmetry (dA) and the relative *amplitude* asymmetry (da) of Stokes V LSD profiles. Statistical error bars are not indicated in the plots whenever they are smaller than the symbol size

neighbouring spectral domain (centred around 847.58 and 870.49 nm) to normalize the index, which is therefore computed as follows:

$$I = \frac{Ca_1 + Ca_2 + Ca_3}{C_1 + C_2}. \tag{1}$$

The temporal evolution of this index is plotted in Fig. 9.3. Significant fluctuations are observed. If a global decrease is visible between March and April, the evolution is not as smooth as the increase of B_l. The Ca II emission is generally expected to be correlated to the local magnetic field strength, so that the (loose) anti-correlation between B_l and the Ca II index might seem paradoxical. We stress, however, that in the case of stars with low $v \sin i$ values, the polarimetric signatures are specifically sensitive to the low-order component of the magnetic topology, while the chromospheric emission retains information about a wide range of spatial scales [25]. The different temporal trend of both tracers is therefore not a surprise in the context of a magnetic field probably shaped by a complex pattern of convective flows.

9.4.2 Radial Velocity and Profile Bisectors

The radial velocities (RV) of Stokes I LSD profiles, derived from the fitting of a Gaussian on the line profiles, are plotted in Fig. 9.3, following previous work by Gray [10]. The wavelength calibration is performed using telluric lines recorded in NARVAL spectra. For solar-type dwarfs, the resulting RV stability is of the order of 15 to 30 m s^{-1} [19]. A similar accuracy can be expected for Betelgeuse, for which LSD profiles benefit from a much larger number of spectral lines. The observed fluctuations are far above this limit, reaching about 1.5 km s^{-1}. The evolution is smooth and anti-correlated to B_l measurements.

The Stokes I LSD profiles of Betelgeuse are also highly asymmetric, with a red wing of the line much wider than the blue wing [4]. Bisectors computed from the LSD profiles confirm this trend, with a clear deviation towards the red near the continuum. The total velocity span of the bisectors (plotted in Fig. 9.3) is comprised between 1 and 2 km s^{-1}. Fluctuations are less organized than those recorded in RV, with a marginal trend to display a lower bisector span in April.

9.4.3 Stokes V Asymmetry and Zero-Crossing of Stokes V Profiles

The double-peaked Stokes V profiles of Betelgeuse all possess a significant level of asymmetry, in the sense that the blue and red lobes of the profiles have different amplitudes and delimit different surfaces. This property of Stokes V signatures is widely observed on the Sun [31] and is generally linked to vertical gradients of both velocity and magnetic field in magnetic elements [17].

To quantify this observation, we use the approach of Petit et al. [24] and call a_b and a_r the amplitudes of the blue and red lobes of the profile. We then derive a relative *amplitude* asymmetry $da = (a_b - a_r)/(a_b + a_r)$. In a similar manner, we define a relative *surface* asymmetry $dA = (A_b - A_r)/(A_b + A_r)$, where A_b and A_r are the areas of the blue and red lobes, respectively. The successive values of dA and da are plotted in Fig. 9.3. We observe a global decrease of both parameters, with a more regular trend in da. Both parameters change sign during the time-series.

The double-lobed shape of Stokes V profiles implies a sign reversal close to the line-core. The zero-crossing radial velocity observed for the Stokes V profiles of Betelgeuse is of about 33 km s^{-1} (with no clear temporal trend), while the core of the Stokes I profile is located at a significantly smaller velocity of about 24 km s^{-1}. These systematic velocity shifts between Stokes I and V can have different origins. A first possible explanation would involve the Stokes V asymmetry reported above, as it can induce a shift of the zero-crossing in the case of a limited spectral resolution [29]; but in this case, the sharp evolution of da and dA should progressively displace the zero-crossing towards lower radial velocities, which is not observed. A second option involves a large-scale toroidal magnetic field [24], but the existence of a toroidal component would be hard to reconcile with the absence of any significant rotation in Betelgeuse (unless the slowly rotating surface is hiding faster rotation in internal layers). As a third option, we propose that the red-shift of Stokes V profiles may be related to the concentration of magnetic elements in the sinking component of the convective mixing, as observed in small-scale solar magnetic elements. If this last interpretation seems more easy to reconcile with the basic properties of Betelgeuse and is also quantitatively consistent with the convective velocity measurements of Gray [10], further investigation is obviously needed before reaching any conclusion.

9.5 A Small-Scale Dynamo in Betelgeuse?

In the past, strong magnetic fields have been detected in fast-rotating giants or subgiants belonging to RS CVn systems or to the FK Com class [22, 23]. More recently, repeated magnetic field detections have been obtained in active, single red giants [3, 14, 15, 28] and in a likely descendant of a strongly magnetic Ap star [2].

In this growing literature on evolved stars, the detection of a weak surface magnetic field at the surface of Betelgeuse [4] is an important observational result, in the sense that the physical interpretations proposed for other objects to account for their magnetic nature cannot be applied here. Firstly, the magnetic field of Betelgeuse has to be generated without the help of a fast, or even moderate stellar rotation, and this specificity should exclude any global dynamo. Secondly, the very large radius implies that any magnetic remnant of a strong magnetic field on the main sequence would be too diluted to be detectable at photospheric level. In this situation, a more natural interpretation would involve the convection alone as the engine of a dynamo, bringing the first strong observational evidence that such a process (proposed by Dorch and Freytag [9]) can be efficient in cool stars.

This exciting result, later confirmed for a larger sample of cool supergiants [11], comes together with a number of additional tracers of magnetic activity and convection (chromospheric emission, radial velocities, line bisectors, Stokes V asymmetries). This wealth of information is a motivation to pursue the spectropolarimetric monitoring of Betelgeuse, in order to investigate longer-term trends that may affect the various measurements at our disposal and study the possible role of the surface magnetic field in the onset of the mass-loss of Betelgeuse and other supergiant stars.

References

1. Aurière, M.: EAS Publ. Ser. **9**, 105 (2003)
2. Aurière, M., et al.: Astron. Astrophys. **491**, 499 (2008)
3. Aurière, M., et al.: Astron. Astrophys. **504**, 231 (2009)
4. Aurière, M., Donati, J.-F., Konstantinova-Antova, R., Perrin, G., Petit, P., Roudier, T.: Astron. Astrophys. **516**, L2 (2010)
5. Cattaneo, F.: Astrophys. J. **515**, L39 (1999)
6. Charbonneau, P.: Living Rev. Sol. Phys. **7**, 3 (2010)
7. Chiavassa, A., Haubois, X., Young, J.S., Plez, B., Josselin, E., Perrin, G., Freytag, B.: Astron. Astrophys. **515**, A12 (2010)

8. Donati, J.-F., Semel, M., Carter, B.D., Rees, D.E., Collier, Cameron A.: Mon. Not. R. Astron. Soc. **291**, 658 (1997)
9. Dorch, S.B.F., Freytag, B.: IAU Symp. **210**, 12P (2003)
10. Gray, D.F.: Astron. J. **135**, 1450 (2008)
11. Grunhut, J.H., Wade, G.A., Hanes, D.A., Alecian, E.: Mon. Not. R. Astron. Soc. **408**, 2290 (2010)
12. Kervella, P., Perrin, G., Chiavassa, A., Ridgway, S.T., Cami, J., Haubois, X., Verhoelst, T.: Astron. Astrophys. **531**, A117 (2011)
13. Kochukhov, O., Makaganiuk, V., Piskunov, N.: Astron. Astrophys. **524**, A5 (2010)
14. Konstantinova-Antova, R., Aurière, M., Iliev, I.K., Cabanac, R., Donati, J.-F., Mouillet, D., Petit, P.: Astron. Astrophys. **480**, 475 (2008)
15. Konstantinova-Antova, R., et al.: Astron. Astrophys. **524**, A57 (2010)
16. Lites, B.W., et al.: Astrophys. J. **672**, 1237 (2008)
17. López Ariste, A.: Astrophys. J. **564**, 379 (2002)
18. Michelson, A.A., Pease, F.G.: Astrophys. J. **53**, 249 (1921)
19. Moutou, C., et al.: Astron. Astrophys. **473**, 651 (2007)
20. Parker, E.N.: Astrophys. J. **122**, 293 (1955)
21. Perrin, G., Ridgway, S.T., Coudé du Foresto, V., Mennesson, B., Traub, W.A., Lacasse, M.G.: Astron. Astrophys. **418**, 675 (2004)
22. Petit, P., et al.: Mon. Not. R. Astron. Soc. **348**, 1175 (2004)
23. Petit, P., et al.: Mon. Not. R. Astron. Soc. **351**, 826 (2004)
24. Petit, P., et al.: Mon. Not. R. Astron. Soc. **361**, 837 (2005)
25. Petit, P., et al.: Mon. Not. R. Astron. Soc. **388**, 80 (2008)
26. Rees, D.E., Semel, M.D.: Astron. Astrophys. **74**, 1 (1979)
27. Schwarzschild, M.: Astrophys. J. **195**, 137 (1975)
28. Sennhauser, C., Berdyugina, S.V.: Astron. Astrophys. **529**, A100 (2011)
29. Solanki, S.K., Stenflo, J.O.: Astron. Astrophys. **170**, 120 (1986)
30. Uitenbroek, H., Dupree, A.K., Gilliland, R.L.: Astron. J. **116**, 2501 (1998)
31. Viticchié, B., Sánchez Almeida, J.: Astron. Astrophys. **530**, A14 (2011)
32. Vögler, A., Schüssler, M.: Astron. Astrophys. **465**, L43 (2007)

Chapter 10
The Formation of Circumstellar Disks Around Evolved Stars

Olivier Chesneau

Abstract The recent high angular resolution observations have shown that the transition between a globally symmetrical giant and a source surrounded by a spatially complex environment occurs relatively early, as soon as the external layers of the stars are not tightly bound to the core of the star anymore. In this review, the emphasis will be put on the delineating the differences between the torus and disk classification through the presentation of many examples of near-IR and mid-IR high angular resolution observations. These examples cover the disks discovered in the core of some bipolar nebulae, post-AGB disks, the dusty environment around born-again stars and recent novae, and also the disks encountered around more massive evolved sources. We discuss the broad range of circumstances and time scales for which bipolar nebulae with disks are observed.

10.1 Direct Detection of Binarity with High Spatial Resolution Imaging

A major breakthrough on the study on asymmetrical Planetary Nebulae (PNs) has been the recognition by a large part of the community of the growing importance of binary systems as main shaping agent of the bipolar and asymmetrical PNs. Companions encompassing a large range of mass, from the stellar objects to Jovian-mass planets are suspected to deeply influence the ejecta when the star reaches the AGBs or even as early as the RGB. This interaction can potentially influence dramatically the fate of the star, leading to poorly known evolutionary paths, influencing deeply the time scales of the different evolutionary stages, eventually bearing only little similarities compared to the time scales involved for the evolution of a single, naked star. Nevertheless, it is still far from excluded that a single star may also provide the conditions for the shaping of an asymmetrical nebula if it expels at one time or another some mass with a significant pole-to-equator density gradient [2].

O. Chesneau (✉)
UMR 6525 Fizeau, CNRS, Obs. de la Côte d'Azur, Univ. Nice Sophia Antipolis, Bvd de l'Obs., BP4229, 06304 Nice Cedex 4, France
e-mail: Olivier.Chesneau@oca.eu

J.-P. Rozelot, C. Neiner (eds.), *The Environments of the Sun and the Stars*,
Lecture Notes in Physics 857,
DOI 10.1007/978-3-642-30648-8_10, © Springer-Verlag Berlin Heidelberg 2013

Many detections of companion of PNs central stars were recently reported, some based on difficult long-term spectral monitoring and many on the recent extensive photometric campaigns from automated telescopes. The high spatial resolution techniques may at first sight represent a large potential for detecting some well-separated (2–100 milli-arcsecond, hereafter mas) companions, but they suffer from intrinsic constraints that do not allow them to currently play a significant role in that domain. Adaptive optics on 8-m-class telescopes are mostly limited in terms of spatial resolution, whilst optical interferometry has strong limitations in terms of contrast and imaging capabilities. The brighter the circumstellar environment and the central source, the harder will be the detectability of a small point-like structure in the close vicinity. The best case is the detection of low T_{eff} companions ($\Delta \sim 5$ mag, $d \sim 10$–100 mas) around hot, and preferentially naked, sources for which it is easy to separate the SED of the stellar components. For a review of the methods used for detecting binarity, the reader is referred to [32].

10.2 Defining the Different Kinds of Equatorial Overdensities

Whatever the origin of the shaping process of an asymmetrical planetary nebula, an equatorial overdensity of material is often involved in the models, and a growing number of observations unveil their presence at many stages of the star evolution. A wealth of new high spatial resolution techniques are now routinely available for the observer, namely the adaptive optics techniques, the speckle and lucky imaging techniques (often called the 'burst modes') in the optical and the infrared, and also the interferometry in the optical, infrared, millimetric and radio wavelengths. All these techniques have their own spatial resolution, sensitivity, contrast and astrometric constraints to the point that it is often quite difficult to compare and put into context the outcome of these observations. Hence, the vocabulary can often be somehow misleading. The detection of disks around evolved sources, as in post-AGB stars, may be closely associated with a high probability to see a binary system. However, this claim is highly dependent on the kind of structure encompassed in the 'disk' denotation, and one shall in the following sections make more precise the equatorial overdensities, dividing them into the 'torus' (or 'outflowing wind') and the 'stratified disks' families. The detection of compact, hot, dust-less accretion disks whose SEDs peak in the UV/B and that are hardly detectable by the above-mentioned infrared techniques is out of the scope of this review.

10.2.1 Dusty Torii: Wind-Related Equatorial Overdensities

An equatorial overdensity is a region of higher density and lower expansion speed compared to the polar circumstellar regions. The kinematics of such a structure is dominantly radial, and the total angular momentum carried by the structure is limited. Equatorial overdensities are short-term structures (these are deflected winds),

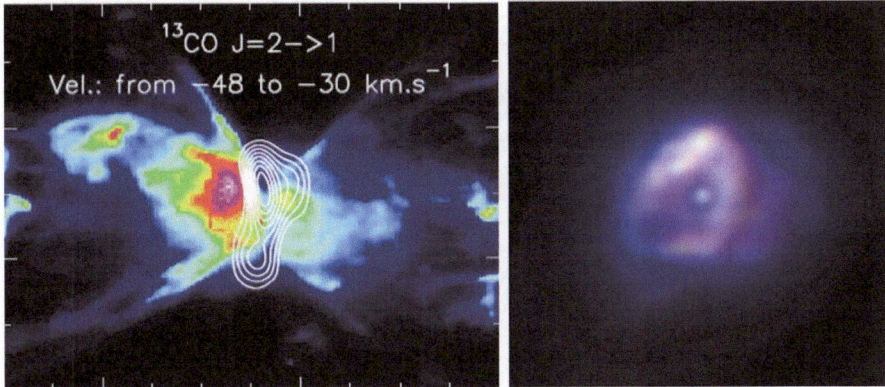

Fig. 10.1 Examples of torus-like overdensities. *Left*: The famous torus of the close and extended bipolar nebula NGC6302 [52]. The millimetric observations are superimposed to the Hα HST image. *Right*: The young Planetary Nebula Hen2-113 [36]. This color image is composed of an HST observation in Hα (in *blue*), an L band image obtained with NACO/VLT (*green*) and an N band image (*red*) obtained using the MIDI/VLTI instrument by direct imaging (i.e. without using the interferometric mode)

e.g. if the mass-loss ceases, then the fate of the material is to rapidly expand and vanish. The increase of density towards the equatorial plane can be ascribed easily by a dependence on the co-latitude of the star in spherical coordinates, and there is no mass nor energy storage in the structure.

To date, the best examples of such torii originate from millimetric interferometry that associates a good spatial resolution ($\sim 1''$) and an excellent spectral resolution (i.e. 1 km s^{-1}). One can cite as a good example of expanding torii [16] or [52] (see Fig. 10.1, left). In the near- or mid-IR, it is more difficult to access the density distribution of the circumstellar material due to opacity effects. Nevertheless, if the scale height of the structure is comparable to its radial extension, without clear sign of marked density increase towards the equatorial plane, the torus hypothesis can be favoured (Fig. 10.1, right). Long slit spectroscopy can provide further evidence for this classification when a significant expansion velocity (i.e. ~ 15–40 km s^{-1}) of the structure can be measured.

10.2.2 Stratified Disks

A disk exhibits a clear vertical stratification whose scale height is governed by the gas pressure only. Its aperture angle is very small (i.e. less than ~ 10 degrees), and its kinematics is dominated by (quasi-)Keplerian velocities, with a small expansion component (≤ 10 km s^{-1}). Thus, its lifetime is much longer than structures blown within a wind, competing or even exceeding the reference time scale of a PN lifetime of a few tens of thousand years.

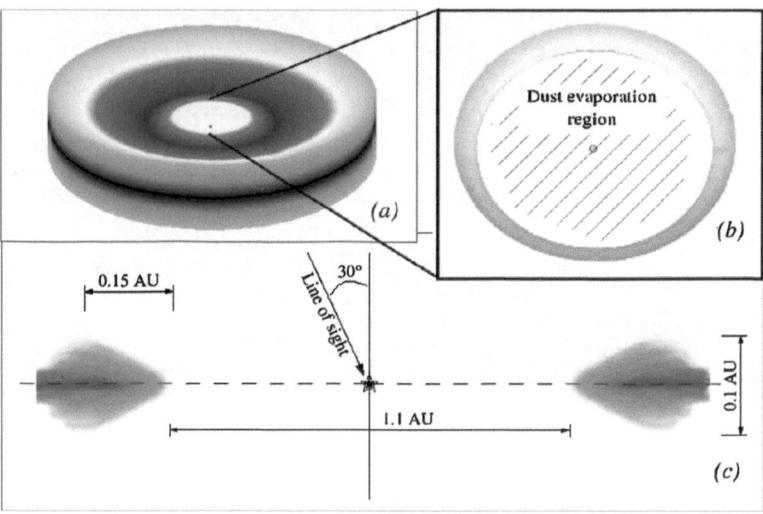

Fig. 10.2 Example of a stratified disk (**a**) with a close-up view of the appearance at inclination $i = 30°$ (**b**) and the vertical structure (**c**) of the inner rim from [30]

The density structure of a Keplerian disk is best ascribed in cylindrical coordinates with a radial law in the equatorial plane and a vertical stratification perpendicular to this plane for the hydrostatic equilibrium.

Such a model is used in a wealth of astronomical contexts and in particular for the formation of Young Stellar Objects (YSOs). Some examples of application of a similar model can be found in [15, 19, 21, 30, 63].

The density law used in this model is

$$\rho(r, z) = \rho_0 \left(\frac{R_*}{r} \right)^{\alpha} \exp\left[-\frac{1}{2} \left(\frac{z}{h(r)} \right)^2 \right], \tag{10.1}$$

$$h(r) = h_0 \left(\frac{r}{R_*} \right)^{\beta}, \tag{10.2}$$

where r is the radial distance in the disk's mid-plane, R_* is the stellar radius, β defines the flaring of the disk, α defines the density law in the mid-plane, and h_0 is the scale height at a reference distance (often 100 AU) from the star (Fig. 10.2). Until recently, such Keplerian disks could only be studied by millimetric interferometry at the highest possible spatial resolution and for the closest targets. One of the best example of such disks is the Red Rectangle [4, 62]. Again, the great advantage of such a technique is to provide both the spatial distribution and the kinematics of the source [5].

In absence of any information on the kinematics in the infrared due to the limited spectral resolution of optical interferometric observations, one has to rely on some indirect evidence to claim for the discovery of a stratified dusty disk, namely one has to prove at least that the disk opening angle is small, or better, that the density

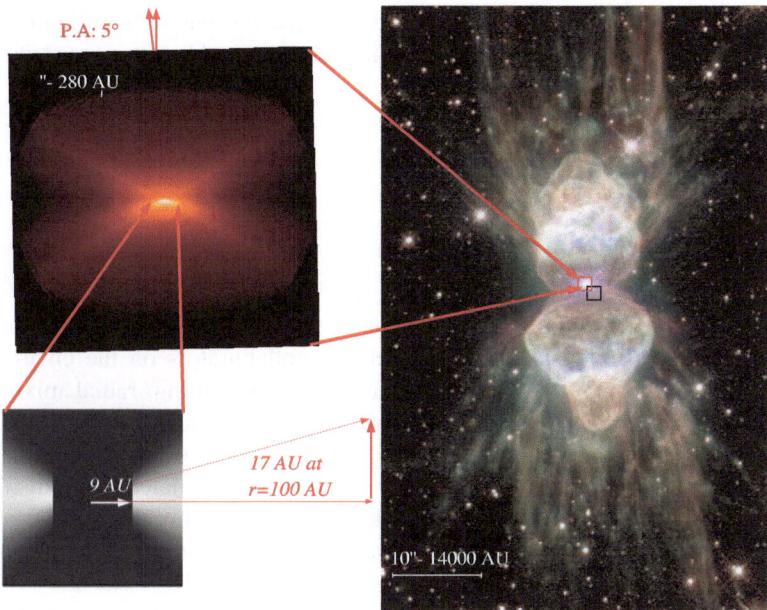

Fig. 10.3 *Right*: Impressive HST multi-color view of the bipolar nebula Menzel 3. *Left*: reconstructed model of the stratified detected in the core using the MIDI/VLTI mid-IR interferometer [6]

law follows radially and vertically the model described above. This is a possible task when the disk is seen close to edge-on (Fig. 10.3). In this case, the radial and vertical directions are well separated on the sky. The best example is the discovery of the edge-on disk in the core of Menzel 3, very well fitted by a stratified disk model [6]. However, our group has conducted some tests trying to invert the disks parameters from artificial interferometric datasets [18, 46]. When the disk is seen at low inclination, many degeneracies appear between the parameters of the vertical density law (such as the flaring parameter β) and the parameters of the radial law. Moreover, the size of the dust grains and their composition affect also critically the fits.

To improve the constraints on the disk temperature law, a good approach is to perform near-IR and mid-IR observations using the AMBER and MIDI instruments of the VLTI respectively. The post-AGB binary IRAS 08544-4431 was studied this way [13], and the same strategy is used for YSOs [35]. It is more difficult to model the near-IR interferometric data due to the potential spatial complexity of the dusty disk's inner rim, and an intense theoretical and observational effort is currently performed, mainly in the YSOs community to better understand the so-called 'puffed-up inner rims' [20, 33, 57]. A better determination of the spatial properties and fine chemical content of the dust forming region is a challenge for the future. Other interesting targets are the double-chemistry sources such as BM Gem, in which a companion has recently been discovered [31, 50, 51]. These sources seem to harbor systemically an equatorially enhanced circumstellar environment, but there is no firm confirmation yet that the structure is best described by a torus or a disk.

10.3 Disk Evolution

10.3.1 Stratified Disk and Binarity

Stratified disks detected in evolved systems are potentially highly correlated with binaries as demonstrated by Van Winckel and collaborators on the environment of binary post-AGBs [11, 25, 60, 61]. Grain growth, settling, radial mixing and crystallization are efficient in such an environment, and the circumbinary disk of these sources seems to be governed by the same physical processes that govern the proto-planetary discs around young stellar objects. It seems that another distinctive character of these long-lived stratified disks as seen in the infrared would be their content in highly processed grains [25, 30, 44]. The key point for the stabilization of the disk is to provide enough angular momentum [1, 55]. A star may (hardly) supply this angular momentum via magnetic fields [22, 59]. But even in this case the formation and stabilization of a Keplerian disk remains a challenge. Of course, this argument does not apply to the accretion disks encountered around YSOs, for which the difficulty in the contrary is to understand how the excess of angular momentum is dissipated. The angular momentum provided by a low-mass stellar or even sub-stellar companion may potentially have a dramatic influence on a growing RGB or AGB star [3, 47–49, 56]. The presence of a stratified disk and the associated Lindblad resonances seems to be a key ingredient in the orbital evolution of the binary system. Without this ingredient, it is difficult to reproduce the observed morphology of the eccentricity-period diagram.

My personal opinion is that the discovery of a stratified disk with proved Keplerian kinematics is directly connected to the influence of a companion, albeit the few exceptions presented above, namely the Young Stellar Objects or the critical velocity rotating massive sources such as Be stars. This hypothesis must be confirmed by further observations.

10.3.2 Time Scale of Formation

When dealing with the theoretical building-up of a disk, time matters, and any constraint on the time scale on which the observed structures were formed is of importance. An interesting study was presented by Huggins [29] on the close time-scale connection between jets and torii, based on many kinematical studies. π^1 Gruis is a good example of a recently formed structure [9, 54]. OH231.8+4.2 is another example of a similar process caught in the act [14, 38].

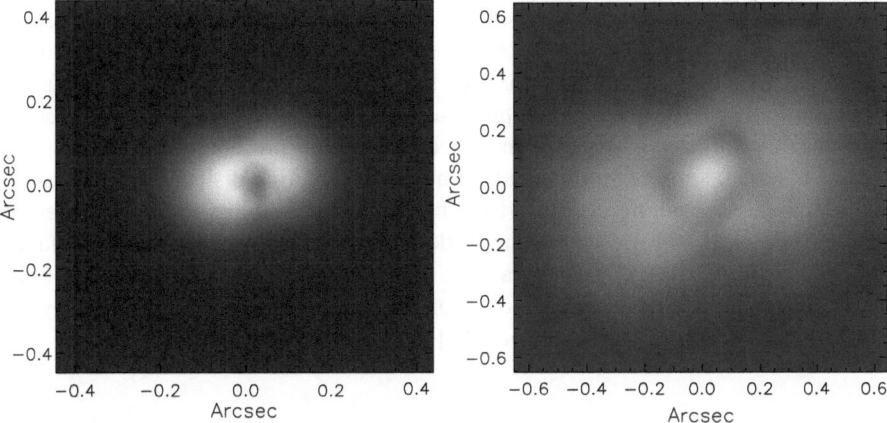

Fig. 10.4 Near-IR view of the fast evolving nebulae formed around the nova V1280 Sco. *Left*: K band image secured two years after the outburst (2009) in the K band using the adaptive optics instrument NACO/VLT. *Right*: J, H, K color image obtained with the same instrument one year later. The stellar source, embedded in the previous observation, is apparent, and the bipolar nebula seems detached [7]

Some recent examples show how fast may be the building up of an equatorial overdensity. The born-again stars, V605 Aql [27] and the Sakurai's Object are surrounded by an equatorial overdensity with a large-scale height that may be described as a torus [8]. In the case of the Sakurai's object, the torus was detected in 2007, about 10 years after the dust began to form. A dense equatorial structure was formed even faster, in less than 2 years, around the slow dust-forming nova V1280 Sco [7]. There is no doubt in this case that the slow ($\sim$500 km s^{-1}) ejecta from the outbursting nova were deeply affected by the common envelope phase that lasted more than tens of orbital period of the companion, leading to the fast formation of a bipolar nebula (Fig. 10.4).

These examples show that as soon as the primary in a binary system gets larger and increases its mass-loss rate when evolving, the influence of a companion can rapidly focus the ejected material onto the equatorial plane of the system, leading to an equatorial overdensity, as already investigated theoretically [37, 58]. This physical process depends on many parameters (mass ratio, orbital parameters, mass-loss rate of the primary, etc.), but the efficiency is such that a large number of targets might be affected at one stage or another [3].

10.3.3 Time Scale of Dissipation and Fate

The observations of the inner circumstellar structures around evolved sources have been to date too scarce to put them into an evolutionary sequence. As written above, the torii are supposed to expand and dissipate much faster than the Keplerian disks.

One may even consider the extreme case in which a binary system surrounded by a stable circumbinary disk continuously replenished by the interaction of the stars may remain virtually unchanged for time scale as long as many 10^5–10^6 years, as proposed for the Red Rectangle. A proposed unified picture of the such an interacting binary evolution involving the presence of a torus is presented in [23].

What is the fate of a stratified disk? An 'old' dissipating disk can see its density and kinematical structure deeply affected by the fast evolving wind of the central star. Gesicki et al. [24] performed an in-depth kinematical study of the dissipating disk in the core of the M2-29 nebula [26]. The torus found around Hen2-113 might also be dissipating under the influence of the fast radiative wind emitted by the Wolf–Rayet star in the core of the nebula [36].

10.4 Stratified Disks Around Massive Stars

There is now a large bunch of evidence that the disks encountered around Be stars and at least some B[e] supergiants are rotating close to Keplerian velocities [39]. The Be stars are proven statistically to represent the tail of the fastest rotators among B stars, and their disks of plasma may be explained, without invoking the influence of a companion, by a combination of an extreme centrifugal force at the equator and some pulsation properties of the star, which leads to erratic ejection of material with high angular momentum [10, 28]. By contrast, there is still no consensus for the more massive supergiant counterparts, the B[e] stars, that are surrounded by dense disks of plasma *and* dust. The dust survives much closer to this hot star than expected so far [17, 40, 41]. The B[e] supergiant's rotation rate is strongly decreased by the increase of their radius while leaving the main sequence, and the rotation alone is probably far from being sufficient to explain the formation of a disk without invoking the influence of a close companion [43]. Note also that dusty circumbinary disks are commonly encountered around interacting binary systems [45].

The example of the A[e] supergiant HD 62623 is informative in this context [40, 42]. HD 62623 is an A supergiant showing the characteristics of the 'B[e]' spectral type, namely a spectrum dominated by strong emission lines and a large infrared excess. Spectrally and spatially resolved observations of AMBER/VLTI in the Brγ line have shown that the supergiant lies in a cavity and is surrounded by a dense disk of plasma. The Brγ line in the location of the central star is *in absorption* showing that the star is not different from a normal member of its class such as Deneb (A3Ia), albeit with a significantly large *vsini* of about 50 km s^{-1}. By contrast, the Balmer and Bracket lines are wider (*vsini* of about 120 km s^{-1}), and the AMBER observations demonstrated that they originate from a disk of plasma, most probably in Keplerian rotation. In absence of any proof of binarity, it is often difficult to understand how such a dense equatorial disk could have been generated. However, HD62623 is a known binary with a stellar companion that orbits close to the supergiant with a period of about 136 days [53]. The mass ratio inferred is very large, and the companion is probably a solar mass star. Plets et al. [53] proposed that an efficient angular momentum transfer occurs near the L2 Lagrangian point of the system, propelling the mass lost from the supergiant by its radiative wind and probably also by strong tides into a stable dense circumbinary disk ([12], in the context of an AGB star). A similar idea was proposed by Dermine et al. [12] in relation with radiation pressure acting on the wind of AGB stars and modifying the Roche lobe geometry, therefore probably easing the formation of a circumbinary disk.

The comparison between the disks encountered, around low and intermediate mass stars and those observed around the B[e] supergiants, a rare spectral type among the zoo of massive stars might not appear relevant at first sight. Yet, recent Spitzer observations of 9 LMC B[e] stars showed an interesting homogeneity of their spectra and a great similarity with the spectra of post-AGBs harboring dense dusty disks [34]. This tightens further the connection between B[e] stars and binarity.

References

1. Akashi, M., Soker, N.: A model for the formation of large circumbinary disks around post AGB stars. New Astron. **13**, 157–162 (2008)
2. Balick, B., Frank, A.: Shapes and shaping of planetary nebulae. Annu. Rev. Astron. Astrophys. **40**, 439–486 (2002)
3. Bear, E., Soker, N.: Spinning-up the envelope before entering a common envelope phase. New Astron. **15**, 483–490 (2010)
4. Bujarrabal, V., Castro-Carrizo, A., Alcolea, J., Neri, R.: The orbiting gas disk in the Red Rectangle. Astron. Astrophys. **441**, 1031–1038 (2005)
5. Bujarrabal, V., van Winckel, H., Neri, R., Alcolea, J., Castro-Carrizo, A., Deroo, P.: The nebula around the post-AGB star 89 Herculis. Astron. Astrophys. **468**, L45–L48 (2007)
6. Chesneau, O., Lykou, F., Balick, B., Lagadec, E., Matsuura, M., Smith, N., Spang, A., Wolf, S., Zijlstra, A.A.: A silicate disk in the heart of the ant. Astron. Astrophys. **473**, L29–L32 (2007)
7. Chesneau, O., Banerjee, D.P.K., Millour, F., Nardetto, N., Sacuto, S., Spang, A., Wittkowski, M., Ashok, N.M., Das, R.K., Hummel, C., Kraus, S., Lagadec, E., Morel, S., Petr-Gotzens, M., Rantakyro, F., Schöller, M.: VLTI monitoring of the dust formation event of the Nova V1280 Scorpii. Astron. Astrophys. **487**, 223–235 (2008)
8. Chesneau, O., Clayton, G.C., Lykou, F., de Marco, O., Hummel, C.A., Kerber, F., Lagadec, E., Nordhaus, J., Zijlstra, A.A., Evans, A.: A dense disk of dust around the born-again Sakurai's object. Astron. Astrophys. **493**, L17–L20 (2009)
9. Chiu, P.-J., Hoang, C.-T., Dinh-V-Trung, et al.: Astrophys. J. **645**, 605 (2006)
10. Cranmer, S.R.: A pulsational mechanism for producing Keplerian disks around Be stars. Astrophys. J. **701**, 396–413 (2009)
11. De Ruyter, S., van Winckel, H., Maas, T., Lloyd Evans, T., Waters, L.B.F.M., Dejonghe, H.: Keplerian discs around post-AGB stars: a common phenomenon? Astron. Astrophys. **448**, 641–653 (2006)
12. Dermine, T., Jorissen, A., Siess, L., Frankowski, A.: Radiation pressure and pulsation effects on the Roche lobe. Astron. Astrophys. **507**, 891–899 (2009)
13. Deroo, P., Acke, B., Verhoelst, T., Dominik, C., Tatulli, E., van Winckel, H.: AMBER and MIDI interferometric observations of the post-AGB binary IRAS 08544-4431: the circumbinary disc resolved. Astron. Astrophys. **474**, L45–L48 (2007)

14. Desmurs, J.-F., Alcolea, J., Bujarrabal, V., Sánchez Contreras, C., Colomer, F.: Astron. Astrophys. **468**, 189 (2007)

15. Di Folco, E., Dutrey, A., Chesneau, O., Wolf, S., Schegerer, A., Leinert, C., Lopez, B.: The flared inner disk of the Herbig Ae star AB Aurigae revealed by VLTI/MIDI in the N-band. Astron. Astrophys. **500**, 1065–1076 (2009)

16. Dinh-V-Trung, B.V., Castro-Carrizo, A., Lim, J., Kwok, S.: Massive expanding torus and fast outflow in planetary nebula NGC 6302. Astrophys. J. **673**, 934–941 (2008)

17. Domiciano de Souza, A., Driebe, T., Chesneau, O., Hofmann, K., Kraus, S., Miroshnichenko, A.S., Ohnaka, K., Petrov, R.G., Preisbisch, T., Stee, P., Weigelt, G., Lisi, F., Malbet, F., Richichi, A.: AMBER/VLTI and MIDI/VLTI spectro-interferometric observations of the B[e] supergiant CPD-57°2874. Size and geometry of the circumstellar envelope in the near- and mid-IR. Astron. Astrophys. **464**, 81–86 (2007)

18. Domiciano de Souza, A., Bendjoya, P., Niccolini, G., Chesneau, O., Borges Fernandes, M., Carciofi, A.C., Spang, A., Stee, P., Driebe, T.: Fast ray-tracing algorithm for circumstellar structures (FRACS). II. Disc parameters of the B[e] supergiant CPD-57°, 2874 from VLTI/MIDI data. Astron. Astrophys. **525**, A22 (2011)

19. Dominik, C., Dullemond, C.P., Waters, L.B.F.M., Walch, S.: Understanding the spectra of isolated Herbig stars in the frame of a passive disk model. Astron. Astrophys. **398**, 607–619 (2003)

20. Dullemond, C.P., Monnier, J.D.: The inner regions of protoplanetary disks. arXiv:1006.3485 (2010)

21. Fedele, D., van den Ancker, M.E., Acke, B., van der Plas, G., van Boekel, R., Wittkowski, M., Henning, T., Bouwman, J., Meeus, G., Rafanelli, P.: The structure of the protoplanetary disk surrounding three young intermediate mass stars, II: spatially resolved dust and gas distribution. Astron. Astrophys. **491**, 809–820 (2008)

22. Frank, A., Blackman, E.G.: Application of magnetohydrodynamic disk wind solutions to planetary and protoplanetary nebulae. Astrophys. J. **614**, 737–744 (2004)

23. Frankowski, A., Jorissen, A.: Binary life after the AGB—towards a unified picture. Balt. Astron. **16**, 104–111 (2007)

24. Gesicki, K., Zijlstra, A.A., Szyszka, C., Hajduk, M., Lagadec, E., Guzman Ramirez, L.: Disk evaporation in a planetary nebula. Astron. Astrophys. **514**, A54 (2010)

25. Gielen, C., van Winckel, H., Min, M., Waters, L.B.F.M., Lloyd Evans, T.: SPITZER survey of dust grain processing in stable discs around binary post-AGB stars. Astron. Astrophys. **490**, 725–735 (2008)

26. Hajduk, M., Zijlstra, A.A., Gesicki, K.: An occultation event in the nucleus of the planetary nebula M 2-29. Astron. Astrophys. **490**, L7–L10 (2008)

27. Hinkle, K.H., Lebzelter, T., Joyce, R.R., Ridgway, S., Close, L., Hron, J., Andre, K.: Imaging ejecta from the final flash star V605 Aquilae. Astron. Astrophys. **479**, 817–826 (2008)

28. Huat, A., Hubert, A., Baudin, F., Floquet, M., Neiner, C., Frémat, Y., Gutiérrez-Soto, J., Andrade, L., de Batz, B., Diago, P.D., Emilio, M., Espinosa Lara, F., Fabregat, J., Janot-Pacheco, E., Leroy, B., Martayan, C., Semaan, T., Suso, J., Auvergne, M., Catala, C., Michel, E., Samadi, R.: The B0.5IVe CoRoT target HD 49330, I: photometric analysis from CoRoT data. Astron. Astrophys. **506**, 95–101 (2009). doi:10.1051/0004-6361/200911928

29. Huggins, P.J.: Jets and tori in proto-planetary nebulae. Astrophys. J. **663**, 342–349 (2007)

30. Isella, A., Testi, L., Natta, A.: Large dust grains in the inner region of circumstellar disks. Astron. Astrophys. **451**, 951–959 (2006)

31. Izumiura, H., Noguchi, K., Aoki, W., Honda, S., Ando, H., Takada-Hidai, M., Kambe, E., Kawanomoto, S., Sadakane, K., Sato, B., Tajitsu, A., Tanaka, W., Okita, K., Watanabe, E., Yoshida, M.: Evidence for a companion to BM gem, a silicate carbon star. Astrophys. J. **682**, 499–508 (2008)

32. Jorissen, A., Frankowski, A.: Detection methods of binary stars with low- and intermediate-mass components. In: Pellegrini, P., Daflon, S., Alcaniz, J.S., Telles, E. (eds.) American Institute of Physics Conference Series, vol. 1057, pp. 1–55 (2008)

33. Kama, M., Min, M., Dominik, C.: The inner rim structures of protoplanetary discs. Astron. Astrophys. **506**, 1199–1213 (2009)
34. Kastner, J.H., Buchanan, C., Sahai, R., Forrest, W.J., Sargent, B.A.: The dusty circumstellar disks of B[e] supergiants in the Magellanic Clouds. Astron. J. **139**, 1993–2002 (2010)
35. Kraus, S., Preibisch, T., Ohnaka, K.: Detection of an inner gaseous component in a Herbig be star accretion disk: near- and mid-infrared spectrointerferometry and radiative transfer modeling of MWC 147. Astrophys. J. **676**, 490–508 (2008)
36. Lagadec, E., Chesneau, O., Matsuura, M., et al..: Astron. Astrophys. **448**, 203 (2006)
37. Mastrodemos, N., Morris, M.: Bipolar preplanetary nebulae: hydrodynamics of dusty winds in binary systems, I: formation of accretion disks. Astrophys. J. **497**, 303 (1998)
38. Matsuura, M., Chesneau, O., Zijlstra, A.A., et al..: Astrophys. J. Lett. **646**, L123 (2006)
39. Meilland, A., Stee, P., Vannier, M., Millour, F., de Domiciano, S.A., Malbet, F., Martayan, C., Paresce, F., Petrov, R.G., Richichi, A., Spang, A.: First direct detection of a Keplerian rotating disk around the Be star α Arae using AMBER/VLTI. Astron. Astrophys. **464**, 59–71 (2007)
40. Meilland, A., Kanaan, S., Borges Fernandes, M., Chesneau, O., Millour, F., Stee, P., Lopez, B.: Resolving the dusty circumstellar environment of the A[e] supergiant HD 62623 with the VLTI/MIDI. Astron. Astrophys. **512**, A73 (2010)
41. Millour, F., Chesneau, O., Borges Fernandes, M., Meilland, A., Mars, G., Benoist, C., Thiébaut, E., Stee, P., Hofmann, K., Baron, F., Young, J., Bendjoya, P., Carciofi, A., Domiciano de Souza, A., Driebe, T., Jankov, S., Kervella, P., Petrov, R.G., Robbe-Dubois, S., Vakili, F., Waters, L.B.F.M., Weigelt, G.: A binary engine fuelling HD 87643's complex circumstellar environment. Determined using AMBER/VLTI imaging. Astron. Astrophys. **507**, 317–326 (2009)
42. Millour, F., Meilland, A., Chesneau, O., Stee, P., Kanaan, S., Petrov, R., Mourard, D., Kraus, S.: Imaging the spinning gas and dust in the disc around the supergiant A[e] star HD 62623. Astron. Astrophys. **526**, A107 (2011)
43. Miroshnichenko, A.S.: Toward understanding the B[e] phenomenon. I. Definition of the galactic FS CMa stars. Astrophys. J. **667**, 497–504 (2007)
44. Murakawa, K., Ueta, T., Meixner, M.: Evidence of grain growth in the disk of the bipolar proto-planetary nebula M 1-92. Astron. Astrophys. **510**, A30 (2010)
45. Netolický, M., Bonneau, D., Chesneau, O., Harmanec, P., Koubský, P., Mourard, D., Stee, P.: The circumbinary dusty disk around the hydrogen-deficient binary star υ Sagittarii. Astron. Astrophys. **499**, 827–833 (2009)
46. Niccolini, G., Bendjoya, P., Domiciano de Souza, A.: Fast ray-tracing algorithm for circumstellar structures (FRACS). I. Algorithm description and parameter-space study for mid-IR interferometry of B[e] stars. Astron. Astrophys. **525**, A21 (2011)
47. Nordhaus, J., Blackman, E.G.: Low-mass binary-induced outflows from asymptotic giant branch stars. Mon. Not. R. Astron. Soc. **370**, 2004–2012 (2006)
48. Nordhaus, J., Blackman, E.G., Frank, A.: Isolated versus common envelope dynamos in planetary nebula progenitors. Mon. Not. R. Astron. Soc. **376**, 599–608 (2007)
49. Nordhaus, J., Spiegel, D.S., Ibgui, L., Goodman, J., Burrows, A.: Tides and tidal engulfment in post-main sequence binaries: period gaps for planets and brown dwarfs around white dwarfs. arXiv:1002.2216 (2010)
50. Ohnaka, K., Driebe, T., Hofmann, K.-H., et al.: Astron. Astrophys. **445**, 1015 (2006)
51. Ohnaka, K., Izumiura, H., Leinert, C., et al.: Astron. Astrophys. **490**, 173 (2008)
52. Peretto, N., Fuller, G., Zijlstra, A., Patel, N.: The massive expanding molecular torus in the planetary nebula NGC 6302. Astron. Astrophys. **473**, 207–217 (2007)
53. Plets, H., Waelkens, C., Trams, N.R.: The peculiar binary supergiant 3 Puppis. Astron. Astrophys. **293**, 363–370 (1995)
54. Sacuto, S., Jorissen, A., Cruzalèbes, P., Chesneau, O., Ohnaka, K., Quirrenbach, A., Lopez, B.: Astron. Astrophys. **482**, 561–574 (2008)
55. Soker, N.: Properties that cannot be explained by the progenitors of planetary nebulae. Astrophys. J. Suppl. Ser. **112**, 487 (1997)

56. Soker, N., Rappaport, S.: The formation of very narrow waist bipolar planetary nebulae. Astrophys. J. **538**, 241–259 (2000)
57. Tannirkulam, A., Harries, T.J., Monnier, J.D.: The inner rim of YSO disks: effects of dust grain evolution. Astrophys. J. **661**, 374–384 (2007)
58. Theuns, T., Jorissen, A.: Wind accretion in binary stars, I: intricacies of the flow structure. Mon. Not. R. Astron. Soc. **265**, 946 (1993)
59. Ud-Doula, A., Owocki, S.P.: Dynamical simulations of magnetically channeled line-driven stellar winds, I: isothermal, nonrotating, radially driven flow. Astrophys. J. **576**, 413–428 (2002)
60. Van Winckel, H.: Post-Agb binaries. Balt. Astron. **16**, 112–119 (2007)
61. Van Winckel, H., Lloyd Evans, T., Reyniers, M., Deroo, P., Gielen, C.: Binary post-AGB stars and their Keplerian discs. Mem. Soc. Astron. Ital. **77**, 943 (2006)
62. Witt, A.N., Vijh, U.P., Hobbs, L.M., Aufdenberg, J.P., Thorburn, J.A., York, D.G.: The Red Rectangle: its shaping mechanism and its source of ultraviolet photons. Astrophys. J. **693**, 1946–1958 (2009)
63. Wolf, S., Padgett, D.L., Stapelfeldt, K.R.: The circumstellar disk of the Butterfly Star in Taurus. Astrophys. J. **588**, 373–386 (2003)

Index

Symbols
3 µm bump, 196
3 µm IR bump, 187
α Col, 151
δ Sco, 154
δ Cen, 150
δ Sco, 150, 153, 157
η Car, 157
υ Sgr, 157

A
aa-index, 58, 74, 87, 96
AATAU objects, 187, 198, 199
AATAU-type stars, 187
Accretion, 163, 164, 168–171, 174–176
Achernar, 150, 153
Active regions, 18, 42
Activity, 231–233, 235, 237
Advection, 61, 64–66, 71, 72, 75, 76, 80
Advection time scale, 64
Alfvén radius, 223, 226, 227, 229
AMBER, 150, 151, 153, 154, 156
Anomalous cosmic rays, 45
Arcades, 43

B
Babcock–Leighton mechanism, 57
Balmer discontinuity, 186
Balmer jump, 185, 186
Balmer line, 156
Betelgeuse, 231–234, 236, 237
Binary, 174
Birthline, 184
Blobs, 31
Brγ, 156
Brackett line, 156

C
CCTR, 17, 20, 21, 23, 26, 27, 48
CHARA, 150, 153
Chromosphere, 17–20, 28, 36, 39, 48
Circumbinary disk, 156, 157
Circumstellar disk, 150, 155, 157
Circumstellar environment, 149, 150, 153, 155, 157
Class 0 object, 183
Class I object, 183
Class II object, 183
Class III object, 183
Classical T Tauri stars, 185
CME (coronal mass ejection), 54, 59, 87, 89, 92, 94, 95, 97, 101
Coherence index, 12
Convection, 163
Convection zone, 55–58, 61, 62, 64–66, 68–72, 74–76, 78, 79, 81, 85, 87, 97, 99, 100, 102
Convective envelope, 55, 82
Cool supergiant, 231, 232, 237
Coplanar discontinuity, 5
Corona, 15, 17–21, 23, 25, 29, 36, 39, 42, 43, 47, 59, 92, 207–209
Coronal, 54, 59, 60, 66, 87, 91–97, 100, 102
Coronal holes, 19, 24, 25, 47
Coronal mass ejections, 15, 36, 40, 43
CTTS, 185, 186
Current sheet, 28, 43, 44, 47
Cycle, 21

D
De Hoffman–Teller frame, 5
Diffusion, 61, 64, 66, 69, 71, 72, 74, 75, 80, 81
Diffusion time scale, 64
Diffusivity, 61, 64, 65, 69, 71, 81, 97, 102

J.-P. Rozelot, C. Neiner (eds.), *The Environments of the Sun and the Stars*,
Lecture Notes in Physics 857,
DOI 10.1007/978-3-642-30648-8, © Springer-Verlag Berlin Heidelberg 2013